Squares and Square Roots

n	n^2	$\sqrt{n}$	$\sqrt{10n}$	n	n^2	$\sqrt{n}$	$\sqrt{10n}$
1	1	1.000	3.162	51	2601	7.141	22.583
2	4	1.414	4.472	52	2704	7.211	22.804
3	9	1.732	5.477	53	2809	7.280	23.022
4	16	2.000	6.325	54	2916	7.348	23.238
5	25	2.236	7.071	55	3025	7.416	23.452
6	36	2.449	7.746	56	3136	7.483	23.664
7	49	2.646	8.367	57	3249	7.550	23.875
8	64	2.828	8.944	58	3364	7.616	24.083
9	81	3.000	9.487	59	3481	7.681	24.290
10	100	3.162	10.000	60	3600	7.746	24.495
11	121	3.317	10.488	61	3721	7.810	24.698
12	144	3.464	10.954	62	3844	7.874	24.900
13	169	3.606	11.402	63	3969	7.937	25.100
14	196	3.742	11.832	64	4096	8.000	25.298
15	225	3.873	12.247	65	4225	8.062	25.495
16	256	4.000	12.649	66	4356	8.124	25.690
17	289	4.123	13.038	67	4489	8.185	25.884
18	324	4.243	13.416	68	4624	8.246	26.077
19	361	4.359	13.784	69	4761	8.307	26.268
20	400	4.472	14.142	70	4900	8.367	26.458
21	441	4.583	14.491	71	5041	8.426	26.646
22	484	4.690	14.832	72	5184	8.485	26.833
23	529	4.796	15.166	73	5329	8.544	27.019
24	576	4,899	15.492	74	5476	8.602	27.203
25	625	5.000	15.811	75	5625	8.660	27.386
26	676	5.099	16.125	76	5776	8.718	27.568
27	729	5.196	16.432	77	5929	8.775	27.749
28	784	5.292	16.733	78	6084	8.832	27.928
29	841	5.385	17.029	79	6241	8.888	28.107
30	900	5.477	17.321	80	6400	8.944	28.284
31	961	5.568	17.607	81	6561	9.000	28.460
32	1024	5.657	17.889	82	6724	9.055	28.636
33	1089	5.745	18.166	83	6889	9.110	28.810
34	1156	5.831	18.439	84	7056	9.165	28.983
35	1225	5.916	18.708	85	7225	9.220	29.155
36	1296	6.000	18.974	86	7396	9.274	29.326
37	1369	6.083	19.235	87	7569	9.327	29.496
38	1444	6.164	19.494	88	7744	9.381	29.665
39	1521	6.245	19.748	89	7921	9.434	29.833
40	1600	6.325	20.000	90	8100	9.487	30.000
41	1681	6.403	20.248	91	8281	9.539	30.166
42	1764	6.481	20.494	92	8464	9.592	30.332
43	1849	6.557	20.736	93	8649	9.644	30.496
44	1936	6.633	20.976	94	8836	9.695	30.659
45	2025	6.708	21.213	95	9025	9.747	30.822
46	2116	6.782	21.448	96	9216	9.798	30.984
47	2209	6.856	21.679	97	9409	9.849	31.145
48	2304	6.928	21.909	98	9604	9.899	31.305
49	2401	7.000	22.136	99	9801	9.950	31.464
50	2500	7.071	22.361	100	10000	10.000	31.623

BEGINNING ALGEBRA

SECOND EDITION

MARGARET L. LIAL / CHARLES D. MILLER

AMERICAN RIVER COLLEGE SACRAMENTO, CALIFORNIA

BEGINNING ALGEBRA

SECOND EDITION

SCOTT, FORESMAN AND COMPANY GLENVIEW, ILLINOIS

DALLAS, TEX. • OAKLAND, N.J. • PALO ALTO, CAL. • TUCKER, GA. • ABINGDON, ENGLAND

Library of Congress Cataloging in Publication Data

Lial, Margaret L
Beginning algebra.

Includes index.
1. Algebra. I. Miller, Charles David, 1942– joint author. II. Title.
QA152.2.L5 1975 512.9′042 75-22460
ISBN 0-673-07933-3

Copyright © 1976, 1971, 1969 Scott, Foresman and Company.
All Rights Reserved.
Printed in the United States of America.

5 6 7 8 9 10 – KPK – 85 84 83 82 81 80 79

PREFACE

This book is for an introductory course in algebra for college students. The only prerequisite we assume is some background in arithmetic. (Even so, much arithmetic, especially of fractions, is discussed and reviewed in the text.) The length is reasonable for one semester or quarter. This second edition contains more exercises and examples than the first, but not more topics. Both a study guide and an instructor's manual are available for this book. The study guide is in a semi-programmed style, and offers a wealth of supplementary exercises over the spectrum of elementary algebra topics. The instructor's manual features answers to exercises not in the book, teaching hints, sample chapter tests, and a sample final.

This second edition has been completely rewritten, along guidelines suggested by our own classroom experience as well as the comments from many users of the first edition. Thus we have made the following changes:

- The number of exercises has been increased by more than 50%. Exercise sets are carefully graded, from the very elementary to the more challenging.
- We have added many worked-out examples—over 60% more. The text almost always features a worked-out example for each type of problem in an exercise set.
- Since fractions require a knowledge of factoring, the chapters on polynomials, factoring, and algebraic fractions (Chapters 3–5) are now grouped together.
- Chapter 8, on roots and radicals, has been moved closer to the end of the book. This is difficult material for many students. Presenting it after a generous exposure to algebra should make it easier to grasp. This chapter is followed by a chapter on quadratic equations, which provides continued use and reinforcement of roots and radicals.
- We have included an appendix introducing the metric system.

This second edition retains the popular features from the first edition:

- An open format enhances the straightforward approach.
- Examples are well set off and can be easily identified.
- Solutions of equations is presented early, and then used throughout the text. Students have direct access to the useful ideas of algebra.

We are grateful to the people who have helped us with the revision of this book: Vern Heeren, Norman Andersen, and Virginia-Jane Gleadall, American River College; Howard Raphael, College of the Sequoias; Harold Oxsen, Diablo Valley College; Lucille Groenke, Maricopa Technical College; Robert Mosher, California State University, Long Beach; James Curl, Modesto Junior College; James Hajek, Lincoln Land Community College; Roy Dubisch, University of Washington; Lawrence Frederick, College of the Desert; G. J. Corley, Carol Hall, Anita Harkness, Stuart Mills, and Barron Tabor, Louisiana State University, Shreveport; Walter Klann, Glendale Community College; Judith Hall, West Virginia University; Ralph Mansfield, Chicago Loop College; Calvin Lathan, Monroe Community College; William Gordon, Roxbury Community College; George Vick, Sam Houston State University.

It seems to us that Scott, Foresman is really interested in producing useful textbooks of high quality. To that end, Robert Runck, Nat Weintraub, Mary Helfrich, and George Lobell have worked closely with us at various stages of the development, editing, and publication of this book.

Margaret Lial

Charles Miller

TO THE STUDENT

Studying algebra is different than studying history, English, or science. Here are some suggestions on how to get the most out of your algebra class and textbook.

1 Before going to class, look over any new topics that the instructor will cover. Read the introductory paragraphs of the new section to see how the new ideas fit in with previous ones. Skim quickly through the section to see if any new formulas or rules are given.

2 Follow through the worked-out examples in the text. This may take some time. First read the statement of the example, and make sure you completely understand what the example asks for. Try to decide how the example fits in the topic of the section. Write out the solution yourself, supplying any steps missing in the book. Work through to the finish, and then check your answer to make sure it is correct.

3 Try a few exercises of each type. Check your answers with those in the back of the book to make sure you're on the right track.

4 There will probably be a few places where you get stuck. Go to class, and listen to the instructor's presentation. You will find that many of your trouble spots are cleared up during class discussion. Don't be afraid to ask questions. You are in class to learn, so go ahead and ask about anything that troubles you.

5 If not all your questions are cleared up in class, ask the instructor for help during office hours. If your school offers a mathematics lab or tutorial services, take advantage of what is offered.

6 That night, work the assigned homework exercises.

7 After you finish the homework, look over the next section of new material.

8 At the end of each chapter, check your understanding of the topics of the chapter by taking the test.

9 If you would like some supplementary exercises and review questions, get a copy of the *Study Guide for Beginning Algebra.* Your bookstore should have copies. If not, they can order one for you.

10 When it comes time for the final, make up your own sample final, as follows. Go through the sections you studied, and write down the statement of some typical worked-out examples. Then work through the ones you have written down. You can then refer to the book to see how to work any items that may have given you trouble.

11 Finally, after all this work, you are ready for your reward—take the next mathematics course you need.

CONTENTS

3 Polynomials

4 Factoring

5 Rational Numbers and Rational Expressions

6 Linear Equations

7 Linear Systems of Equations

Roots and Radicals

Quadratic Equations

Appendix

Tables

Answers

DIAGNOSTIC PRETEST

Each problem in the following diagnostic pretest is worked out in this book, on the given page.

	page
1. Simplify: (a) $2(5+6)+7\cdot 3$	**3**
(b) $\dfrac{4(5+3)+3}{2(3)-1}$	**4**
2. Simplify: (a) $\lvert -7\rvert$	**15**
(b) $-\lvert -14\rvert$	**15**
3. Add: (a) $-2+(-9)$	**18**
(b) $-8+(-12)$	**18**
4. Simplify: $-6-[2-(8+3)]$	**22**
5. Solve: $-1+2x+16+(-x)=12$	**48**
6. Solve: $4(k-3)-k=k-6$	**58**
7. The perimeter of a square is 96 inches. Find the length of a side.	**67**
8. Solve: $3z+2-5>-z+7+2z$	**79**
9. Evaluate 5^4. Name the base and exponent.	**87**
10. Simplify: $(2x)^3(2x)^2$	**91**
11. Subtract: $6x^3-4x^2+2$ from $11x^3+2x^2-8$	**99**
12. Multiply: $3p-5$ and $2p+6$	**103**
13. Divide: $2x+3$ into $8x^3-4x^2-14x+15$	**112**
14. Simplify: $\left(\dfrac{3x^{-2}}{4^{-1}y^3}\right)^{-3}$	**117**
15. Factor: $m^2+9m+14$	**127**
16. Factor: $12a^2-ab-20b^2$	**132**
17. Solve the quadratic equation: $x^2-x-20=0$	**138**
18. The product of two consecutive odd integers is 1 less than 5 times their sum. Find the integers.	**143**
19. Find the product: $\dfrac{x^2+3x}{x^2-3x-4}\cdot\dfrac{x^2-5x+4}{x^2+2x-3}$	**165**

20. Add the rational expressions: $\dfrac{x}{x^2-1}+\dfrac{x}{x+1}$ **169**

21. Solve: $\dfrac{2}{x^2-x}=\dfrac{1}{x^2-1}$ **174**

22. Graph: $4x-5y=20$ **201**

23. Graph: $x=3$ **204**

24. Graph: $2x-5y\geq 10$ **207**

25. Let $P(x)=5x^2-4x+3$. Find: (a) $P(0)$ **213**

(b) $P(-2)$ **213**

26. Solve the system: $2x+3y=-15$
$5x+2y=1$ **226**

27. Solve the system: $3x-\ \ y=4$
$-9x+3y=-12$ **230**

28. Find each of the following: (a) $-\sqrt{144}$ **249**

(b) $\sqrt{10{,}000}$ **249**

(c) $-\sqrt{10{,}000}$ **249**

29. Simplify: $\sqrt{72}$ **252**

30. Find the product: $\sqrt{25}\cdot\sqrt{75}$ **253**

31. Simplify: $2\sqrt{12}+3\sqrt{75}$ **257**

32. Rationalize the denominator: $\sqrt{\dfrac{27}{5}}$ **259**

33. Solve: $3\sqrt{x}=\sqrt{x+8}$ **271**

34. Find the solutions of the quadratic equation:
$x^2-2x-1=0$ **286**

35. Solve the equation: $\dfrac{1}{10}t^2=\dfrac{2}{5}-\dfrac{1}{2}t$ **287**

36. Simplify: $2\sqrt{-96}+\sqrt{-54}$ **289**

37. Solve: $x^2-2x-2=0$ **291**

38. Graph the parabola: $y=(x-2)^2$ **299**

1 OPERATIONS WITH THE REAL NUMBERS

1.1 USING SYMBOLS

There is a natural human tendency to simplify things that must be done often. In mathematics, symbols are used to simplify mathematical statements. For example, the problem "three plus five" is easier and faster to work by using the symbols 3, $+$, and 5 to write

$$3 + 5 = 8.$$

Four symbols familiar from arithmetic are the symbols for the basic operations:

$+$ for addition	$\times$ for multiplication
$-$ for subtraction	$\div$ for division

A common symbol is the equals sign, $=$, used in algebra, as well as in arithmetic, to represent the idea of *equality*. For example, the statement $8 = 6 + 2$ expresses the fact that 8 and $6 + 2$ both represent the same number.

The symbol $\neq$ expresses the fact that two numbers are not equal. For example, the statement $7 \neq 8$ is read "7 is not equal to 8."

If two numbers are not equal, then one of the numbers must be smaller than the other. We know that $7 \neq 8$; we also know that 7 is less than 8. The symbol $<$ represents "is less than," so that "7 is less than 8" can be written $7 < 8$. Also, "6 is less than 9" is written $6 < 9$.

The symbol $>$ means "is greater than." Since 8 is greater than 2, we have $8 > 2$. The statement "17 is greater than 11" is written $17 > 11$.

One way to keep the symbols $<$ and $>$ straight is to remember that the symbol always points to the smaller number. To write "8 is less than 15," point the symbol toward the 8:

$$8 < 15.$$

Just as we write "is not equal to" as $\neq$, we write "is not less than" as $\not<$. The statement "5 is not less than 4" is written $5 \not< 4$. "Is not greater than" is written $\not>$. For example, $6 \not> 9$ is read as "6 is not greater than 9."

Example 1 Write each word statement below in symbols.

(a) Twelve equals ten plus two.
In symbols: $12 = 10 + 2$.

(b) Nine is less than ten.
$9 < 10$.

(c) Fifteen is not equal to eighteen.
$15 \neq 18$.

(d) Seven is not greater than ten.
$7 \not> 10$.

Two other symbols also represent the idea of *inequality*, the symbols $\leq$ and $\geq$. The symbol $\leq$ means "less than or equal to."

$$5 \leq 9.$$

It is true that 5 is less than or equal to 9 because the part about 5 less than 9 is true. If one part of the statement is true, then the whole statement is true.

The symbol $\geq$ means "greater than or equal to." Again, $9 \geq 5$ is true because $9 > 5$ is true—that makes the whole statement true.

In the same way, $8 \geq 8$ is a true statement since $8 = 8$. But it is not a true statement that $13 \leq 9$. Neither part is true.

In arithmetic, the symbol $\times$ is used to show that two numbers should be multiplied. In algebra, the symbol $\times$ looks too much like the letter x, which is used for another purpose. To avoid the look-alike symbol $\times$, we use a dot for multiplication. For example, $5 \cdot 8$ tells us to multiply 5 and 8. Thus,

$$5 \cdot 8 = 40.$$

The number 40 is called the **product** of 5 and 8. We can also write the product of 5 and 8 using parentheses, so that

$$5(8), \quad (5)8, \quad \text{and} \quad (5)(8)$$

all mean 5 times 8, or 40.

Many of the problems that we will work will have combinations of more than one symbol. For example, let us find the value of

$$2 \cdot 3 + 5.$$

We have a choice—we could first multiply 2 and 3, or we could first add 3 and 5. If we first multiply 2 and 3, the result is

$$2 \cdot 3 + 5 = 6 + 5 = 11.$$

If we first add 3 and 5, and then multiply by 2, the result is

$$2 \cdot 3 + 5 = 2 \cdot 8 = 16.$$

To make sure that everyone who works a problem like this always gets the same answer, the following **order of operations** has been agreed upon as being the most useful.

(1) Do any operations inside parentheses.
(2) Do any multiplications or divisions, working from left to right.
(3) Do any additions or subtractions, working from left to right.
(4) If the problem involves a fraction bar, do all work above the bar and below the bar separately, and then simplify, if possible, to get the final answer.

Example 2 Find the value of $2 \cdot 3 + 5$.

Using the order of operations given above, we first multiply.

$$2 \cdot 3 + 5 = 6 + 5.$$

Then add.

$$6 + 5 = 11.$$

Therefore, $2 \cdot 3 + 5 = 11$. By following the order of operations given above, there is only one possible answer for this problem.

Example 3 Find the simplest possible answer for each of the following.

(a) $3 \cdot 8$

This represents the product of 3 and 8, or $3 \cdot 8 = 24$.

(b) $6 \cdot 8 + 5 \cdot 2$

Do any multiplications, working from left to right, and then add.

$$\begin{aligned} 6 \cdot 8 + 5 \cdot 2 &= 48 + 10 && \text{Multiply} \\ &= 58 && \text{Add} \end{aligned}$$

(c) $9(6 + 11)$

This expression can be simplified by first working inside the parentheses, using rule 1.

$$\begin{aligned} 9(6 + 11) &= 9(17) && \text{Work inside parentheses} \\ &= 153 && \text{Multiply} \end{aligned}$$

(d) $$\begin{aligned} 2(5 + 6) + 7 \cdot 3 &= 2(11) + 7 \cdot 3 && \text{Work inside parentheses} \\ &= 22 + 21 && \text{Multiply} \\ &= 43 && \text{Add} \end{aligned}$$

(e) $$\frac{4(5 + 3) + 3}{2(3) - 1}$$

Simplify the top and bottom of the fraction, separately, by rule (4).

$$\frac{4(5+3)+3}{2(3)-1} = \frac{4(8)+3}{2(3)-1} \quad \text{Work first inside parentheses}$$

$$= \frac{32+3}{6-1} \quad \text{Multiply}$$

$$= \frac{35}{5} \quad \text{Add or subtract}$$

$$= 7 \quad \text{Simplify}$$

An expression with double parentheses, such as $2(8 + 3(6 + 5))$, can be confusing. To eliminate this, square brackets, [and], are used instead of one pair of parentheses.

$$2[8 + 3(6 + 5)].$$

To work out this expression, the brackets are treated just like parentheses.

$$\begin{aligned} 2[8 + 3(6 + 5)] &= 2[8 + 3(11)] \\ &= 2[8 + 33] \\ &= 2[41] \\ &= 82. \end{aligned}$$

EXERCISES

In Exercises 1–16, answer *true* or *false* for each statement.

1. $8 + 2 = 10$

2. $9 + 12 = 21$

3. $6 \neq 5 + 1$

4. $8 \neq 9 - 1$

5. $9 < 12$

6. $15 < 21$

7. $12 \geq 10$

8. $25 \geq 19$

9. $11 < 11$

10. $45 < 45$

11. $18 \not< 20$

12. $29 \not< 30$

13. $0 < 15$

14. $9 < 0$

15. $8 \not> 0$

16. $0 \not> 4$

In Exercises 17–36, carry out the indicated operations. Then decide whether the given statement is *true* or *false*.

17. $3 \cdot 4 + 7 < 10$

18. $8 \cdot 2 - 5 > 12$

19. $9 \cdot 3 - 11 \leq 16$

20. $6 \cdot 5 - 12 \leq 18$

21. $5 \cdot 11 + 2 \cdot 3 \leq 60$

22. $9 \cdot 3 + 4 \cdot 5 \geq 48$

23. $9 \cdot 2 - 6 \cdot 3 \geq 2$

24. $8 \cdot 3 - 4 \cdot 6 < 1$

25. $12 \cdot 3 - 6 \cdot 6 \leq 0$

26. $13 \cdot 2 - 15 \cdot 1 \geq 10$

27. $6 \cdot 5 + 3 \cdot 10 \leq 0$

28. $5 \cdot 8 + 10 \cdot 4 \geq 0$

29. $4[2 + 3(4)] \geq 50$

30. $3[5(2) - 3] < 20$

31. $60 < 5[8 + (2 + 3)]$

32. $80 < 9[(14 + 5) - 10]$

33. $2[2 + 3(2 + 5)] \leq 45$

34. $3[4 + 3(4 + 1)] \leq 55$

35. $\dfrac{2(5 + 1) - 3(1 + 1)}{5(8 - 6) - 4 \cdot 2} \leq 3$

36. $\dfrac{3(8 - 3) + 2(4 - 1)}{9(6 - 2) - 11(5 - 2)} \geq 7$

In Exercises 37–50, insert parentheses in each expression so that the resulting statement is true.

Example The statement $9 - 3 - 2 = 8$ will be true if parentheses are inserted around $3 - 2$. Thus

$$9 - (3 - 2) = 8.$$

It is not true that $(9 - 3) - 2 = 8$, because $6 - 2 \neq 8$.

37. $10 - 7 - 3 = 6$

38. $3 \cdot 5 + 7 = 36$

39. $3 \cdot 5 + 7 = 22$

40. $3 \cdot 5 - 4 = 3$

41. $3 \cdot 5 - 4 = 11$

42. $3 \cdot 5 + 2 \cdot 4 = 23$

43. $3 \cdot 5 + 2 \cdot 4 = 84$

44. $3 \cdot 5 + 2 \cdot 4 = 68$

45. $3 \cdot 5 - 2 \cdot 4 = 36$

46. $3 \cdot 5 - 2 \cdot 4 = 7$

47. $100 \div 20 \div 5 = 1$

48. $360 \div 18 \div 4 = 5$

49. $100 \div 20 \div 5 = 25$

50. $4096 \div 256 \div 4 = 4$

1.2 VARIABLES, EQUATIONS, AND INEQUALITIES

A **variable** is a letter, such as x, y, or z, that represents an unknown number. For example, a problem might contain the words "the sum of an unknown number and 5." To translate this statement into mathematical symbols, first choose a letter, such as x, to represent the unknown number.

Words The sum of an unknown number and 5.

The sum of x and 5.

Symbols $x + 5$.

Any combination of variables, numbers, and symbols for operations is called an **algebraic expression**: for example, $x + 5$. The expression $x + 5$ takes on different numerical values as the variable x takes on different numerical values.

For example, if x represents 12, we can find the value of $x + 5$ by substituting 12 for x. If $x = 12$, then $x + 5$ takes on the value $12 + 5$, or 17. However, if $x = 3$, then $x + 5$ takes on the value $3 + 5$, or 8. For $x = 7$, the value of $x + 5$ is 12.

Example 1 Find the value of the algebraic expression $x + 9 + x + 5$ when $x = 1$; when $x = 10$.

If $x = 1$, then $x + 9 + x + 5 = 1 + 9 + 1 + 5 = 16$.

If $x = 10$, then $x + 9 + x + 5 = 10 + 9 + 10 + 5 = 34$.

Multiplication is indicated with a dot or with parentheses. For example, $9 \cdot 2$ and $9(2)$ both mean 9×2, or 18. The product of 8 and x can also be written as $8 \cdot x$, or $8(x)$, or $(8)x$. However, for simplicity, the product of a number and a variable is written without special symbols. The product of 8 and x is written $8x$, and the product of 12 and z is written $12z$. The product of x and z is xz.

Example 2 Find the value of each of the following expressions when $x = 5$ and $y = 3$.

(a) $2x + 5y$

Replace x with 5 and y with 3. Do the multiplications first, and then add.

$$\begin{aligned} 2x + 5y &= 2 \cdot 5 + 5 \cdot 3 && \text{Let } x = 5 \text{ and } y = 3 \\ &= 10 + 15 && \text{Multiply} \\ &= 25. && \text{Add} \end{aligned}$$

(b) $\dfrac{9x - 8y}{2x - y}$

Replace x with 5 and y with 3.

$$\begin{aligned} \frac{9x - 8y}{2x - y} &= \frac{9 \cdot 5 - 8 \cdot 3}{2 \cdot 5 - 3} && \text{Let } x = 5 \text{ and } y = 3 \\ &= \frac{45 - 24}{10 - 3} && \text{Multiply} \\ &= \frac{21}{7} && \text{Subtract} \\ &= 3. && \text{Simplify} \end{aligned}$$

An **equation** is a sentence which says that two algebraic expressions are equal. Examples of equations include

$$x + 4 = 11, \qquad 2y = 16, \qquad 4p + 1 = 25 - p.$$

To solve an equation, it is necessary to find all values of the variable that make the equation true. The values of the variable that make the equation true are called the **solutions** of the equation.

In this chapter, whenever you are given an equation to solve, you will also be given a set of numbers which might be solutions. You must then decide which, if any, of these numbers are really solutions of the equation.

For example, which numbers in the set $\{6, 7, 8\}$ are solutions of the equation $x + 4 = 11$?* To decide which of the numbers in the set $\{6, 7, 8\}$ are solutions of $x + 4 = 11$, take each number in turn, and substitute it for x.

If $x = 6$,	If $x = 7$,	If $x = 8$,
$x + 4 = 11$	$x + 4 = 11$	$x + 4 = 11$
$6 + 4 = 11$	$7 + 4 = 11$	$8 + 4 = 11$
$10 = 11$	$11 = 11$	$12 = 11$
False	True	False

The only true result is $11 = 11$. Therefore, the only number from the set $\{6, 7, 8\}$ which leads to a true statement is the number 7. Therefore, 7 is the only solution from the given set for the equation $x + 4 = 11$. This can be expressed by saying that the **solution set** of the equation $x + 4 = 11$ is $\{7\}$. However, we usually say only that *the solution of* $x + 4 = 11$ *is* 7.

The set of possible replacements for the variable, $\{6, 7, 8\}$ in the example above, is called the **replacement set**, or **domain**, of the variable.

Example 3 Solve $3x - 6 = 24$. The domain of the variable is $\{9, 10, 11\}$.

To find the solution, try each number from the domain.

If $x = 9$,	If $x = 10$,	If $x = 11$
$3x - 6 = 24$	$3x - 6 = 24$	$3x - 6 = 24$
$3 \cdot 9 - 6 = 24$	$3 \cdot 10 - 6 = 24$	$3 \cdot 11 - 6 = 24$
$27 - 6 = 24$	$30 - 6 = 24$	$33 - 6 = 24$
$21 = 24$	$24 = 24$	$27 = 24$
False	True	False

A true statement is obtained only when the number 10 replaces the variable. Therefore, the solution of $3x - 6 = 24$ is 10.

Sometimes we want to say that one expression is less than or greater than another expression. Remember that "less than" is written using the symbol $<$. "Greater than" is written using the symbol $>$.

If we want to say that an expression is either less than or equal to a second expression, we use the symbol $\leq$. If we want to say that the first expression is greater than or equal to the second expression, we use the symbol $\geq$.

Such statements are called **inequalities**. Examples of inequalities include

$$x + 2 \leq 6, \qquad 3y \geq 18, \qquad 2x + 1 < 9, \qquad 5x - 2 > 13.$$

To solve an inequality, we must find all numbers from the given domain that make the inequality true.

* For us, a *set* is a collection of numbers.

Example 4 Solve the inequality $2x + 1 < 9$. The domain is $\{1, 2, 3, 4, 5\}$.

Replace the variable x, in turn, by the numbers 1, 2, 3, 4, and 5. The inequality $2x + 1 < 9$ is true for 1, 2, and 3, but not true for 4 and 5. Therefore, the solutions of $2x + 1 < 9$ are 1, 2, and 3. The solution set is $\{1, 2, 3\}$.

The solutions for the equations and inequalities in this section are found by inspection. This means that you try all possible numbers from the given domain, and see which ones work.

It is often helpful to draw a diagram that shows the solution of an equation or inequality. To do this we begin with a **number line**. First draw a straight line. Choose any point on the line, and label it 0. (See Figure 1.1.) Then choose another point to the right of 0, not too far from it. Label your second point 1.

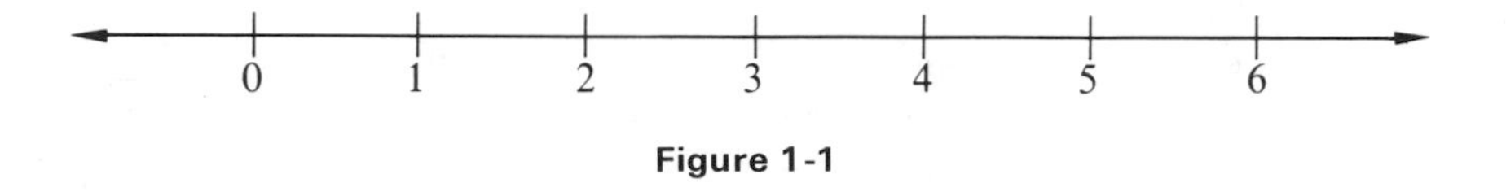

Figure 1-1

The distance between 0 and 1 sets up a unit measure that you use to locate more points, labeled 2, 3, 4, 5, 6, and so on. Other points could be located by dividing the units into halves, thirds, and so on. Each number is called the **coordinate** of the point it labels. For example, 2 is the coordinate of the point labeled A in Figure 1.2.

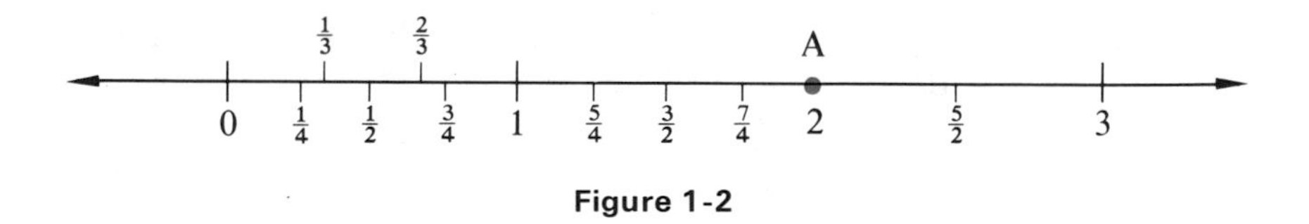

Figure 1-2

In Example 4, we found that if the set $\{1, 2, 3, 4, 5\}$ is the domain for the inequality $2x + 1 < 9$, then the solutions from the domain are the numbers 1, 2, and 3. To **graph** these solutions, draw dots at the points labeled 1, 2, and 3. See Figure 1.3.

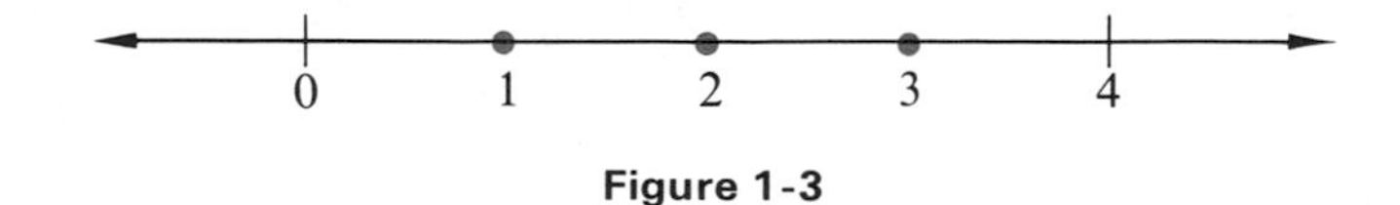

Figure 1-3

Example 5 Draw a graph of the solutions of $5r - 2 \geq 18$. The domain of the variable is the set $\{2, 3, 4, 5, 6, 7\}$.

By trying each of the numbers in the given domain, you should find that the solutions of the given inequality are the numbers 4, 5, 6, and 7. The graph of these solutions is shown in Figure 1.4.

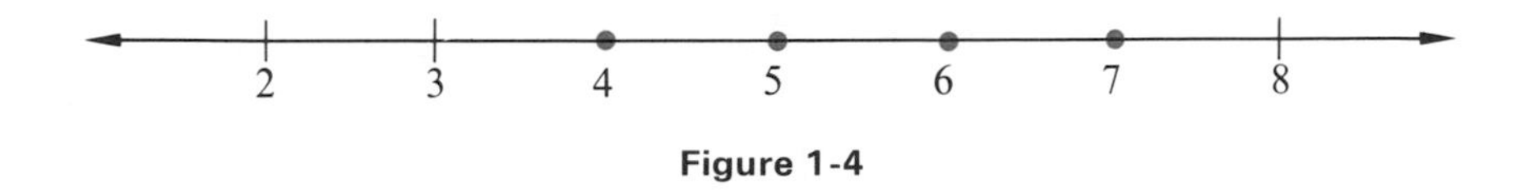

Figure 1-4

Example 6 Graph the solutions of the equation $2p + 1 = 19$. The domain of the variable is $\{7, 8, 9, 10, 11\}$.

By trying each number in the domain, you should find that 9 is the only solution. The graph therefore includes only one point, as in Figure 1.5.

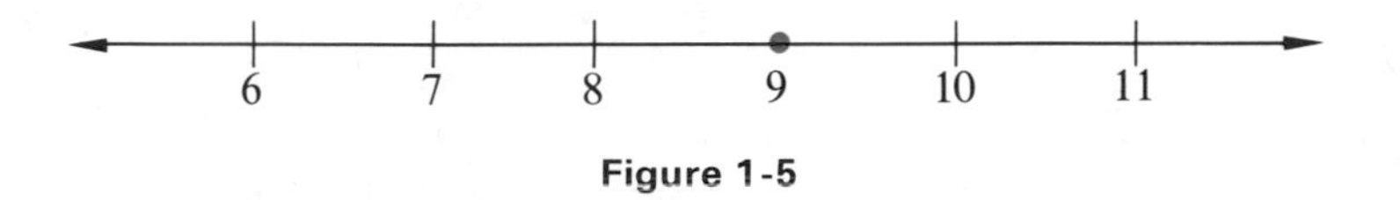

Figure 1-5

Example 7 Graph the solutions of the equation $x + 1 = 10$. The domain of the variable is $\{0, 2, 4, 6, 8, 10\}$.

None of the numbers in the domain make the given equation true. (The only solution to the given equation is 9, which is not in the domain.) Thus, we must say that there are no solutions *from the given domain*, or that the solution set contains no numbers. Any set containing no numbers or other elements is called the **empty set**, or **null set**. The symbol for the empty set is $\varnothing$. The graph of the solution here is a number line with no points marked on it, as in Figure 1.6.

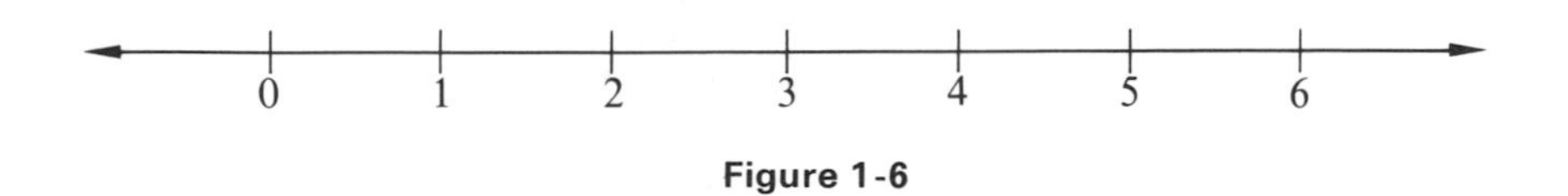

Figure 1-6

EXERCISES

In Exercises 1–6, find the numerical value of the given expression when $x = 3$.

1. $x + 9$

2. $x - 2$

3. $5x$

4. $7x$

5. $2x + 8$

6. $9x - 5$

In Exercises 7–18, find the numerical value of the given expression when $x = 4$ and $y = 2$.

7. $x + y$
8. $x - y$
9. $8x + 3y + 5$
10. $4x + 2y + 2$
11. $3(x + 2y)$
12. $2(2x - y)$
13. $5(4x - 7y)$
14. $8(5x - 9y)$
15. $\frac{2x + 3y}{x + y + 1}$
16. $\frac{5x - 3y + 1}{2x}$
17. $\frac{2x + 4y - 6}{5y + 2}$
18. $\frac{4x + 3y - 1}{x}$

In Exercises 19–28 an equation is given, along with a number. Decide whether or not the number is a solution for the given equation.

19. $x + 6 = 15$; 9
20. $p - 5 = 12$; 17
21. $3r + 5 = 8$; 2
22. $5m + 2 = 7$; 2
23. $6a + 2(a + 3) = 14$; 1
24. $2y + 3(y - 2) = 14$; 4
25. $2x + 3x + 8 = 38$; 6
26. $6x + 4x - 9 = 11$; 2
27. $2 + 3y + 4y = 20$; 3
28. $6 + 8r - 3r = 11$; 5

In Exercises 29–48, an equation or inequality is given, along with a domain. Write the solutions. Graph the solutions on a number line.

29. $2 + x = 4$; $\{2, 4, 6\}$
30. $3 + p = 6$; $\{1, 2, 3\}$
31. $y + 8 = 12$; $\{1, 3, 4, 5\}$
32. $r + 2 = 10$; $\{7, 8, 9\}$
33. $2k = 12$; $\{5, 6, 7\}$
34. $3m = 15$; $\{4, 5, 6\}$
35. $4a + 3 = 19$; $\{3, 4, 5\}$
36. $9z + 1 = 19$; $\{2, 3, 4\}$
37. $11k - 5 = 17$; $\{4, 5, 6\}$
38. $7s - 2 = 12$; $\{2, 3, 4\}$
39. $y + 1 \le 6$; $\{3, 4, 5, 6\}$
40. $k - 3 \ge 5$; $\{6, 7, 8, 9, 10\}$
41. $3r - 2 \le 7$; $\{1, 2, 3, 4, 5\}$
42. $5m - 1 \le 9$; $\{1, 2, 3, 4\}$
43. $2z + 3 < 12$; $\{2, 3, 4, 5\}$
44. $4k - 5 > 8$; $\{2, 3, 4, 5, 6\}$
45. $4r - 3 \ge 13$; $\{0, 1, 2, 3\}$
46. $5p + 1 \le 14$; $\{3, 4, 5, 6\}$
47. $2x + 1 \ne 9$; $\{2, 3, 4, 5, 6\}$
48. $5x - 2 \ne 13$; $\{0, 1, 2, 3, 4, 5, 6\}$

In Exercises 49–58, write each word phrase using mathematical symbols. Use x to represent the variable.

Example "A number plus 18" would be written $x + 18$. "Four times a number, subtracted from 19" is $19 - 4x$.

49. Twice a number
50. Four added to a number

51. 6 added to a number

52. Five times a number

53. A number subtracted from 8

54. 9 subtracted from a number

55. 8 added to three times a number

56. 6 subtracted from 5 times a number

57. Twice a number, subtracted from 15

58. 8 times a number, added to 52

In Exercises 59–68, write the given word statement in mathematical symbols. Use x as the variable. Find the solution for the statement if the domain is $\{0, 2, 4, 6, 8, 10\}$.

Example The sum of a number and four is six.

First write this sentence in symbols as $x + 4 = 6$. Then find the solution from the domain. The solution is the number 2.

59. The sum of a number and 8 is 12.

60. A number minus 3 equals 7.

61. The sum of a number and 2 is less than 11.

62. The sum of twice a number and 6 is less than 10.

63. Five more than twice a number is less than ten.

64. The sum of a number and 2, divided by 4, is 3.

65. The product of a number and three is greater than eight.

66. Three times a number is equal to one more than twice the number.

67. Twelve divided by a number equals three times that number.

68. Six times a number, minus three, is greater than twenty-four.

1.3 NEGATIVE NUMBERS

Set up a number line, and let the distance between 0 and 1 be the unit measure. This time go to the left of zero. Mark off points one unit to the *left* of 0, two units to the *left*, three units to the *left*, and so on. (See Figure 1.7.) The points to the left of zero are labeled with a minus sign:

$$\ldots, -4, -3, -2, -1.$$

The three dots indicate that the numbers continue without end to the left.

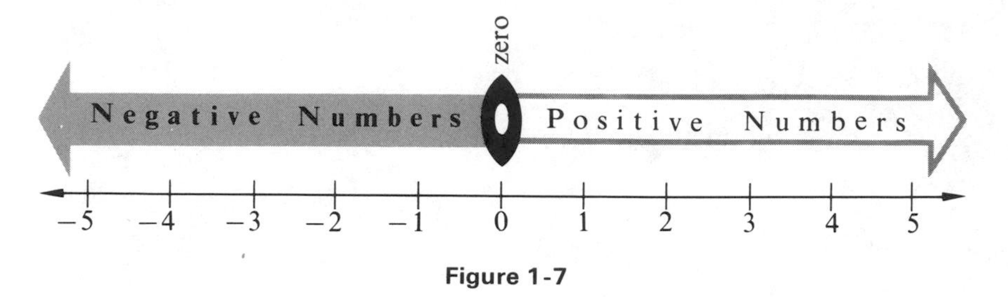

Figure 1-7

The coordinates of points to the left of 0 on the number line are called **negative numbers**. To the right of 0, the coordinates of points are called **positive numbers**. The number 0 itself is neither positive nor negative.

There are many examples of negative numbers used in practical applications. For example, the altitude of Badwater in Death Valley, California, is -282 feet (282 feet below sea level, which is taken to be 0). Some land in certain river deltas is at an elevation of -12 feet when compared to the river—the land is 12 feet lower than the river, which is held in levees. A temperature on a cold January day can be written as -10°, or 10 degrees below 0. A business which spends more than it takes in has a negative profit.

The set of numbers which corresponds to *all* the points on the number line is the set of **real numbers**. The set of real numbers is made up of negative numbers, positive numbers, and zero. The set of real numbers includes all fractions, decimals, and square roots, and many more numbers besides. One of the most common sets of numbers within the real numbers is the set of **whole numbers**

$$\{0, 1, 2, 3, 4, 5, 6, \ldots\} \qquad \text{Whole numbers}$$

The three dots show that the whole numbers continue in the same way.

The real numbers have *properties*, that is, features that describe the set. One property is this: If we select any real number x (except 0), we can find exactly one number on the number line the same distance from 0 as x, but on the opposite side of 0.

For example, Figure 1.8 shows that the numbers 3 and -3 are each the same distance from 0, but are on opposite sides of 0. The numbers 3 and -3 are **additive inverses**, or **negatives**, of each other.

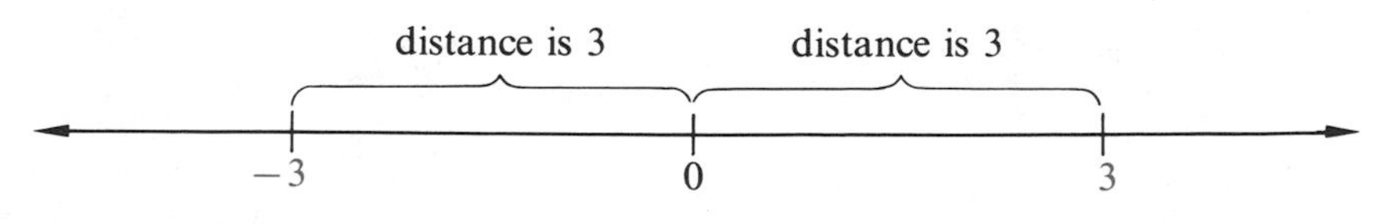

Figure 1-8

In general, the additive inverse of a number x is that number which is the same distance from 0 on the number line as x, but on the opposite side of 0.

The additive inverse of the number 0 is 0 itself. This makes 0 the only real number that is its own additive inverse.

Except for the number 0, additive inverses occur in pairs. For example, 4 and -4, 3 and -3, and 5 and -5 are additive inverses of each other. Several pairs of additive inverses are shown in Figure 1.9.

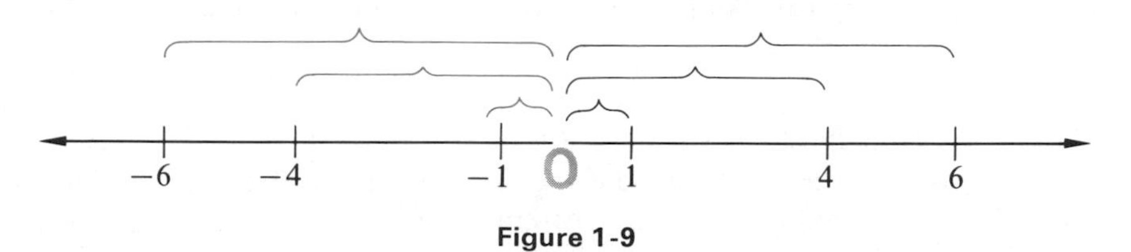

Figure 1-9

The additive inverse of a number can be indicated by writing the symbol $-$ in front of the number. With this symbol, the additive inverse of 7 is written -7. The additive inverse of -3 can be written $-(-3)$. We know that 3 is an additive inverse of -3. Since a number can have only one additive inverse, the symbols 3 and $-(-3)$ must really represent the same number, which means that

$$-(-3) = 3.$$

In general, for any real number x, it is true that $-(-x) = x$.

With the symbol $-$, the additive inverse of 0 is written -0. Since 0 is its own additive inverse, then $0 = -0$.

Example 1

Number	Additive inverse
-3	3
-4	$-(-4)$, or 4
0	-0, or 0
-2	$-(-2)$, or 2
5	-5
19	-19

We have defined the set of whole numbers:

$$\{0, 1, 2, 3, 4, 5, 6, \ldots\}.$$

If we include with these numbers the additive inverses (or negatives) of the whole numbers, we get the set of **integers.**

$$\{\ldots, -4, -3, -2, -1, 0, 1, 2, 3, 4, \ldots\} \qquad \text{Integers}$$

If you are given any two whole numbers, you probably can tell which number is the smaller of the two. What happens when we look at negative numbers, such as those in the set of integers? Positive numbers increase as the corresponding points on the number line go to the right. For example, $8 < 12$, and 8 is to the left of 12 on the number line. We can extend this idea to all real numbers.

> The smaller of any two different real numbers is the one farther to the left on the number line.

Then any negative number is smaller than 0, and any negative number is smaller than any positive number. Also, 0 is smaller than any positive number.

Example 2 Is it true that $-3 < -1$?

To decide whether the statement $-3 < -1$ is true, locate both numbers -3 and -1 on a number line, as in Figure 1.10. Since -3 is to the left of -1 on the number line, -3 is smaller than -1. The statement $-3 < -1$ is true.

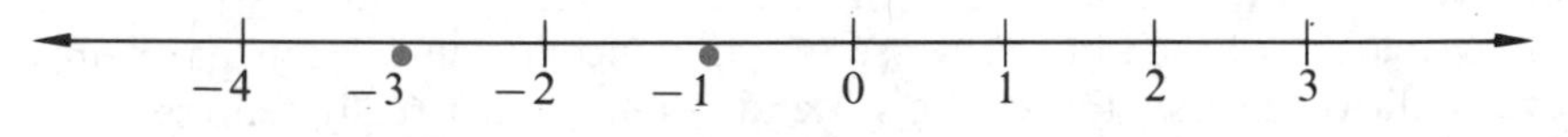

Figure 1-10 -3 is to the left of -1, so that $-3 < -1$

Two numbers, each of which is the additive inverse of the other, are the same distance from 0 on the number line. For example, 2 and -2 are additive inverses of each other, and 2 and -2 are each the same distance from 0 on the number line.

To express this, we say that 2 and -2 have the same **absolute value.** The absolute value of a number is defined as the number of units of distance from the number to 0 on the number line.

The symbol for the absolute value of the number x is $|x|$, read "the absolute value of x." For example, the distance from 2 to 0 on the number line is 2 units, so that

$$|2| = 2.$$

Also, the distance from -2 to 0 on the number line is 2 units, so that

$$|-2| = 2.$$

Since distance is a physical measurement which is never negative, the absolute value of a number can never be negative. For example, $|12| = 12$ and $|-12| = 12$, since both 12 and -12 lie 12 units from 0 on the number line. Since 0 is 0 units from 0, we have $|0| = 0$.

We formally define absolute value as follows:

$$|x| = x \text{ if } x \geq 0, \qquad \text{and} \qquad |x| = -x \text{ if } x < 0.$$

The second part of this definition is a little tricky. Since x is just a variable, x can be either positive or negative, and we do not know which. In the second part of the definition we say that $|x| = -x$ if $x < 0$. If $x < 0$, then x is already negative by definition, so that $-x$ will be *positive.*

Figure 1.11 shows a negative number x and its additive inverse, $-x$, which is a positive number.

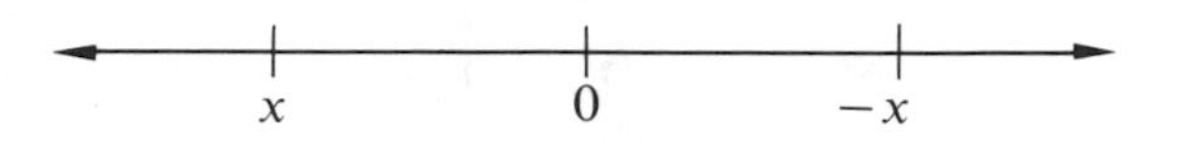

Figure 1-11 x is negative, and $-x$ is positive

Example 3 **(a)** $|7| = 7$
(b) $|-7| = 7$
(c) $|-5| = 5$
(d) $-|5| = -(5) = -5$
(e) $-|-14| = -(14) = -14$
(f) $-|-8| = -(8) = -8$

Example 4 Find the solutions of each of the following equations or inequalities. The domain of the variable is the set

$$\{-4, -3, -2, -1, 0, 1, 2, 3, 4\}.$$

(a) $|x| = 2$

The equation $|x| = 2$ is true if x is replaced with 2, since $|2| = 2$. The equation is also true if x is replaced with -2, since $|-2| = 2$. Therefore, the numbers 2 and -2 are the solutions of the equation $|x| = 2$.

(b) $|x| \leq 2$

By substitution from the domain above, check that -2, -1, 0, 1, and 2 are the solutions of this inequality.

(c) $|x| \geq 3$

This inequality is true if x takes on any of the values -4, -3, 3, or 4.

EXERCISES

For Exercises 1–12, give the additive inverse of each number. Identify any answers which turn out to be negative numbers.

1. 8
2. 12
3. -9
4. -11
5. -2
6. -3
7. $|15|$
8. $|3|$
9. $|-5|$
10. $|-2|$
11. $|-4|$
12. $|-8|$

In Exercises 13–24, select the smaller of the two given numbers.

13. $-5, \quad 5$

14. $9, \quad -3$

15. $-12, \quad -4$

16. $-9, \quad -14$

17. $-8, \quad -1$

18. $-15, \quad -16$

19. $3, \quad |-4|$

20. $5, \quad |-2|$

21. $|-3|, \quad |-4|$

22. $|-8|, \quad |-9|$

23. $-|-6|, \quad -|-4|$

24. $-|-2|, \quad -|-3|$

In Exercises 25–42, write *true* or *false* for each statement.

25. $-2 < -1$

26. $-8 < -4$

27. $-3 \geq -7$

28. $-9 \geq -12$

29. $-15 \leq -20$

30. $-21 \leq -27$

31. $-8 \leq -(-4)$

32. $-9 \leq -(-6)$

33. $8 \leq -(-9)$

34. $6 > -(-2)$

35. $-8 > -(-2)$

36. $-4 < -(-5)$

37. $|-6| < |-9|$

38. $|-12| < |-20|$

39. $-|8| > |-9|$

40. $-|12| > |-15|$

41. $-|-5| \geq -|-9|$

42. $-|-12| \leq -|-15|$

In Exercises 43–48, graph each group of numbers on a separate number line.

43. $0, 3, -5, -6$

44. $2, 6, -2, -1$

45. $-2, -6, |-4|, 3, -|4|$

46. $-5, -3, -|-2|, -0, |-4|$

47. $|3|, -|3|, -|-4|, -|-2|$

48. $|6|, -|6|, -|-8|, -|-3|$

In Exercises 49–66, solve each equation or inequality. Assume that the domain of the variable is

$$\{-4, -3, -2, -1, 0, 1, 2, 3, 4\}.$$

49. $|x| = 3$

50. $|x| = 4$

51. $|x| < 2$

52. $|x| \leq 4$

53. $|x| \leq 1$

54. $|x| < 3$

55. $|x| > 1$

56. $|x| > 2$

57. $|x| < 1$

58. $|x| \geq 4$

59. $|x| \geq 0$

60. $|x| \leq 0$

61. $|x| = -2$

62. $|x| = -4$

63. $|x| > -2$

64. $|x| > -4$

65. $|x| \leq -1$

66. $|x| \leq -3$

1.4 ADDITION OF REAL NUMBERS

We use the number line to extend addition of whole numbers to addition of real numbers, which include both negatives and positives. Consider the sum $2 + 3 = 5$. To work this sum on the number line, first draw an arrow starting at 0 and going two units to the right. This arrow represents the number 2 in the sum $2 + 3$. Then, from the right end of this arrow, draw another arrow of length 3 units, as shown in Figure 1.12. The coordinate at the end of this second arrow is 5, the sum $2 + 3$.

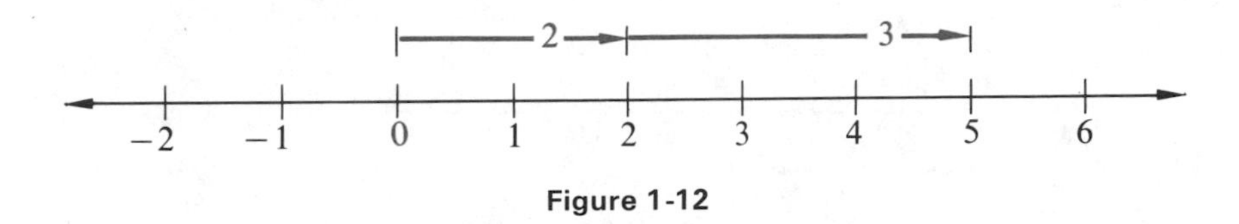

Figure 1-12

To find the sum of two positive numbers, x and y, on the number line, start at 0 and draw an arrow of length x to the right. From the end of that arrow draw another arrow of length y to the right. The coordinate at the end of this second arrow gives the sum $x + y$.

We can find the sum of two negative numbers, such as $-2 + (-4)$, on the number line in much the same way. Here we placed parentheses around the -4 in the sum $-2 + (-4)$ to avoid the confusing use of $+$ and $-$ next to each other. To add the negative numbers -2 and -4, start at 0 and draw an arrow two units to the *left*. Draw the arrow to the left to represent the addition of a *negative* number. From the end of this first arrow, draw a second arrow four units to the left, as shown in Figure 1.13. The coordinate at the end of this second arrow gives the sum $-2 + (-4) = -6$.

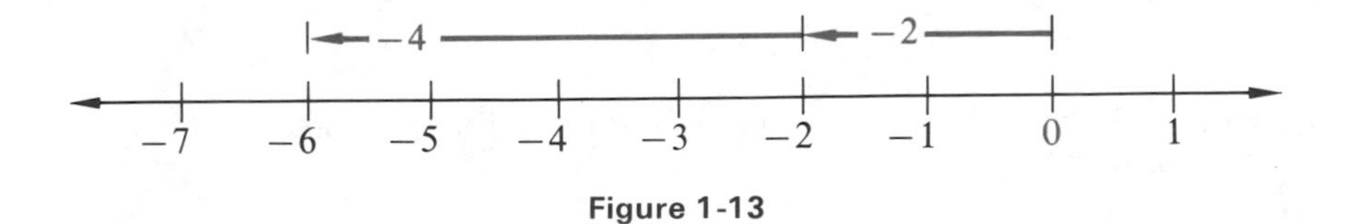

Figure 1-13

The sum of two negative numbers x and y is a negative number whose distance from 0 is the sum of the distance of x from 0 and the distance of y from 0. That is, the sum of two negative numbers is the negative of the sum of their absolute values.

If both x and y are negative numbers, then

$$x + y = -(|x| + |y|).$$

Example 1 **(a)** $-2 + (-9) = -11$.
(b) $-8 + (-12) = -20$.
(c) $-15 + (-3) = -18$.

What about the sum of a positive number and a negative number? For example, what is the sum $-2 + 5$? To find the sum $-2 + 5$ on the number line, start at 0, and draw an arrow two units to the left. See Figure 1.14. From the end of this arrow, shift direction, and draw a second arrow five units to the right. The coordinate at the end of the second arrow gives the sum $-2 + 5 = 3$.

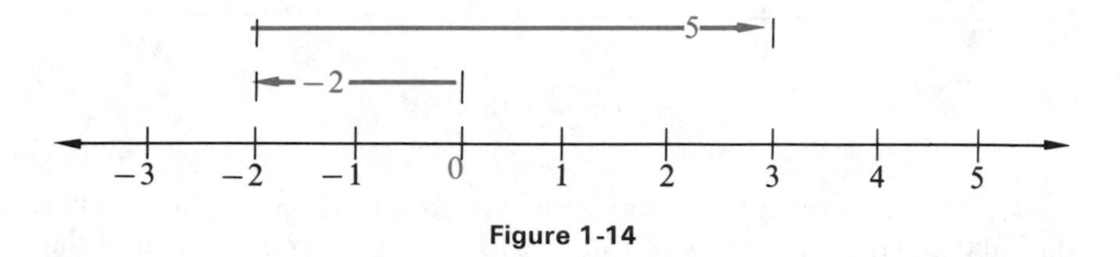

Figure 1-14

To find the sum of 4 and -6, draw arrows as in Figure 1.15. The coordinate at the end of the second arrow gives the sum $4 + (-6) = -2$.

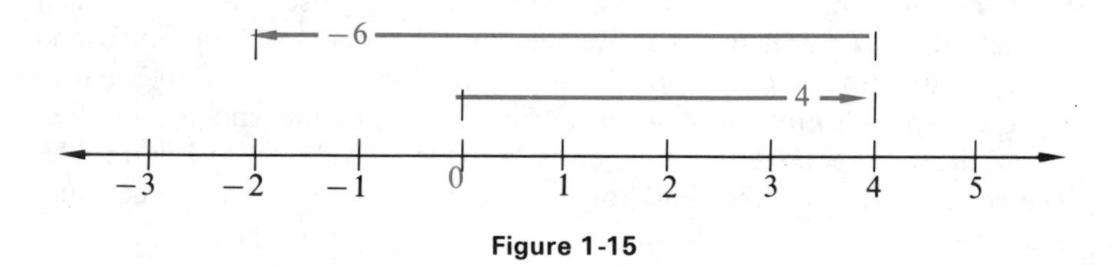

Figure 1-15

Example 2 Use the number line to find the following sums:
(a) $6 + (-3)$

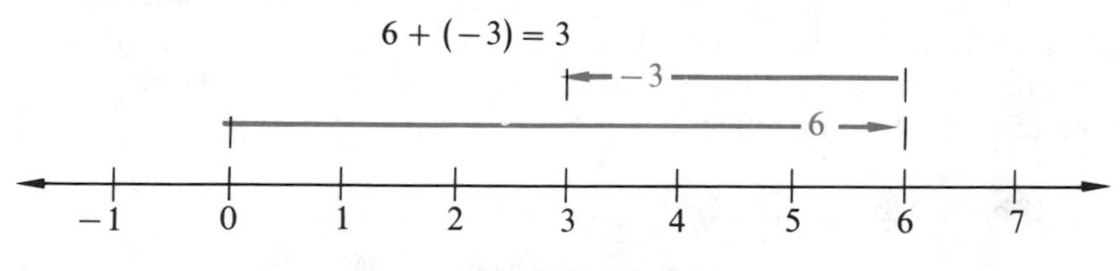

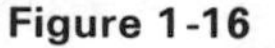

Figure 1-16

(b) $(-7) + 2$

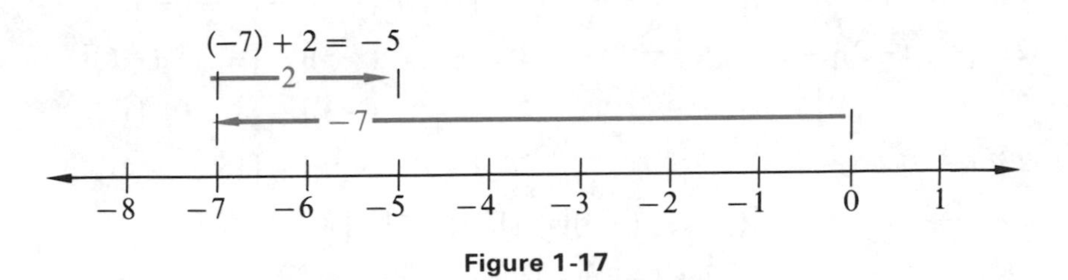

Figure 1-17

Example 3 Add each of the following. Try to work the addition in your head. If you get stuck, work it on a number line.

(a) $7 + (-4)$
(b) $-8 + 12$
(c) $-11 + 15$
(d) $-8 + 2$
(e) $-15 + 4$

Answers: (a) 3 (b) 4 (c) 4 (d) -6 (e) -11

Example 4 To work the problem

$$-3 + [4 + (-8)],$$

first work inside the brackets. Follow the rules for the order of operations from the first section of this chapter.

$$\begin{aligned} -3 + [4 + (-8)] &= -3 + (-4) \\ &= -7. \end{aligned}$$

Example 5 $8 + [(-2 + 6) + (-3)] = 8 + [4 + (-3)]$

$$\begin{aligned} &= 8 + 1 \\ &= 9. \end{aligned}$$

EXERCISES

In Exercises 1–30, find each sum.

1. $5 + (-3)$

2. $11 + (-8)$

3. $6 + (-8)$

4. $3 + (-7)$

5. $-6 + (-2)$

6. $-8 + (-3)$

7. $-9 + (-2)$

8. $-15 + (-6)$

9. $-3 + (-9)$

10. $-11 + (-5)$

11. $12 + (-8)$

12. $10 + (-2)$

13. $4 + [13 + (-5)]$

14. $6 + [2 + (-13)]$

15. $8 + [-2 + (-1)]$

16. $12 + [-3 + (-4)]$

17. $-2 + [5 + (-1)]$

18. $-8 + [9 + (-2)]$

19. $-6 + [6 + (-9)]$

20. $-3 + [4 + (-8)]$

21. $[9 + (-2)] + 6$

22. $[8 + (-14)] + 10$

23. $[(-9) + (-14)] + 12$

24. $[(-8) + (-6)] + 10$

25. $[-3 + (-4)] + [5 + (-6)]$

26. $[-8 + (-3)] + [-7 + (-6)]$

27. $[-4 + (-3)] + [8 + (-1)]$

28. $[-5 + (-9)] + [16 + (-2)]$

29. $[-4 + (-6)] + [(-3) + (-8)] + [12 + (-11)]$

30. $[-2 + (-11)] + [12 + (-2)] + [18 + (-6)]$

In Exercises 31–44, write *true* or *false* for each statement.

31. $-4 + 0 = -4$

32. $-6 + 5 = -1$

33. $-8 + 12 = 8 + (-12)$

34. $15 + (-8) = 8 + (-15)$

35. $-(4 - 2) = -2$

36. $-(6 - 3) = 3$

37. $-3 + 5 = 5 + (-3)$

38. $11 + (-6) = -6 + 11$

39. $[4 + (-6)] + 6 = 4 + (-6 + 6)$

40. $[(-2) + (-3)] + (-6) = 12 + (-1)$

41. $-7 + [-5 + (-3)] = [(-7) + (-5)] + 3$

42. $6 + [-2 + (-5)] = [(-4) + (-2)] + 5$

43. $-5 + (-|-5|) = -10$

44. $|-3| + (-5) = -2$

In Exercises 45–54, find all solutions for each equation. Let the domain of x be the set $\{-3, -2, -1, 0, 1, 2, 3\}$.

45. $x + 2 = 0$

46. $x + 3 = 0$

47. $x + 1 = -2$

48. $x + 2 = -1$

49. $14 + x = 12$

50. $8 + x = 7$

51. $x + (-4) = -6$

52. $x + (-2) = -5$

53. $-8 + x = -6$

54. $-2 + x = -1$

In Exercises 55–60, find the answer to each word problem.

55. Joann has \$15. She then spends \$6. What is her balance at that time?

56. An airplane is at an altitude of 6000 feet. It then descends 4000 feet. What is the final altitude?

57. Chuck is standing 15 feet below sea level in Death Valley. He then goes down another 120 feet. What is his final altitude?

58. Hiram has \$11, and spends \$19. What is his final balance?

59. One measure of Frieda's blood pressure is 120. It then changes by -30. Find the new blood pressure.

60. The temperature is $-14°$. It then changes by $-12°$. Find the new temperature.

1.5 SUBTRACTION OF REAL NUMBERS

We already know how to subtract a positive number from a larger positive number, such as the subtraction $7 - 4$. Let us draw a number line and see how we could find the answer to this subtraction problem. Since addition of a positive number on the number line is shown by an arrow to the right, it is reasonable to represent subtraction of a positive number by an arrow going to the left.

To find the answer to the subtraction problem $7 - 4$, begin at 0 and draw an arrow 7 units to the right. See Figure 1.18. From the end of this arrow, draw an arrow 4 units to the left. The result is the number 3, so that $7 - 4 = 3$.

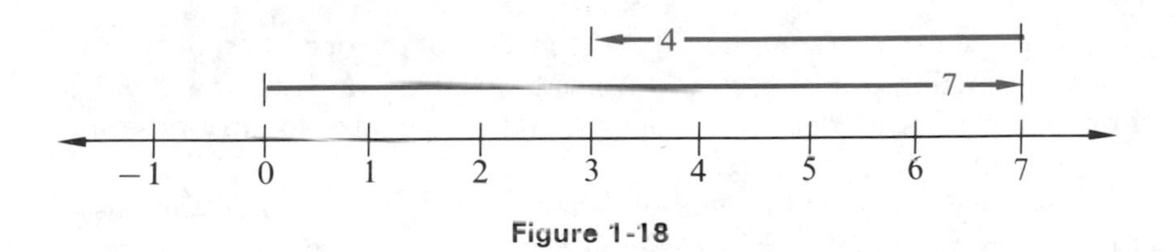

Figure 1-18

The procedure used above to find $7 - 4$ is exactly the same procedure that would be used to find $7 + (-4)$, so that

$$7 - 4 = 7 + (-4).$$

Example 1 To find $5 - 1$ on the number line, draw arrows as in Figure 1.19. From these arrows, we see that $5 - 1 = 4$. The arrows in Figure 1.19 are exactly the same that would be drawn to get the sum $5 + (-1)$. Therefore,

$$5 - 1 = 5 + (-1).$$

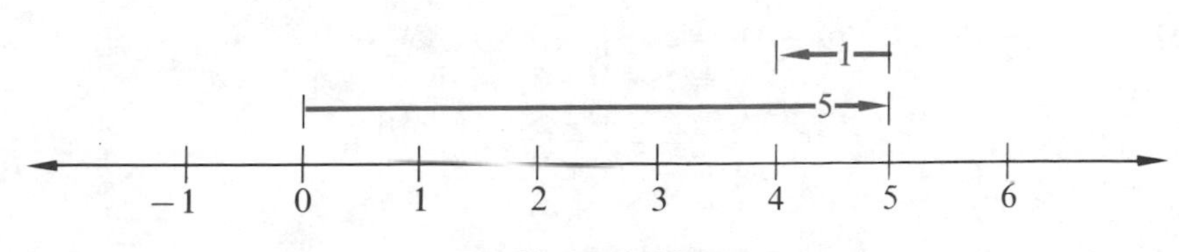

Figure 1-19

Based on these examples, as well as others that we could work, it seems that subtraction of one positive number from a larger positive number is the same as adding the inverse of the smaller to the larger. We extend this definition to all real numbers.

For any two real numbers x and y,

$$x - y = x + (-y).$$

(To subtract y from x, *add* the additive inverse of y to x.)

Example 2 **(a)** $12 - 3 = 12 + (-3) = 9$
(b) $5 - 7 = 5 + (-7) = -2$
(c) $8 - 15 = 8 + (-15) = -7$
(d) $-3 - (-5) = -3 + (+5) = 2$
(e) $-6 - (-9) = -6 + (+9) = 3$
(f) $8 - (-5) = 8 + (+5) = 13$

The operation of subtraction is the inverse of the operation of addition since subtraction can be used to reverse the result of an addition problem. For example, if 4 is added to a number, and then subtracted from the sum, the original number is the result.

We have now used the symbol $-$ for a variety of purposes. One use is to represent subtraction, as in $9 - 5$. The symbol $-$ is also used to represent negative numbers, such as -10, -2, -3, and so on. Finally, the symbol $-$ is used to represent the additive inverse of a number. More than one use may appear in the same problem, such as $-6 - (-9)$, where -9 is subtracted from -6. The meaning of the symbol is usually clear from the context.

Example 3 When working problems involving parentheses and brackets, first do any operations inside the parentheses or brackets, from the inside out.

(a)
$$\begin{aligned} -6 - [2 - (8 + 3)] &= -6 - [2 - 11] \\ &= -6 - (-9) \\ &= -6 + (+9) \\ &= 3. \end{aligned}$$

(b)
$$\begin{aligned} 5 - [(-3 - 2) - (4 - 1)] &= 5 - [(-3 + (-2)) - 3] \\ &= 5 - [(-5) - 3] \\ &= 5 - [(-5) + (-3)] \\ &= 5 - (-8) \\ &= 5 + (+8) \\ &= 13. \end{aligned}$$

EXERCISES

In Exercises 1–20, work each subtraction problem.

1. $3 - 6$

2. $5 - 9$

3. $11 - 12$

4. $8 - 13$

5. $-6 - 2$

6. $-11 - 4$

7. $-9 - 5$

8. $-12 - 15$

9. $6 - (-3)$

10. $8 - (-5)$

11. $5 - (-12)$

12. $12 - (-2)$

13. $-6 - (-2)$

14. $-7 - (-5)$

15. $-8 - (-12)$

16. $-10 - (-15)$

17. $2 - (3 - 5)$

18. $5 - (6 - 13)$

19. $-2 - (5 - 8)$

20. $-3 - (4 - 11)$

In Exercises 21–40, perform the indicated operations in each exercise.

21. $(4 - 6) + 12$

22. $(3 - 7) + 4$

23. $(8 - 1) - 12$

24. $(9 - 3) - 15$

25. $6 - (-8 + 3)$

26. $8 - (-9 + 5)$

27. $-6 - (-4 + 3)$

28. $-12 - (-9 + 12)$

29. $2 + (-4 - 8)$

30. $6 + (-9 - 2)$

31. $(-5 + 6) + (9 - 2)$

32. $(-4 - 8) - (6 - 1)$

33. $(-6 + 9) - (3 + 5)$

34. $(-2 + 11) - (4 + 1)$

35. $(-8 - 2) - (-9 - 3)$

36. $(-4 - 2) - (-8 - 1)$

37. $-9 - [(3 - 2) - (-4 - 2)]$

38. $-8 - [(-4 - 1) - (9 - 2)]$

39. $-3 + [(-5 - 8) - (-6 + 2)]$

40. $-4 + [(-12 + 1) - (-1 - 9)]$

In Exercises 41–50, write the given problem in symbols (no variables are needed). Then solve the problem.

41. Subtract -6 from 12.

42. Subtract -8 from 15.

43. From -25, subtract -4.

44. What number is 6 less than -9?

45. The number -24 is how much greater than -27?

46. How much greater is 8 than -5?

47. The temperature dropped 10° below the previous temperature of $-5°$. Find the new temperature.

48. Bill owed his brother \$10. He repaid \$6, and later borrowed \$7. What positive or negative number represents his present financial status?

49. The bottom of Death Valley is 282 feet below sea level. The top of Mt. Whitney, visible from Death Valley, and in the same county, has an altitude of 14,494 feet above sea level. Using sea level as zero, find the difference in elevations.

50. Harriet has \$15, while Thomasina is \$12 in debt. Find the difference of their financial positions.

1.6 PROPERTIES OF ADDITION

The first property of addition that we consider is perhaps the most basic of all the properties about addition. Called the **closure property of addition,** it says that the sum of any two real numbers is a real number.

If x and y are any two real numbers, then their sum, $x + y$, is a real number.

Although the closure property may seem obvious, it is not true for all operations or for all sets of numbers.

For example, the operation of subtraction using only the positive numbers does not always produce a number which is positive. While the numbers 4 and 6 are both positive, the difference, $4 - 6$, is not positive. Because of this, subtraction with the positive numbers does not have the closure property. Another way of saying this is to say that the set of positive numbers is not *closed* under subtraction. In the example $4 - 6$, it is necessary to go outside the set of positive numbers for the answer, -2.

We know that adding 8 to 5 gives the same answer as adding 5 to 8.

$$8 + 5 = 5 + 8.$$

The same is true when adding two negative numbers, or a positive and a negative number.

$$-2 + (-5) = -5 + (-2) \qquad \text{and} \qquad 9 + (-2) = -2 + 9.$$

Reversing the order of the two numbers to be added in a sum produces no change in the result. This fact is known as the **commutative property of addition.**

If x and y are any two real numbers, then

$$x + y = y + x.$$

Starting with the sum $6 + (-3) + 5$, we could add in either of two ways.

First add 6 and -3, then add 5 to the sum.	First add -3 and 5, then add 6 to the sum.
$6 + (-3) + 5$	$6 + (-3) + 5$
$3 + 5$	$6 + 2$
8	8

The results are the same. To express the two ways in which we added, brackets can be used. Therefore,

$$[6 + (-3)] + 5 = 6 + [-3 + 5].$$

In the same way,

$$5 + (8 + 2) = (5 + 8) + 2.$$

$$[-4 + (-6)] + (-2) = -4 + [-6 + (-2)].$$

These examples suggest that in finding the sum of three numbers, the order in which we group the numbers in pairs does not change the result. This property of real numbers is called the **associative property of addition.**

If x, y, and z are any real numbers, then

$$x + (y + z) = (x + y) + z.$$

By the associative property, the sum of three numbers will be the same, however we "associate" the numbers in groups. For this reason, parentheses can be left out in many addition problems. For example, we can write

$$2 + 3 + (-1)$$

instead of

$$(2 + 3) + (-1) \qquad \text{or} \qquad 2 + [3 + (-1)].$$

Example 1 Is $(2 + 4) + 5 = 2 + (4 + 5)$ an example of the associative property?

The order of the three numbers is the same on both sides of the equals sign. The only change is in the grouping of the numbers. On the left, $2 + 4$ is grouped. On the right, $4 + 5$ is grouped. This example, therefore, illustrates the associative property.

Example 2 Is $6 + (3 + 10) = 6 + (10 + 3)$ an example of the associative property or the commutative property?

Here the same numbers, 3 and 10, are grouped. However, on the left the 3 appears first $(3 + 10)$. On the right, the 10 appears first. Since the only change involves the order of the numbers, this is an example of the commutative property.

Example 3 Is $(8 + 1) + 7 = 8 + (7 + 1)$ an example of the associative property or the commutative property?

In the statement both the order and the grouping are changed. On the left the order of the three numbers is 8, 1, and 7. On the right it is 8, 7, and 1. On the left the 8 and 1 are grouped, while on the right the 7 and 1 are grouped. Therefore, in this example both the associative and the commutative properties are used.

The sum of 0 and any real number equals that real number. For example, $8 + 0 = 8$, $0 + (-9) = -9$, and so on. The number 0 leaves the identity of any real number unchanged by addition. For this reason, 0 is called the **identity element** for addition. This special property of 0 is called the **identity property of addition.**

For any real number x,

$$x + 0 = x \quad \text{and} \quad 0 + x = x.$$

By the discussion of the previous section, $4 + (-4) = 0$ and $-6 + 6 = 0$. In general, $x + (-x) = 0$ and $-x + x = 0$ for any real number x. The pairs of numbers 4 and -4, and -6 and 6, are additive inverses. The fact that every real number has an additive inverse and that the sum of the number and its additive inverse is 0 is the **additive inverse property**.

For every real number x, there is exactly one real number $-x$ such that

$$x + (-x) = 0 \quad \text{and} \quad -x + x = 0.$$

Summary Properties of the real numbers for addition

For any real numbers x, y, and z:

$x + y$ is a real number	Closure property
$x + y = y + x$	Commutative property
$(x + y) + z = x + (y + z)$	Associative property
$x + 0 = x$ and $0 + x = x$	Identity property
$x + (-x) = 0$ and $-x + x = 0$	Inverse property

EXERCISES

In Exercises 1–20, label each statement as an example of the commutative, associative, closure, identity, or inverse property.

1. $6 + 15 = 15 + 6$

2. $9 + (11 + 4) = (9 + 11) + 4$

3. $5 + (15 + 8) = (5 + 15) + 8$

4. $23 + 9 = 9 + 23$

5. $12 + (-8) = -8 + 12$

6. $-9 + [6 + (-2)] = (-9 + 6) + (-2)$

7. $-8 + [-4 + (-3)] = [-8 + (-4)] + (-3)$

8. $-4 + (-12) = -12 + (-4)$

9. $-6 + 12$ is a real number

10. $-9 + (-11)$ is a real number

11. $6 + (-6) = 0$

12. $-8 + 8 = 0$

13. $-4 + 0 = -4$

14. $0 + (-9) = -9$

15. $(m + n) + 4 = (n + m) + 4$

16. $2 + (p + r) = (p + r) + 2$

17. $k + [-2 + (-r)] = [k + (-2)] + (-r)$

18. $[y + (-8)] + 4 = y + (-8 + 4)$

19. $a + (-a) + k = 0 + k$

20. $m + y + (-y) = m + 0$

In Exercises 21–30, solve the given equation. Let the domain of the variable be the set $\{0, 2, 4, 6, 8, 10\}$. Use the properties to help you find the solution.

21. $4 + 8 = x + 4$

22. $6 + x = 8 + 6$

23. $y + (-2) = 0$

24. $-10 + k = 0$

25. $x + (2 + 5) = (4 + 2) + 5$

26. $8 + (x + 3) = (8 + 2) + 3$

27. $(5 + x) + 4 = 4 + (5 + 6)$

28. $3 + (9 + 10) = (9 + x) + 3$

29. $6 + (9 + x) = 6 + (6 + 9)$

30. $(14 + 12) + 2 = x + (14 + 12)$

In Exercises 31–38, determine whether or not the given set of numbers has the closure property for (a) addition, and for (b) subtraction.

31. $\{0, 1, 2, 3, 4, 5, 6, 7, 8, \ldots\}$

32. $\{0, -1, -2, -3, -4, -5, \ldots\}$

33. $\{\ldots, -3, -2, -1, 0, 1, 2, 3, \ldots\}$

34. $\{\ldots, -4, -2, 0, 2, 4, 6, \ldots\}$

35. $\{0, 5, 10, 15, 20, 25, \ldots\}$

36. $\{0, 3, 6, 9, 12, 15, 18, \ldots\}$

37. $\{0, 1\}$

38. $\{0\}$

In Exercises 39–41, write down the property which justifies each step.

39. $(6+8)+(-8) = 6+[8+(-8)]$
$= 6+0$
$= 6.$

40. $(-4+9)+4 = -4+(9+4)$
$= -4+(4+9)$
$= (-4+4)+9$
$= 0+9$
$= 9.$

41. $[(-2+4)+2]+6 = [-2+(4+2)]+6$
$= [-2+(2+4)]+6$
$= [(-2+2)+4]+6$
$= (0+4)+6$
$= 4+6.$

1.7 MULTIPLICATION OF REAL NUMBERS

Any rules we develop for multiplication of real numbers ought to be consistent with the usual multiplication of positive numbers. For example, we would want the product of 0 and *any* real number (positive or negative) to be 0.

$$x \cdot 0 = 0.$$

How can we define the product of a positive and a negative number so that the result is consistent with past work? Look at the following pattern of products.

$$3 \cdot 6 = 18$$
$$3 \cdot 5 = 15$$
$$3 \cdot 4 = 12$$
$$3 \cdot 3 = 9$$
$$3 \cdot 2 = 6$$
$$3 \cdot 1 = 3$$
$$3 \cdot 0 = 0$$
$$3 \cdot (-1) = ?$$

What number should we assign as the product $3(-1)$ so that the pattern is maintained? The numbers on the left of the equals sign decrease by 1 each time. The products, on the right, decrease by 3 each time. To maintain the pattern, the number on the right in the bottom row must be 3 less than 0, which is -3. Therefore, we must have

$$3(-1) = -3.$$

The pattern continues with

$$3(-2) = -6$$
$$3(-3) = -9$$
$$3(-4) = -12,$$

and so on. In the same way, we could find that

$$(-3)2 = -6$$
$$(-3)3 = -9$$
$$(-3)4 = -12,$$

and so on.

In general, if x and y both represent positive numbers, then

$$x(-y) = -(xy)$$

or

$$(-x)y = -(xy).$$

(In other words, the product of a positive number and a negative number is negative.)

Example 1 **(a)** $8(-5) = -40$
(b) $5(-4) = -20$
(c) $(-7)(2) = -14$
(d) $(-9)(3) = -27$
(e) $(-6)(4) = -24$

Example 2 To simplify the expression

$$(-9)(2) - (-3)(2),$$

we first multiply from left to right:

$$\begin{aligned}(-9)(2) - (-3)(2) &= -18 - (-6)\\ &= -18 + (+6)\\ &= -12.\end{aligned}$$

The product of two positive numbers is positive, and we have seen that the product of a positive and a negative number is negative. What about the product of two negative numbers? Consider another pattern.

$$(-5)(4) = -20$$
$$(-5)(3) = -15$$
$$(-5)(2) = -10$$
$$(-5)(1) = -5$$
$$(-5)(0) = 0$$
$$(-5)(-1) = ?$$

The numbers on the left of the equals sign decrease by 1 each time. The products on the right increase by 5 each time. To maintain the pattern, we will have to agree that $(-5)(-1)$ is 5 more than $(-5)(0)$. Therefore, we must have

$$(-5)(-1) = 5.$$

Continuing this same pattern, we have

$$\begin{aligned}(-5)(-2) &= 10\\(-5)(-3) &= 15\\(-5)(-4) &= 20\\(-5)(-5) &= 25,\end{aligned}$$

and so on.

In general, if x and y both represent positive numbers, then

$$(-x)(-y) = xy.$$

(The product of two negative numbers is positive.)

Example 3 **(a)** $(-9)(-2) = 18$
(b) $(-6)(-12) = 72$
(c) $(-8)(-1) = 8$
(d) $(-15)(-2) = 30$

Example 4 **(a)** $(-6)(-2) - (3)(-4) = 12 - (-12)$
$= 12 + (+12)$
$= 24$

(b) $-5(-2 - 3) = -5(-5) = 25$

EXERCISES

In Exercises 1–20, find each product.

1. $(-3)(-4)$
2. $(-3)(4)$
3. $3(-4)$
4. $-2(-8)$
5. $(-1)(-5)$
6. $(-9)(-5)$
7. $(-4)(-11)$
8. $(-5)(7)$
9. $(-10)(-12)$
10. $9(-5)$
11. $(8)(-6)$
12. $(13)(-2)$
13. $(-6)(5)$
14. $(-9)0$
15. $(13)(-5)$
16. $(12)(5)$
17. $0(-11)$
18. $3(-15)$
19. $(15)(-11)$
20. $(-9)(-4)$

In Exercises 21–40, carry out the operations.

21. $9(6 - 10)$

22. $5(12 - 15)$

23. $-6(2 - 4)$

24. $-9(5 - 8)$

25. $(4 - 9)(2 - 3)$

26. $(6 - 11)(3 - 6)$

27. $(2 - 5)(3 - 7)$

28. $(5 - 12)(2 - 6)$

29. $(-4 - 3)(-2) + 4$

30. $(-5 - 2)(-3) + 6$

31. $5(-2) - 4$

32. $9(-6) - 8$

33. $3(-4) - (-2)$

34. $5(-2) - (-9)$

35. $(-8 - 2)(-4) - (-5)$

36. $(-9 - 1)(-2) - (-6)$

37. $|-4(-2)| + |-4|$

38. $|8(-5)| + |-2|$

39. $|2|(-4) + |6| \cdot |-4|$

40. $|-3|(-2) + |-8| \cdot |5|$

In Exercises 41–50, evaluate the given expression for $x = -2$, $y = 3$, and $a = -4$.

41. $2x + 7y$

42. $3x + 5y$

43. $5x - 2y + 3a$

44. $6x - 5y + 4a$

45. $(2x + y)(3a)$

46. $(5x - 2y)(-2a)$

47. $(3x - 4y)(-5a)$

48. $(6x + 2y)(-3a)$

49. $(-5 + x)(-3 + y)(2 - a)$

50. $(6 - x)(5 + y)(3 + a)$

In Exercises 51–60, find the solution for each equation. The domain of the variable is the set $\{-3, -2, -1, 0, 1, 2, 3\}$.

51. $2x = -4$

52. $3x = -6$

53. $-4x = 0$

54. $-9x = 0$

55. $-8x = 16$

56. $-9x = 27$

57. $2x + 1 = -3$

58. $3x + 3 = -3$

59. $-4x + 2 = 10$

60. $-5x + 6 = 11$

1.8 PROPERTIES OF MULTIPLICATION, AND THE DISTRIBUTIVE PROPERTY

For each of the five properties of addition in Section 1.6, there is a corresponding property for multiplication. The sum of any two real numbers is a real number. This is true for the product of any two real numbers, as stated by the **closure property of multiplication.**

If x and y are real numbers, then their product xy is a real number.

We know that two numbers can be added in any order without affecting the answer. The same is true for multiplication. The product of the numbers x and y is the same whether we calculate xy or yx. This is the **commutative property of multiplication.**

If x and y are any two real numbers, then

$$xy = yx.$$

Example 1 **(a)** $(-6)(4) = (4)(-6)$
(b) $(-12)(-3) = (-3)(-12)$
(c) $(-4)(6-2) = (6-2)(-4)$

When adding three numbers x, y, and z, we can first get the sum $x + y$ and then add z. Or we can find $y + z$ and then add x. The result is the same in either case. There is a similar **associative property for multiplication.**

For any real numbers x, y, and z,

$$x(yz) = (xy)z.$$

Example 2 **(a)** $4(6 \cdot 3) = (4 \cdot 6)(3)$
(b) $[(-2)(-3)]4 = -2[(-3) \cdot 4]$

If 0 is added to any real number, the result is that real number. For this reason, 0 is called the identity element for addition. The number 1 plays a similar role in multiplication. The product of any real number and the number 1 is that real number. The number 1 is called the **identity element** for multiplication. The **identity property for multiplication** is:

For any real number x,

$$x \cdot 1 = x \qquad \text{and} \qquad 1 \cdot x = x.$$

For any real number x there is an additive inverse, $-x$, such that $x + (-x) = 0$. We also have a multiplicative inverse for real numbers. For example,

$$2 \cdot \frac{1}{2} = 1,$$

where 1 is the identity element for multiplication. Since the product of 2 and $\frac{1}{2}$ is 1, which is the identity element, the numbers 2 and $\frac{1}{2}$ are **multiplicative inverses,** or **reciprocals,** of each other.

Example 3

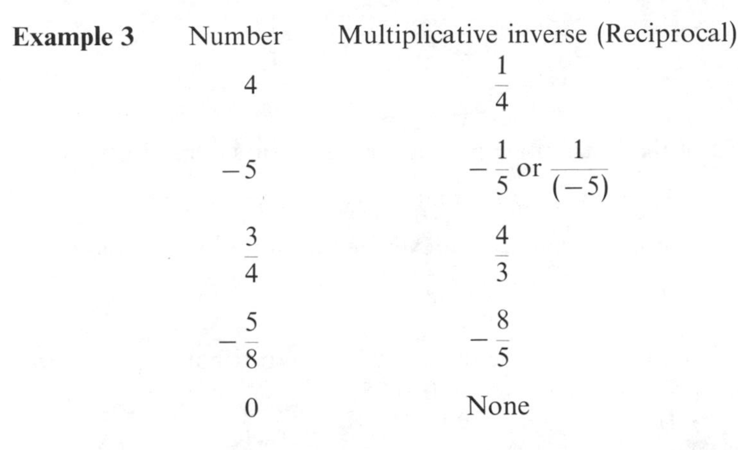

Number	Multiplicative inverse (Reciprocal)
4	$\frac{1}{4}$
-5	$-\frac{1}{5}$ or $\frac{1}{(-5)}$
$\frac{3}{4}$	$\frac{4}{3}$
$-\frac{5}{8}$	$-\frac{8}{5}$
0	None

Why is there no multiplicative inverse for the number 0? To answer this, suppose the number k is to be a multiplicative inverse for 0. Then $0 \cdot k = 1$. However, for every real number k, we know that $0 \cdot k = 0$. So there is no value of k which is a solution of the equation $0 \cdot k = 1$. For this reason, the number 0 has no multiplicative inverse. The **multiplicative inverse property** is:

For every real number x (except 0) there is a multiplicative inverse, or reciprocal, $1/x$, such that

$$x \cdot \frac{1}{x} = 1 \quad \text{and} \quad \frac{1}{x} \cdot x = 1.$$

The distributive property is different from the other properties of the real numbers. It involves both addition and multiplication. For example, to find the product $3(5 + 9)$, first add within parentheses, and then multiply the result by 3.

$$3(5 + 9) = 3(14) = 42.$$

To find the sum $3 \cdot 5 + 3 \cdot 9$, first multiply, and then add the products.

$$3 \cdot 5 + 3 \cdot 9 = 15 + 27 = 42.$$

Based on these results, we can say that

$$3(5 + 9) = 3 \cdot 5 + 3 \cdot 9.$$

The number 3 on the left was "distributed" over the numbers inside the parentheses, giving the result on the right.

Example 4 **(a)** $6(9 + 4) = 6 \cdot 9 + 6 \cdot 4$

(b) $-5(-8 + 2) = (-5)(-8) + (-5)(2)$

(c) $(4 + 7)9 - 4 \cdot 9 + 7 \cdot 9$

On the basis of these examples, we have the **distributive property.**

For all real numbers x, y, and z,

$$x(y + z) = xy + xz \qquad \text{and} \qquad (y + z)x = yx + zx.$$

Another form of the distributive property is valid for subtraction:

$$x(y - z) = xy - xz \qquad \text{and} \qquad (y - z)x = yx - zx.$$

The distributive property can also be extended to more than two numbers.

$$a(b + c + d) = ab + ac + ad, \qquad \text{and so on.}$$

Example 5 Simplify each of the following by the distributive property.

(a) $5(9 + 6) = 5 \cdot 9 + 5 \cdot 6 = 45 + 30 = 75$
(b) $4(x + 5) = 4x + 4 \cdot 5 = 4x + 20$
(c) $-2(x + 3) = -2x + (-2)(3) = -2x - 6$
(d) $3(k - 9) = 3k - 3 \cdot 9 = 3k - 27$
(e) $6 \cdot 8 + 6 \cdot 2 = 6(8 + 2)$
(f) $4x - 4m = 4(x - m)$
(g) $8(3r + 5z) = 8(3r) + 8(5z)$
$= (8 \cdot 3)r + (8 \cdot 5)z$ Associative property
$= 24r + 40z$

EXERCISES

In Exercises 1–10, give the reciprocal of each number that has a reciprocal.

1. 9
2. 8
3. -4
4. -10
5. $2/3$
6. $3/4$
7. $-9/10$
8. $-4/5$
9. 0
10. $6 - 6$

In Exercises 11–30, use the indicated property to write an expression which is equal to the given expression. Simplify the new expression if possible.

Example $x + 5$; commutative property
By the commutative property, $x + 5$ is equal to $5 + x$.

11. $9 + k$; commutative
12. $z + 5$; commutative
13. $m + 0$; identity
14. $(-9) + 0$; identity
15. $3(r + m)$; distributive
16. $11(k + z)$; distributive
17. $8 \cdot \frac{1}{8}$; inverse
18. $\frac{1}{6} \cdot 6$; inverse
19. $12 + (-12)$; inverse
20. $-8 + 8$; inverse
21. $5 + (-5)$; commutative
22. $-9 + 9$; commutative
23. $-3(r + 2)$; distributive

24. $4(k-5)$; distributive

25. $9 \cdot 1$; identity

26. $1(-4)$; identity

27. $(k+5)+(-6)$; associative

28. $(m+4)+(-2)$; associative

29. $(4z+2r)+3k$; associative

30. $(6m+2n)+5r$; associative

In Exercises 31–54, use the distributive property to rewrite each expression.

31. $5(m+2)$

32. $6(k+5)$

33. $-4(r+2)$

34. $-3(m+5)$

35. $-8(k-2)$

36. $-4(z-5)$

37. $-9(a+3)$

38. $-4(m-6)$

39. $(r+8)4$

40. $(m+12)6$

41. $(8-k)(-2)$

42. $(9-r)(-3)$

43. $2(5r+6m)$

44. $5(2a+4b)$

45. $-4(3x-4y)$

46. $-9(5k-12m)$

47. $5 \cdot 8 + 5 \cdot 9$

48. $4 \cdot 3 + 4 \cdot 9$

49. $7 \cdot 2 + 7 \cdot 8$

50. $6x + 6m$

51. $9p + 9q$

52. $8(2x) + 8(3y)$

53. $5(7z) + 5(8w)$

54. $11(2r) + 11(3s)$

Each of Exercises 55–68 gives a true statement. Name the property which justifies the statement.

55. $2 \cdot 1 = 1 \cdot 2$

56. $2 \cdot 1 = 2$

57. $3(2y) = (3 \cdot 2)y$

58. $4 + m = m + 4$

59. $(-2) + 2 = 0$

60. $(3+r)+(-2) = 3+[r+(-2)]$

61. $(6+y)+5 = 6+(y+5)$

62. $9 \cdot \frac{1}{9} = 1$

63. $[3+(-3)]+4 = 0+4$

64. $9(-2)$ is a real number

65. $a+[5+(-5)] = a+0$

66. $3(2+x) = 3 \cdot 2 + 3x$

67. $6(5-2x) = 6 \cdot 5 - 6(2x)$

68. $5(2m)+5(7n) = 5(2m+7n)$

1.9 DIVISION OF REAL NUMBERS

In Section 1.5 we saw that a subtraction problem can be worked by *adding* the additive inverse of the second number. That is, if x and y are any two real numbers, then

$$x - y = x + (-y).$$

For example,

$$3 - (-5) = 3 + (+5) = 8$$
$$-6 - 9 = -6 + (-9) = -15.$$

Division problems are worked in a similar manner. To subtract, we use the additive inverse—to divide we use the multiplicative inverse. The multiplicative inverse (or reciprocal) of 8 is $\frac{1}{8}$, and of $\frac{5}{4}$ is $\frac{4}{5}$. We define the **quotient** of the real numbers x and y (that is, division of the real number x by the number y) as multiplication of x by the multiplicative inverse of y.

For any real numbers x and y (where $y \neq 0$), the quotient of x and y is

$$\frac{x}{y} = x \cdot \frac{1}{y}.$$

(In other words, the quotient of x and y equals the product of x and the reciprocal of y.)

Example 1 **(a)** $\frac{12}{3} = 12 \cdot \frac{1}{3} = 4$

(b) $\frac{-10}{2} = -10 \cdot \frac{1}{2} = -5$

(c) $\frac{8}{-4} = 8 \cdot \left(\frac{1}{-4}\right) = -2$

(d) $\frac{-14}{-7} = -14\left(\frac{1}{-7}\right) = 2$

(e) $\frac{-100}{-20} = -100\left(\frac{1}{-20}\right) = 5$

In practice, division is often done directly, with the answer being positive if both numbers have like signs, and negative if both numbers have unlike signs. For example,

$$\frac{8}{-2} = -4 \quad \text{and} \quad \frac{-45}{-9} = 5.$$

In the definition of x divided by y, we said that the number we divide by, namely y, cannot be 0. The reason is that 0 is the only number with no multiplicative inverse, so that 1/0 is not a number. For this reason, division by 0 is meaningless, and is never permitted. If a proposed division problem turns out to involve division by 0, write "no such number."

From the definitions of multiplication and division of real numbers, we see that

$$\frac{-40}{8} = -40 \cdot \frac{1}{8} = -5,$$

and

$$\frac{40}{-8} = 40\left(\frac{1}{-8}\right) = -5,$$

so that

$$\frac{-40}{8} = \frac{40}{-8}.$$

In general, if $y \neq 0$, then three forms are possible.

$$\frac{-x}{y} = \frac{x}{-y} = -\frac{x}{y}.$$

We usually use the forms $\frac{-x}{y}$ or $-\frac{x}{y}$. The form $\frac{x}{-y}$ is seldom used.

In general,

$$\frac{-x}{-y} = \frac{x}{y}.$$

Because x/y is simpler than $-x/-y$, the form $-x/-y$ is seldom used.

If one integer is exactly divisible by another (that is, if the quotient is an integer, and not a fraction), the second integer is a **factor** of the first. For example, 12 is exactly divisible by 3, so that 3 is a factor of 12. The factors of 12 are the numbers −12, −6, −4, −3, −2, −1, 1, 2, 3, 4, 6, and 12.

Example 2

Number	Factors
18	−18, −9, −6, −3, −2, −1, 1, 2, 3, 6, 9, 18
20	−20, −10, −5, −4, −2, −1, 1, 2, 4, 5, 10, 20
15	−15, −5, −3, −1, 1, 3, 5, 15
7	−7, −1, 1, 7
1	−1, 1
0	0

EXERCISES

In Exercises 1–30, find each quotient.

1. $\frac{-10}{5}$

2. $\frac{-12}{3}$

3. $\frac{-15}{5}$

4. $\frac{-20}{2}$

5. $\frac{18}{-3}$

6. $\frac{24}{-6}$

7. $\frac{100}{-20}$

8. $\frac{250}{-25}$

9. $\frac{-12}{-6}$

10. $\frac{-25}{-5}$

11. $\frac{-150}{-10}$

12. $\frac{-280}{-20}$

13. $\frac{-180}{-5}$

14. $\frac{-350}{-7}$

15. $\frac{0}{-2}$

16. $\frac{0}{12}$

17. $\frac{12}{2-5}$

18. $\frac{15}{3-8}$

19. $\frac{50}{2-7}$

20. $\frac{30}{5-8}$

21. $\frac{-30}{2-8}$

22. $\dfrac{-50}{6-11}$

23. $\dfrac{-40}{8-(-2)}$

24. $\dfrac{-72}{6-(-2)}$

25. $\dfrac{-120}{-3-(-5)}$

26. $\dfrac{-200}{-6-(-4)}$

27. $\dfrac{-15-3}{3}$

28. $\dfrac{16-(-2)}{-6}$

29. $\dfrac{-30+(+8)}{-11}$

30. $\dfrac{-17-(-12)}{5}$

In Exercises 31–44, simplify each expression. Simplify the numerator and denominator separately. Then find the quotient.

31. $\dfrac{-8(-2)}{3-(-1)}$

32. $\dfrac{-12(-3)}{-15-(-3)}$

33. $\dfrac{-15(2)}{-7-3}$

34. $\dfrac{-20(6)}{-5-1}$

35. $\dfrac{-2(6)+3}{2-(-1)}$

36. $\dfrac{3(-8)+4}{-6+1}$

37. $\dfrac{-5(2)+3(-2)}{-3-(-1)}$

38. $\dfrac{4(-1)+3(-2)}{-2-3}$

39. $\dfrac{2-4(2)}{4-1}$

40. $\dfrac{-4-3(-2)}{5-3}$

41. $\dfrac{-9(-2)-(-4)(-2)}{-2(3)-2(2)}$

42. $\dfrac{5(-2)-3(4)}{-2[3-(-2)]-1}$

43. $\dfrac{4(-2)-5(-3)}{2[-1+(-3)]-(-8)}$

44. $\dfrac{5(-3)-(-2)(-4)}{5[-4+(-2)]+3(10)}$

In Exercises 45–54, find all integer factors of each of the numbers.

45. 36

46. 32

47. 25

48. 14

49. 40

50. 50

51. 17

52. 13

53. 29

54. 37

In Exercises 55–64, find the solution of each of the equations. The domain of the variable is the set $\{-8, -6, -4, -2, 0, 2, 4, 6, 8\}$.

55. $\dfrac{x}{4}=-2$

56. $\dfrac{x}{2}=-1$

57. $\dfrac{x}{-2}=3$

58. $\dfrac{x}{-2}=-2$

59. $\dfrac{x}{-3}=0$

60. $\dfrac{x}{5}=0$

61. $\frac{x}{-2} = -4$

62. $\frac{x}{-1} = 2$

63. $\frac{x}{4} = -1$

64. $\frac{x}{3} = -2$

In Exercises 65–74, write each sentence in symbols, and find the solution. The domain is the set of all real numbers.

65. Six times a number is -42.

66. Four times a number is -32.

67. When a number is divided by 5, the answer is 15.

68. When a number is divided by 6, the result is -3.

69. When a number is divided by 3, the result is -9.

70. When a number is divided by -3, the answer is -4.

71. The quotient of a number and 2 is -6. (Write the quotient as $x/2$.)

72. The quotient of a number and -9 is 2.

73. The quotient of 6 and one more than a number is 3.

74. When 12 is divided by twice a number, the quotient is 2.

1.10 THE SET OF REAL NUMBERS AND ITS SUBSETS

The set of real numbers contains many kinds of numbers. We have already defined the set of whole numbers.

$$\{0, 1, 2, 3, 4, 5, 6, \ldots\} \qquad \text{Whole numbers}$$

The three dots show that the whole numbers continue indefinitely without stopping. Every whole number is a real number. For this reason, the set of whole numbers is a *subset* of the set of real numbers.

In general, a set of elements A is a **subset** of a set of elements B if every element in set A is also in set B.

The set of positive numbers is also a subset of the real numbers. So is the set of negative numbers. Also, the set of positive whole numbers (the *counting numbers*) is a subset of the set of whole numbers.

The set made up of both the whole numbers and the additive inverses of the whole numbers is the set of integers.

$$\{\ldots, -3, -2, -1, 0, 1, 2, 3, \ldots\} \qquad \text{Integers}$$

The set of integers is a subset of the set of real numbers.

The set of all numbers which can be written as the quotient of two integers with denominator not 0 is called the set of **rational numbers**. The rational numbers include the numbers

$$\frac{3}{4}, \frac{5}{8}, \frac{-2}{3}, -\frac{11}{7}, \frac{15}{4}, 9\left(=\frac{9}{1}\right), -15\left(=\frac{-15}{1}\right), 0\left(=\frac{0}{1}\right),$$

and so on. Since division by 0 is not permitted, 6/0 or $-8/0$ are not rational numbers (or any other kind of number). The set of integers is a subset of the set of rational numbers. The set of rational numbers, in turn, is a subset of the set of real numbers.

It seems reasonable to think that the set of rational numbers would include enough numbers to correspond to every point on the number line, but this is not true. There are many points on the number line whose coordinates are *not* rational numbers.

An example is the real number $\sqrt{2}$. (The number $\sqrt{2}$ is that number which can be multiplied by itself to give 2. That is, $\sqrt{2} \cdot \sqrt{2} = 2$.) The real number $\sqrt{2}$ measures the length of the longest side of a right triangle whose other sides each measure 1 unit. If such a triangle is constructed on a number line, the location of the point corresponding to $\sqrt{2}$ on the number line can be found. This is shown in Figure 1.20.

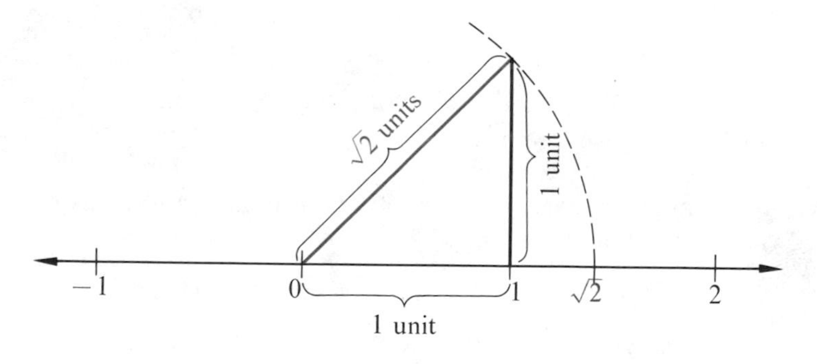

Figure 1-20

No matter how fine the divisions on a ruler, the segment labeled $\sqrt{2}$ units is not measured by any of them. There is no rational number exactly equal to $\sqrt{2}$, although it can be shown mathematically that $\sqrt{2}$ is between the two rational numbers 1.414 and 1.415. That is,

$$1.414 < \sqrt{2} < 1.415.$$

Many numbers exist that are not rational numbers—infinitely many. The set of irrational numbers includes $\sqrt{2}$, $\sqrt{3}$, $\sqrt{5}$, $\sqrt{7}$, π, and many more.

The real numbers can be divided into two sets, the set of rational numbers and the set of irrational numbers. All real numbers belong to one of these two sets—no real number belongs to both. Figure 1.21 diagrams the relationship between the set of real numbers and its subsets that we have discussed.

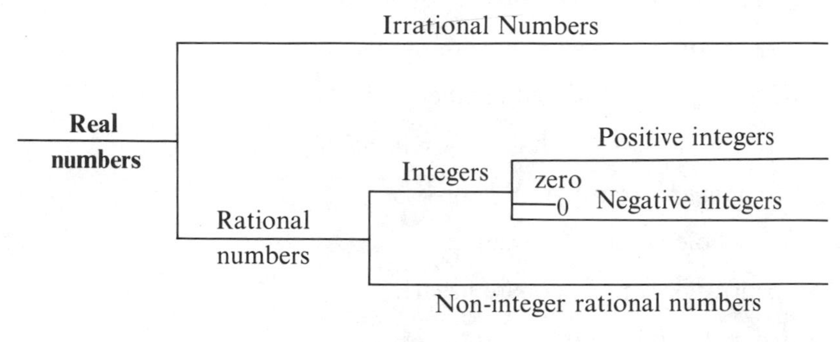

Figure 1-21

EXERCISES

In Exercises 1–20, first write the number in simplest form. Then select all words from the list that apply to the given expressions.

Real number	Integer
Rational number	Whole number
Irrational number	Not a real number

1. 23

2. -46

3. $5/8$

4. $-2/3$

5. -12

6. $\sqrt{19}$

7. $-\sqrt{14}$

8. $0/3$

9. $0/(-2)$

10. $-15/5$

11. $35/(-7)$

12. $39/2$

13. $-\sqrt{6}$

14. $-\sqrt{15}$

15. $8/0$

16. $-4/0$

17. $\sqrt{16}$

18. $\sqrt{25}$

19. $\dfrac{-4}{2-2}$

20. $\dfrac{15}{3+(-3)}$

In Exercises 21–26, give three examples of numbers which satisfy the given condition.

21. Positive real numbers but not integers.

22. Real numbers but not positive.

23. Real numbers but not whole numbers.

24. Rational numbers but not integers.

25. Real numbers but not rational numbers.

26. Rational numbers but not negative numbers.

In Exercises 27–36, answer *true* or *false* for each statement.

27. Every rational number is a real number.

28. Every integer is a rational number.

29. There are integers which are not real numbers.

30. Every integer is positive.

31. Every whole number is positive.

32. Some irrational numbers are negative.

33. Some real numbers are not rational.

34. Not every rational number is positive.

35. Some whole numbers are not integers.

36. The number 0 is irrational.

CHAPTER 1 SUMMARY

Key Words

(Look in the Index for the pages where these words are defined.)

Algebraic expression
(or Expression)
Variable
Numerical value
(or Value)
Absolute value
Additive inverse
(or Negative)
Product
Quotient
Factor
Multiplicative inverse
(or Reciprocal)
Identity element

Equation
Inequality
Domain (or
Replacement set)
Solution set
Solution

Number line
Coordinate
Graph

Set
Subset
Empty set
(or Null set)

Real numbers
Irrational numbers
Rational numbers
Integers
Negative numbers
Positive numbers
Zero
Whole numbers

Closure property
Commutative
property
Associative property
Identity property
Inverse property
Distributive property

Symbols introduced in this chapter

(A complete listing for the book is inside the front cover.)

Symbol	Meaning
$=$	Equals sign; equals, is equal to
$\neq$	Is not equal to
$<$	Is less than
$\nless$	Is not less than
$>$	Is greater than
$\ngtr$	Is not greater than
$\leq$	Is less than or equal to
$\geq$	Is greater than or equal to
$\{a, b\}$	Set containing the elements a and b
$\lvert x \rvert$	Absolute value of x
$\varnothing$	Empty set
$a \cdot b$	Product of a times b; also $a(b)$, ab
[ ]	Brackets

Order of Operations

1. Perform all operations within parentheses or brackets, from inside out.
2. Perform all multiplications and divisions, from left to right.
3. Perform all additions and subtractions, from left to right.
4. If the problem involves a fraction bar, do all work above and below the bar separately. Then perform the division if possible.

Definitions

For all real numbers x and y:

Subtraction $x - y = x + (-y)$.

Division $\frac{x}{y} = x \cdot \frac{1}{y}, \ y \neq 0.$

Absolute value $|x|$ is the distance from x to 0, or

$|x| = x$ if x is positive or zero, and $|x| = -x$ if x is negative.

Sets of Numbers

Whole numbers	$\{0, 1, 2, 3, 4, 5, 6, \ldots\}$
Integers	$\{\ldots, -3, -2, -1, 0, 1, 2, 3, \ldots\}$
Rational numbers	{all numbers that can be represented as the quotient of two integers, with denominator not zero}
Irrational numbers	{real numbers that are not rational numbers}
Real numbers	{numbers corresponding to all points on the number line}

Properties of the Real Numbers

For all real numbers x, y, and z:

	Addition	*Multiplication*
Closure	$x + y$ is a real number	xy is a real number
Commutative	$x + y = y + x$	$xy = yx$
Associative	$(x + y) + z = x + (y + z)$	$(xy)z = x(yz)$
Identity	$x + 0 = x$ and $0 + x = x$	$x \cdot 1 = x$ and $1 \cdot x = x$
Inverse	$x + (-x) = 0$ and $-x + x = 0$	$x \cdot \frac{1}{x} = 1$ and $\frac{1}{x} \cdot x = 1$, provided that $x \neq 0$
Distributive	$x(y + z) = xy + xz$ and $(y + z)x = yx + zx$	

CHAPTER 1 TEST

Perform each of the following operations (whenever possible).

1. $-6 + 10$
2. $-2 + (-7)$
3. $12 + (-13)$
4. $27 - (-5)$
5. $10 - 18$
6. $-1 - 3$
7. $-5 + 15$
8. $-2(-3)$
9. $12(-6)$
10. $-1(3)$
11. $\dfrac{27}{-3}$
12. $\dfrac{-9}{9}$
13. $\dfrac{-8}{-4}$
14. $\dfrac{5}{0}$
15. $\dfrac{0}{-3}$

Find the value of each variable expression.

16. $3 + m; \quad m = -5$
17. $4x - 3y; \quad x = -1, y = 3$
18. $\dfrac{p - 8}{3p - 4}; \quad p = 3$
19. $\dfrac{27 + 3t}{-5t - 1}; \quad t = 2$

Graph the solution from the given domain for each inequality.

20. $x > 5$; $\{2, 4, 6, 8, 10\}$

21. $3k \le 9$; $\{-2, -1, 0, 1, 2, 3, 4\}$

22. $4y + 1 > 5$; $\{0, 1, 2, 3, 4\}$

Select the smaller number from each pair.

23. $-3, -5$

24. $6, -|-8|$

25. $|-4|, 0$

26. $3, |-5|$

27. $-9, 9$

28. $|4|, |-6|$

Match the property in Column I with all examples of it from Column II.

Column I	*Column II*
29. Commutative	**A.** $-2 + 2 = 0$
30. Associative	**B.** $3 + (7 + x) = (3 + 7) + x$
31. Closure	**C.** $8 + 0 = 8$
32. Identity	**D.** $17 \cdot 1 = 17$
33. Inverse	**E.** $3(x + y) = 3x + 3y$
34. Distributive	**F.** $-7 + \sqrt{2}$ is a real number
	G. $3(-2)$ is a real number
	H. $8 + m = m + 8$
	I. $-5\left(\frac{1}{-5}\right) = 1$
	J. $mn = nm$

List all items from the set $\{-5, 0/2, \sqrt{3}, 9/4, 15/3, 12, -6, 3/0\}$ **which ar**

35. Whole numbers

36. Integers

37. Rational numbers

38. Irrational numbers

39. Real numbers

40. Not a real number

SOLVING LINEAR EQUATIONS AND INEQUALITIES IN ONE VARIABLE

An equation or inequality cannot be solved without a domain for the variable. In Chapter 1, domains were given with the equation or inequality. But from now on, the domain of any variable will be the set of *all real numbers.* This means that when you have an equation or inequality to solve, you must find all real number solutions for it.

2.1 SIMPLIFYING EQUATIONS

In solving an equation, the first step is to simplify the expressions on the left-hand side and on the right-hand side of the equation as much as possible. You can do this by adding, or multiplying, or both.

Example 1 Simplify the equation $x + 8 + 9 = 3 + 20$.

Add the 8 and 9 on the left of the equals sign. Then add the 3 and 20 on the right. The result is a simpler equation.

$$x + 17 = 23.$$

Example 2 Simplify the equation $(5 + x) + (-2) = (-3) + [11 + (-1)]$.

Here you need to rearrange numbers and variables. On the left, use the commutative property and the associative property to reverse the order of x and -2.

$$\begin{aligned} (5 + x) + (-2) &= (-3) + [11 + (-1)] \\ [5 + (-2)] + x &= (-3) + 10 \\ 3 + x &= 7. \end{aligned}$$

The final equation, $3 + x = 7$, is simpler than the given equation. Find the solution by inspection. (The domain is the set of all real numbers.) You should have found that the solution is 4.

To check that the solution is correct, substitute 4 for the variable in the given equation. Does this result in a true statement?

$$(5 + x) + (-2) = (-3) + [11 + (-1)]$$
$$(5 + 4) + (-2) = (-3) + [11 + (-1)] \quad \text{Let } x = 4.$$
$$9 + (-2) = (-3) + 10$$
$$7 = 7 \quad \text{True}$$

Since this result is true, 4 is the solution.

To simplify an equation containing a sum such as $3x + 5x$, use the distributive property.

$$3x + 5x = (3 + 5)x$$
$$= 8x.$$

In the same way, $9m + 5m = (9 + 5)m = 14m$, and $6r + 3r + 2r = 11r$. Also, $4x + x = 4x + 1x = 5x$.

The distributive property can be used to simplify a difference.

$$16x - 9x = (16 - 9)x$$
$$= 7x.$$

Example 3 Simplify the equation $12m - 3m - 27$.

Use the distributive property on the left-hand side.

$$12m - 3m = (12 - 3)m = 9m.$$

The equation $12m - 3m = 27$ is simplified to

$$9m = 27.$$

By inspection, the solution of this equation is 3.

Example 4 Simplify the equation $4x + 2 + 3x - 6x = 9$.

Use the associative, commutative, and distributive properties.

$$4x + 2 + 3x - 6x = 9$$
$$4x + 3x - 6x + 2 = 9$$
$$(4 + 3 - 6)x + 2 = 9$$
$$1x + 2 = 9$$
$$x + 2 = 9. \quad 1x = x$$

By inspection, the solution of the equation is 7.

Rearranging numbers within each side of an equation is called *collecting terms*. Carrying out the operations of addition or subtraction is called *combining terms*. A **term** is either a single number or the product of a number and a variable.

Example 5 The terms in the equation $4x + 2 + 3x - 6x = 9$ are as follows.

$$\left.\begin{matrix} 2 \\ 9 \end{matrix}\right\} \quad \text{Single numbers (in this case, integers)}$$

$$\left.\begin{matrix} 4x \\ 3x \\ 6x \end{matrix}\right\} \quad \text{Products of a number and a variable}$$

Example 6 Simplify the equation $3x + 7 - 2x - 5 + 1 = 2$.

First rewrite the equation so that subtractions are written as additions of inverses.

$$3x + 7 + (-2x) + (-5) + 1 = 2.$$

Then collect all terms with variables.

$$3x + (-2x) + 7 + (-5) + 1 = 2.$$

Combine terms.

$$\begin{aligned} (3 - 2)x + 3 &= 2 \\ 1x + 3 &= 2 \\ x + 3 &= 2. \qquad 1x = x \end{aligned}$$

The solution by inspection is -1.

Example 7 Solve the equation $-1 + 2x + 16 - x = 12$.

First rewrite all subtractions as additions of inverses. Then collect the terms on the left side of the equation into two groups—those with a variable and those without. Do not forget that $1x$ is the same as x.

$$\begin{aligned} -1 + 2x + 16 + (-x) &= 12 \\ 2x + (-1x) + (-1) + 16 &= 12 \\ 1x + 15 &= 12 \\ x + 15 &= 12. \end{aligned}$$

By inspection, the solution is -3.

Here is a list of the equations we have worked in this section. The simplified equations for each are given on the right.

$x + 8 + 9 = 3 + 20$	$x + 17 = 23$
$(5 + x) + (-2) = (-3) + [11 + (-1)]$	$3 + x = 7$
$12m - 3m = 27$	$9m = 27$
$4x + 2 + 3x - 6x = 9$	$x + 2 = 9$
$3x + 7 - 2x - 5 + 1 = 2$	$x + 3 = 2$
$-1 + 2x + 16 - x = 12$	$x + 15 = 12$

What do all these equations have in common? The terms of each equation are either numbers or products of a number and a variable. There are no products of variables, such as $x \cdot x$ or $x \cdot y$. Equations having only terms like this are called *linear equations.*

In general, a **linear equation** can be expressed in the form

$$ax + b = c,$$

where a, b, and c represent any real numbers, except that a cannot be 0.

Check that each of the above equations has this form. For example, in $x + 17 = 23$, take $a = 1$, $b = 17$, and $c = 23$. The equation is linear, and so is the equation before it was simplified, $x + 8 + 9 = 3 + 20$. In the equation $9m = 27$, you can take $b = 0$, that is, $9m + 0 = 27$. This, too, is a linear equation.

EXERCISES

In Exercises 1–10, an equation is given, along with a proposed solution. Decide whether the proposed solution is the correct solution.

Example $5r - 3r + 1 = 13$; 6. Substitute the proposed solution, 6, for r in the equation: $5 \cdot 6 - 3 \cdot 6 + 1 = 13$. Is this true? $30 - 18 + 1 = 13$, then $13 = 13$. It is true. Therefore, 6 is the correct solution.

1. $2x + 3 = 13;\quad 5$

2. $4x - 1 = 11;\quad 3$

3. $5k + 1 = 20;\quad 4$

4. $3z + 2 = 21;\quad 6$

5. $3h - 2h + 1 + 4 = 9;\quad 4$

6. $5m - 2m + 4 - 3 = 13;\quad 4$

7. $4y + 1 - 5y + 3y - 5 = 6;\quad 5$

8. $3s - 5s + 6s + 5 - 2 = 17;\quad 4$

9. $4x + 1 = 4x + 6 - 3x + 2x - 5;\quad -2$

10. $8j - 4j + 3 - 5 + 6j = 11j + 1;\quad -3$

In Exercises 11–20, simplify each expression by using the associative, commutative, and distributive properties, as needed.

11. $2k + 9 + 5k + 6$

12. $2 + 17z + 1 + 2z$

13. $m + 1 - m + 2 + m - 4$

14. $12 - 13x - 27 + 2x - x$

15. $-5y + 3 - 1 + 5 + y - 7$

16. $2k - 7 - 5k + 5k - 3 - k$

17. $-2x + 3 + 4x - 17 + 20$

18. $r - 6 - 12r - 4 + 6r$

19. $16 - 5m - 4m - 2 + 2m + 6$

20. $6 - 3z - 2z - 5 + z - 3z - 3$

In Exercises 21–32, simplify the given equation until the solution can be found by inspection. Then write the solution.

21. $3x + 4x - 14$

22. $x + 5 + 6 = 18 + 2$

23. $6k - 5k + 5 = 10 + 3$

24. $5m - 4m - 7 = 2 + 3$

25. $-3z + 7 - 3z + 7z = 31 - 18$

26. $15 - 2 + 4 = -4 + 1 - 5k + 7k - k$

27. $-10 + x + 4x - 7 - 4x = 21 - 19$

28. $-x + 10x - 3x - 4 - 5x = 2 + 10$

29. $1 + 7x - 11x - 1 + 5x = 2 - 6$

30. $-r + 2 - 5r + 3 + 4r + 3r = -8$

31. $2m + m + 4 + 2 - 5 = 16$

32. $4z + 2z + 5 - 8 + 4 = 25$

In Exercises 33–36, write an equation using the given information. Then solve the equation.

33. The sum of five times a number, three times the number, and 1 is 49. Find the number.

34. The difference of four times a number and three times a number, added to 8 is 11. Find the number.

35. On a recent trip to McDonald's, one of the authors noticed a total of five employees. The number of managers was one less than the number of other employees. Find the number of managers.

36. Joann is three times as old as Marilyn. The sum of their ages is 24. Find Joann's age.

2.2 THE ADDITION PROPERTY OF EQUALITY

Our goal in solving an equation is to go through a series of steps, ending up with an equation where only the variable x is on one side of the equals sign, and a number is on the other side. That is, we want to reduce the given equation to the form

$$x = \text{number}.$$

Think about the equation

$$x + 5 = 12.$$

We know that $x + 5$ and 12 represent the same number, since this is the meaning of the $=$ sign. We want to change the left-hand side from $x + 5$ to x. We could do this by adding -5 to $x + 5$.

$$x + 5 + (-5) = x.$$

To keep the equality between $x + 5$ and 12, we must add -5 to 12.

$$12 + (-5) = 7.$$

By adding -5 to both sides of the equation $x + 5 = 12$, we get the simpler equation

$$x = 7.$$

The solution is 7.

The **addition property of equality** is the property of the real numbers that lets us add the same number to both sides of an equation.

If A, B, and C represent algebraic expressions, then the equations

$$A = B$$

and

$$A + C = B + C$$

both have the same solutions.

Example 1 Solve the equation $x - 16 = 7$.

If x were alone on the left, we would have the solution. So use the addition property of equality to add 16 to both sides.

$$\begin{aligned} x - 16 &= 7 \\ (x - 16) + 16 &= 7 + 16 \\ x &= 23. \end{aligned}$$

To check, substitute 23 for x in the original equation.

$$\begin{aligned} x - 16 &= 7 & & \\ 23 - 16 &= 7 & & \text{Let } x = 23 \\ 7 &= 7. & & \text{True} \end{aligned}$$

Since the check results in a true statement, 23 is the correct solution.

In this example, how did we know to add 16 to both sides of the equation $x - 16 = 7$? We want one side of the equation to contain only the variable term and the other side to contain only a number. We know that $x + 0 = x$, so we need to get $x + 0$. What number must be added to $x - 16$ so that $x + 0$ results? Since the sum of any number and its additive inverse is 0, we must add the additive inverse of -16, which is 16, to both sides of the equation.

Example 2 Solve the equation $3x + 17 = 4x$.

We could find the solution if the terms $3x$ and $4x$ were alone on the same side of the equation. One way to do this is to use the addition property of equality, and add $-3x$ to both sides.

$$\begin{aligned} 3x + 17 &= 4x \\ 3x + 17 + (-3x) &= 4x + (-3x) \\ 17 &= x. \end{aligned}$$

The solution is 17.

The equation $3x + 17 = 4x$ could also be solved by first adding $-4x$ to both sides as follows.

$$3x + 17 = 4x$$
$$3x + 17 + (-4x) = 4x + (-4x)$$
$$17 - x = 0.$$

Now add -17 to both sides.

$$17 - x + (-17) = 0 + (-17)$$
$$-x = -17.$$

This result gives the value of $-x$, but not x itself. However, we know that the additive inverse of x is -17. Then x must equal 17.

$$\text{If } -x = -17, \text{ then } x = 17.$$

This answer agrees with the first one.

Example 3 Solve the equation $4r + 5r - 3 + 8 - 3r - 5r = 12 + 8$.
First simplify the equation by combining terms.

$$4r + 5r - 3 + 8 - 3r - 5r = 12 + 8$$
$$r + 5 = 20.$$

Then add -5 to both sides.

$$r + 5 + (-5) = 20 + (-5)$$
$$r = 15.$$

The solution of the given equation is 15.

EXERCISES

In Exercises 1–20, solve each equation by first using the addition property of equality.

1. $x - 3 = 7$
2. $x + 5 = 13$
3. $7 + k = 5$
4. $9 + m = 4$
5. $3r - 10 = 2r$
6. $2p = p + 3$
7. $7z = -8 + 6z$
8. $4y = 3y - 5$
9. $m + 5 = 0$
10. $k - 7 = 0$
11. $2 + 3x = 2x$
12. $10 + r = 2r$
13. $2p + 6 = 10 + p$
14. $5r + 2 = -1 + 4r$
15. $2k + 2 = -3 + k$
16. $6 + 7x = 6x + 3$
17. $x - 5 = 2x + 6$
18. $-3r + 7 = -4r - 19$
19. $6z + 3 = 5z - 3$
20. $6t + 5 = 5t + 7$

In Exercises 21–30, solve each equation. First simplify both sides of the equation as much as possible.

21. $4x + 3 + 2x - 5x = 2 + 8$
22. $3x + 2x - 6 + x - 5x = 9 + 4$

23. $9r + 4r + 6 - 8 = 10r + 6 + 2r$

24. $-3t + 5t - 6t + 4 - 3 = -3t + 2$

25. $11z + 2 + 4z - 3z = 5z - 8 + 6z$

26. $2k + 8k + 6k - 4k - 8 + 2 = 3k + 2 + 10k$

27. $4m + 8m - 9m + 2 - 5 = 4m + 6$

28. $15y - 4y + 8 - 2 + 7 - 4 = 4y + 2 + 8y$

29. $-9p + 4p - 3p + 2p - 6 = -5p - 6$

30. $5x - 2x + 3x - 4x + 8 - 2 + 4 = 5x + 10 - 4x$

In Exercises 31–34, write an equation using the given information, and then solve it.

31. Three times a number is seventeen more than twice the number. Find the number.

32. The sum of two consecutive integers is 13 less than three times the smaller. Find the integers. (Hint: Let x represent the first of the integers. Then $x + 1$ represents the second.)

33. If five times a number is added to three times the number, the result is the sum of seven times the number and nine. Find the number.

34. If six times a number is subtracted from seven times a number, the result is -9. Find the number.

2.3 THE MULTIPLICATION PROPERTY OF EQUALITY

The addition property of equality by itself is not sufficient for solving an equation like $3x + 2 = 17$.

$$3x + 2 = 17$$
$$3x + 2 + (-2) = 17 + (-2)$$
$$3x = 15.$$

We do not end up with the variable x alone on one side of the equation, but with $3x$. We need another property similar to the addition property.

If $3x = 15$, then $3x$ and 15 are names for the same number. This is what equality means. Multiplying both $3x$ and 15 by the same number will also result in equality. Thus the **multiplication property of equality** states that both sides of an equation can be multiplied by the same term.

If A, B, and C represent algebraic expressions, the equations

$$A = B$$

and

$$AC = BC$$

have exactly the same solution. (Assume that $C \neq 0$.)

Now we go back and solve $3x = 15$. On the left, we have $3x$. We need $1x$, or x, instead of $3x$. To get x, multiply both sides of the equation by $\frac{1}{3}$. This works because $3 \cdot \frac{1}{3} = 1$.

$$3x = 15$$

$$\frac{1}{3}(3x) = \frac{1}{3} \cdot 15$$

$$\left(\frac{1}{3} \cdot 3\right)x = \frac{1}{3} \cdot 15$$

$$1x = 5 \qquad \frac{1}{3} \cdot 3 = 1$$

$$x = 5. \qquad 1x = x$$

The solution of the equation is 5.

Example 1 Solve the equation $5x = 30$.

To get x alone on the left, use the multiplication property of equality, and multiply both sides of the equation by $\frac{1}{5}$.

$$5x = 30$$

$$\frac{1}{5}(5x) = \frac{1}{5} \cdot 30$$

$$x = 6.$$

The solution is 6.

Example 2 Solve the equation $x/4 = 3$.

We can replace $x/4$ by $\frac{1}{4}x$, since division by the nonzero number a is the same as multiplication by $1/a$. (This idea was first discussed in Section 1.9.) To get x alone on the left, multiply both sides by 4.

$$\frac{x}{4} = 3$$

$$\frac{1}{4}x = 3$$

$$4\left(\frac{1}{4}x\right) = 4 \cdot 3 \qquad \frac{x}{4} = \frac{1}{4}x$$

$$x = 12. \qquad 4 \cdot \frac{1}{4} = 1$$

Check the answer: $\frac{x}{4} = 3$

$$\frac{12}{4} = 3 \qquad \text{Let } x = 12$$

$$3 = 3. \qquad \text{True}$$

The solution, 12, is correct.

Example 3 Solve the equation $\frac{3}{4}h = 6$.

To get h alone on the left, use the multiplication property of equality. Multiply both sides of the equation by 4/3.

$$\frac{3}{4}h = 6$$

$$\frac{4}{3}\left(\frac{3}{4}h\right) = \frac{4}{3} \cdot 6$$

$$h = \frac{4}{3} \cdot \frac{6}{1}$$

$$h = \frac{24}{3}$$

$$h = 8.$$

The solution is 8.

Example 4 Solve the equation $5m + 6m = 33$.

First use the distributive property to combine terms.

$$5m + 6m = 33$$

$$11m = 33$$

$$\frac{1}{11}(11m) = \frac{1}{11} \cdot 33$$

$$m = 3.$$

Example 5 Solve the equation $-r = 4$.

To find the solution for this equation, we need to get r itself, and not $-r$, on one side of the equals sign. If we multiply both sides of $-r = 4$ by the number -1, we get the solution.

$$-r = 4$$

$$-1 \cdot r = 4 \qquad -1 \cdot r = -r$$

$$-1 \cdot (-1 \cdot r) = -1 \cdot 4$$

$$(-1)(-1)r = -1 \cdot 4$$

$$1r = -4$$

$$r = -4.$$

The solution of the equation $-r = 4$ is thus -4.

EXERCISES

In Exercises 1–36, solve each equation.

1. $5x = 25$

2. $7x = 28$

3. $2m = 50$

4. $6y = 72$

5. $3a = -24$

6. $5k = -60$

7. $8s = -56$

8. $9t = -36$

9. $-4x = 16$

10. $-6x = 24$

11. $-12z = 108$

12. $-11p = 77$

13. $5r = 0$

14. $2x = 0$

15. $-y = 6$

16. $-m = 2$

17. $-n = -4$

18. $-p = -8$

19. $-x = 0$

20. $-x = -0$

21. $2x + 3x = 20$

22. $3k + 4k = 14$

23. $5m + 6m - 2m = 72$

24. $11r - 5r + 6r = 84$

25. $k + k + 2k = 80$

26. $4z + z + 2z = 28$

27. $\frac{m}{2} = 16$

28. $\frac{p}{5} = 3$

29. $\frac{x}{7} = 7$

30. $\frac{k}{8} = 2$

31. $\frac{2}{3}t = 6$

32. $\frac{3}{4}m = 18$

33. $\frac{5}{2}z = 20$

34. $\frac{9}{5}r = 18$

35. $\frac{3}{4}p = -60$

36. $\frac{5}{8}z = -40$

In Exercises 37–40, write an equation for each problem. Then solve the equation.

37. When a number is divided by 4, the result is 6. Find the number.

38. The quotient of a number and -5 is 2. Find the number.

39. Chuck decides to divide a sum of money equally among four relatives, Dennis, Mike, Ed, and Joyce. Each relative received \$62. Find the original sum that was divided up.

40. If twice a number is divided by 5, the result is 4. Find the number.

2.4 A GENERAL METHOD FOR SOLVING LINEAR EQUATIONS

We now summarize the methods we have used so far to simplify equations so that solutions can be found.

Step 1 Combine terms to simplify. Use the commutative, associative, and distributive properties as needed.

Step 2 Use the addition property of equality to further simplify, so that the variable term is on one side of the equals sign, and the number term is on the other.

Step 3 Use the multiplication property of equality to simplify further. This gives an equation of the form $x =$ number.

Step 4 Check the solution by substituting into the original equation.

Example 1 Solve the equation $2x + 3x + 3 = 38$. Follow the four steps of the summary above.

Step 1 Combine terms.

$$2x + 3x + 3 = 38$$
$$5x + 3 = 38.$$

Step 2 Use the addition property of equality. Add -3 to both sides.

$$5x + 3 + (-3) = 38 + (-3)$$
$$5x = 35.$$

Step 3 Use the multiplication property of equality. Multiply both sides by $\frac{1}{5}$:

$$\frac{1}{5} \cdot 5x = \frac{1}{5} \cdot 35$$
$$x = 7.$$

Step 4 Check the solution. Substitute 7 for x in the original equation.

$$2x + 3x + 3 = 38$$
$$2(7) + 3(7) + 3 = 38 \qquad \text{Let } x = 7$$
$$14 + 21 + 3 = 38$$
$$38 = 38. \qquad \text{True}$$

Since the final statement is true, 7 is the correct solution.

Example 2 Solve the equation $3r + 4 - 2r - 7 = 4r + 3$.

Step 1
$$3r + 4 - 2r - 7 = 4r + 3$$
$$r - 3 = 4r + 3 \qquad \text{Combine terms}$$

Step 2
$$r - 3 + 3 = 4r + 3 + 3 \qquad \text{Add}$$
$$r = 4r + 6$$
$$r + (-4r) = 4r + 6 + (-4r)$$
$$-3r = 6$$

Step 3
$$\left(-\frac{1}{3}\right)(-3r) = \left(-\frac{1}{3}\right)6 \qquad \text{Multiply}$$
$$r = -2.$$

Step 4 Substitute -2 for r in the original equation.

$$3r + 4 - 2r - 7 = 4r + 3$$
$$3(-2) + 4 - 2(-2) - 7 = 4(-2) + 3$$
$$-6 + 4 + 4 - 7 = -8 + 3$$
$$-5 = -5. \qquad \text{True}$$

The correct solution for the equation is -2.

Example 3 Solve the equation $4(k - 3) - k = k - 6$.

Step 1 Before combining terms, use the distributive property to simplify $4(k - 3)$.

$$4(k - 3) = 4k - 4 \cdot 3 = 4k - 12.$$

Now combine terms.

$$4(k - 3) - k = k - 6$$
$$4k - 12 - k = k - 6$$
$$3k - 12 = k - 6$$

Step 2

$$3k - 12 + 12 = k - 6 + 12$$
$$3k = k + 6$$
$$3k + (-k) = k + 6 + (-k)$$
$$2k = 6$$

Step 3

$$\frac{1}{2}(2k) = \frac{1}{2} \cdot 6$$
$$k = 3.$$

Step 4 Substitute 3 for k in the original equation. Remember to do all work inside parentheses first.

$$4(k - 3) - k = k - 6$$
$$4(3 - 3) - 3 = 3 - 6$$
$$4(0) - 3 = 3 - 6$$
$$0 - 3 = 3 - 6$$
$$-3 = -3. \qquad \text{True}$$

The correct solution to the equation is 3.

Example 4 Solve the equation $3(m + 5) - 5 + 2m = 2(m - 10)$.

Step 1 Use the distributive property to simplify both $3(m + 5)$ and $2(m - 10)$.

$$
\begin{aligned}
3(m+5)-5+2m &= 2(m-10)\\
3m+3\cdot 5-5+2m &= 2m-2\cdot 10\\
3m+15-5+2m &= 2m-20\\
5m+10 &= 2m-20.
\end{aligned}
$$

Step 2 Add -10 to both sides, and then add $-2m$ to both sides.

$$
\begin{aligned}
5m+10+(-10) &= 2m-20+(-10)\\
5m &= 2m-30\\
5m+(-2m) &= 2m-30+(-2m)\\
3m &= -30.
\end{aligned}
$$

Step 3 Multiply both sides by $\frac{1}{3}$.

$$\frac{1}{3}(3m)=\frac{1}{3}(-30)$$

$$m=-10.$$

Step 4 Substitute -10 for m in the original equation.

$$
\begin{aligned}
3(m+5)-5+2m &= 2(m-10)\\
3(-10+5)-5+2(-10) &= 2(-10-10)\\
3(-5)-5-20 &= 2(-20)\\
-15-5-20 &= -40\\
-40 &= -40. \qquad \text{True}
\end{aligned}
$$

The solution, -10, is correct.

Example 5 Solve $4(8-3t)=32-8(t+2)$.

Step 1 Simplify the equation as much as possible.

$$
\begin{aligned}
4(8-3t) &= 32-8(t+2)\\
32-12t &= 32-8t-16\\
32-12t &= 16-8t.
\end{aligned}
$$

Step 2 First add -32 to both sides, and then add $8t$ to both sides.

$$
\begin{aligned}
32-12t+(-32) &= 16-8t+(-32)\\
-12t &= -16-8t\\
-12t+8t &= -16-8t+8t\\
-4t &= -16.
\end{aligned}
$$

Step 3 Multiply both sides by $-\frac{1}{4}$:

$$-\frac{1}{4}(-4t)=-\frac{1}{4}(-16)$$

$$t=4.$$

Step 4 Substitute 4 for t in the original equation.

$$4(8 - 3t) = 32 - 8(t + 2)$$
$$4(8 - 3 \cdot 4) = 32 - 8(4 + 2)$$
$$4(8 - 12) = 32 - 8(6)$$
$$4(-4) = 32 - 48$$
$$-16 = -16. \qquad \text{True}$$

The correct solution is 4.

EXERCISES

In Exercises 1–16, simplify the expressions. Use the distributive property and combine terms.

1. $3(k - 6)$
2. $5(m + 4)$
3. $6(5t + 11)$
4. $3(2x + 4)$
5. $-3(n + 5)$
6. $-4(v - 8)$
7. $-2(3x - 4)$
8. $-5(4t + 6)$
9. $7(r + 2) - 3r$
10. $3(m - 6) + 5m$
11. $-5(2v - 3) + 2(5v + 3)$
12. $-4(5y - 7) + 3(2y - 5)$
13. $8(2k - 1) - (4k + 5)$
14. $6(3p - 2) - (5p + 1)$
15. $-2(-3k + 2) + 2(5k - 6) - 3k - 5$
16. $-2(3r - 4) - (6 - r) + 2r - 5$

In Exercises 17–30, solve the equation. Check the solution.

17. $4(h + 2) = 16$
18. $3(x - 5) = 9$
19. $3(2k + 4) = -12$
20. $2(m - 3) = -6$
21. $6(2p + 3) = 14p$
22. $5(2m - 3) = 7m$
23. $3x + 9 = -3(2x + 3)$
24. $4z + 2 = -2(z + 2)$
25. $3k - 5 = 2(k + 6) + 1$
26. $4a - 7 = 3(2a + 5) - 2$
27. $2(2r - 1) = -3(r + 3)$
28. $3(3k + 5) = 2(5k + 5)$
29. $2(3x + 4) = 8(2 + x)$
30. $4(3p + 3) = 3(3p - 1)$

In Exercises 31–50, combine terms as necessary. Then solve the equation.

31. $-4 - 3(2x + 1) = 11$
32. $8 - 2(3x - 4) = 2x$
33. $5(2m - 1) = 4(2m + 1) + 7$
34. $3(3k - 5) = 4(2k - 5) + 7$
35. $5(4t + 3) = 6(3t + 2) - 1$
36. $7(2y + 6) = 9(y + 3) + 5$
37. $5(x - 3) + 2 = 5(2x - 8) - 3$
38. $6(2v - 1) - 5 = 7(3v - 2) - 24$

39. $-2(3s + 9) - 6 = -3(3s + 11) - 6$

40. $-3(5z + 24) + 2 = 2(3 - 2z) - 10$

41. $6(2p - 8) + 24 = 3(5p - 6) - 6$

42. $2(5x + 3) - 3 = 6(2x - 3) + 15$

43. $3(m - 4) - (2m - 3) = -4$

44. $4(2a + 6) - (7a - 5) = 2$

45. $-(4m + 2) - (-3m - 5) = 3$

46. $-(6k - 5) - (-5k + 8) = -4$

47. $2x + 6x - 9 + 4 = 3x - 9 + 3$

48. $3(z - 2) + 4z = 8 + z + 1 - z$

49. $2(r - 3) + 5(r + 4) = 9$

50. $-4(m - 8) + 3(2m + 1) = 6$

In Exercises 51–54, use the information to write an equation. Then solve it.

51. If 17 is subtracted from a number, and the result is multiplied by 3, the product is 102. What is the number?

52. A teacher says, "If I had three times as many students in my class as I do have, I would have 46 more than I now have." How many students are now in the class?

53. If three times the sum of a number and 4 is subtracted from 8, the result is 2. Find the number.

54. If five times the sum of a number and 3 is added to 5, the result is -5. Find the number.

2.5 FROM WORD PROBLEMS TO EQUATIONS

To get the answer to a word problem, you must read it and sort out the facts. Then take the following steps.

(a) Choose a variable to represent the numerical value that is not given—the unknown number.
(b) Translate the problem into an equation.
(c) Solve the equation.
(d) Check your solution by using the original words of the problem.

Step (b) is the focus of this section. To translate the problem, go from facts stated in words to mathematical expressions. The problem also gives you some fact about equality, which leads to an equation.

What about the words? You are likely to see certain words again and again in problems. These words are listed here, along with the mathematical symbols that represent them. These words state facts about addition, subtraction, multiplication, or division.

The unknown quantity is x here, but any letter will do as the variable.

	Addition
5 **plus** a number	$5 + x$
add 20 to a number	$x + 20$
1 **added** to a number	$x + 1$
the **sum** of a number and 11	$x + 11$
7 **more than** a number	$x + 7$
a quantity is **increased by** 3	$x + 3$
10 feet **longer than** x	$x + 10$
warmer than x by 10 degrees	$x + 10$
total of x and two other values	$x + a + b$

	Subtraction
3 **less than** a number	$x - 3$
a number **decreased by** 14	$x - 14$
subtract 6 **from** a number	$x - 6$
16 **minus** a number	$16 - x$
the **difference between** x and 9	$x - 9$
take 4 **away** from a number	$x - 4$
ten **fewer than** x	$x - 10$
shorter by 2 feet than x	$x - 2$
your age three years **ago**	$x - 3$

	Multiplication
the **product of** the number and 3	$x \cdot 3$, or $3x$
three **times** a number	$3 \cdot x$, or $3x$
a number is **tripled**	$3x$
a number **multiplied by** 7	$x \cdot 7$, or $7x$
twice the amount	$2x$
two-thirds **of** a number	$\frac{2}{3} \cdot x$, or $\frac{2}{3}x$ (fractions only)
four **at** price x	$4x$

	Division (use a fraction bar instead of $\div$)
the **quotient** of a number and 2	$\frac{x}{2}$

half the number	$\frac{x}{2}$ or $\frac{1}{2}x$
two-thirds **of** a number	$\frac{2}{3}x$, or $\frac{2x}{3}$
a number **divided by** 5	$\frac{x}{5}$
5 **is divided by** some amount	$\frac{5}{x}$
the **reciprocal** of a number	$\frac{1}{x}$
the **ratio** of some number to 4	$\frac{x}{4}$

	Combinations
four times **a number minus 7**	$4(x - 7)$
four times a number, minus 7	$4x - 7$
the sum of two consecutive integers	$x + (x + 1)$
a number **plus** its reciprocal	$x + \frac{1}{x}$
the sum of a number and 2, multiplied by 5	$(x + 2) \cdot 5$, or $5(x + 2)$
a number divided by 4, plus the number	$\frac{x}{4} + x$
a number divided by **4 plus the number**	$\frac{x}{4 + x}$
three times the quotient of a number and 2	$3\left(\frac{x}{2}\right)$

Since equal mathematical statements are names for the same number, any words that mean equality or sameness translate as =.

	Equality
Four times a number decreased by 7 **is** 100.	$4(x - 7) = 100$
The quotient of a number plus 7, and 2, **is** 4.	$\frac{x + 7}{2} = 4$
When 3 times a number was added to 9, the **answer was** 12.	$9 + 3x = 12$
If you add 10 to a number, the **result is** 20.	$x + 10 = 20$
Twice a number, increased by three, **equals** 17.	$2x + 3 = 17$
Increasing a number by 10 **is the same as** three times the number.	$x + 10 = 3x$

In translating word problems about numbers into mathematical statements, begin by representing the unknown number as a variable. Write down exactly what unknown number the variable is to represent. Then, when you translate, the variable represents the unknown number each time it is mentioned in the problem.

Example 1 If three times the sum of a number and 4 is decreased by twice the number, the result is -6. Find the number.

Let x represent the unknown number. "Three times the sum of a number and 4" translates into symbols as $3(x + 4)$. "Twice the number" is $2x$. Now write an equation, using the information of the problem.

three times the sum of a number and 4	decreased by	twice the number	the result is	-6
↓	↓	↓	↓	↓
$3(x + 4)$	$-$	$2x$	$=$	-6

We can now solve the equation.

$$3(x + 4) - 2x = -6$$
$$3x + 12 - 2x = -6$$
$$x + 12 = -6$$
$$x = -18.$$

The number is -18.

Example 2 In a given amount of time, Alice drove 40 miles more than Fred. The total distance that both of them traveled was 204 miles. Find the number of miles driven by Fred.

Let x represent the number of miles driven by Fred. Since Alice drove 40 miles more than Fred, the number of miles she drove is $x + 40$. The problem gives you the total number of miles, 204.

miles (Fred)		miles (Alice)	is	total miles
x	$+$	$x + 40$	$=$	204

Solve the equation.

$$x + (x + 40) = 204$$
$$2x + 40 = 204$$
$$2x = 164$$
$$x = 82.$$

Fred drove 82 miles. Alice drove 40 miles more, or $82 + 40 = 122$ miles. Their total miles is $82 + 122 = 204$ miles. This checks with the information in the problem.

EXERCISES

For each word problem, follow these steps:

(a) Choose a variable to represent the unknown quantity.
(b) Translate the problem into an equation.
(c) Solve the equation.
(d) Check your solution by using the original words of the problem.

1. If three times a number is decreased by 2, the result is 22. Find the number.

2. When six is added to four times a number, the result is 42. Find the number.

3. The sum of a number and 3 is multiplied by 4, giving 36 as a result. Find the number.

4. Five is multiplied by the sum of a number and 8, giving 60. Find the number.

5. Carl has a board 44 inches long. He wishes to cut it into two pieces so that one piece will be three inches longer than the other. How long should the shorter piece be?

6. Johnson and Smith were opposing candidates in the school board election. Johnson received 30 more votes than Smith, and 516 total votes were cast. How many votes did Smith receive?

7. On an algebra test, the highest grade was 42 points more than the lowest grade. The sum of the two grades was 138. Find the lowest grade.

8. In a physical fitness test, Nat did 25 more pushups than Chuck did. The total number of pushups for both men was 173. Find the number of pushups that Chuck did.

9. Pharmacist Mary found that at the end of a day she had 12 more prescriptions for antibiotics than she had for tranquilizers. She had 84 prescriptions altogether for these two types of drugs. How many did she have for tranquilizers?

10. A. L. Gator gives glass-bottom boat rides in Old Creepy Swamp. One day he noticed that the boat contained 17 more men (counting himself) than women, with a total of 165 people on the boat. How many women were on the boat?

11. Madelene Guernsey runs a dairy farm. Last year, her cow Bessie gave 238 more gallons of milk than one of her other cows, Bossie. Between them, the two cows gave 1464 gallons of milk. How many gallons of milk did Bossie give?

12. Myna Byrd, the zookeeper, noticed that her zoo has 82 more flamingos than parrots. If the zoo has a total of 224 of these two types of birds, how many parrots does the zoo have?

13. A farmer raises only chickens and cows. Among the animals, there are a total of 564 legs. If there are 142 chickens, how many cows are there?

14. Joe is three times as old as Dick. Three years ago the sum of their ages was 22 years. How old is each now? (Hint: First write an expression for the age of each now, then for the age of each three years ago.)

15. A bridge has three spans, one of which is 100 feet longer than each of the other two. If the bridge is 2500 feet long, how long is each of the shorter spans?

16. Ms. Emerson is employed at an annual salary of \$6000, with an annual increase of \$300. Ms. White started at the same time, with a salary of \$7000 and an annual increase of \$200. After how many years will the two people be earning the same salary?

17. A store has 39 quarts of milk, some in pint cartons and some in quart cartons. There are six times as many quart cartons as pint cartons. How many quart cartons are there?

18. A table is three times as long as it is wide. If it were three feet shorter and three feet wider, it would be square (with all sides equal). How long and how wide is it?

2.6 FORMULAS

Many word problems can be solved if you know a formula stating the connections among certain dimensions or amounts or quantities. Formulas exist for geometric figures such as squares and circles. Other formulas exist for distance, and for money earned on bank savings.

Suppose a word problem talks about fencing in a rectangular piece of land. You have to find how much fence is needed to enclose it. The problem gives you the measurements of the piece of land, the length and the width.

A formula exists for the perimeter of a rectangle:

$$P = 2l + 2w.$$

In the formula, P stands for perimeter (of a rectangle), l stands for the long side (length), and w stands for the short side (width). The perimeter of a rectangle equals the sum of twice the length and twice the width.

After deciding on the correct formula, insert the given information, as shown in the next example.

Example 1 The perimeter of a rectangle is 80, and the length is 25. (See Figure 2.1.) Find the width.

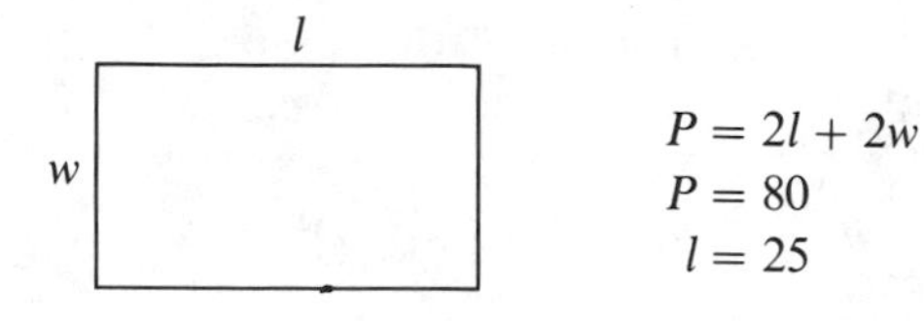

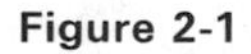
Figure 2-1

You can find the width by substituting 80 for P and 25 for l in the formula $P = 2l + 2w$.

$$P = 2l + 2w$$
$$80 = 2 \cdot 25 + 2w.$$

Solve the second equation to find w. First simplify the equation.

$$80 = 50 + 2w.$$

Then add -50 to both sides.

$$80 + (-50) = 50 + 2w + (-50)$$
$$30 = 2w$$
$$15 = w.$$

The width of the rectangle is 15.

Example 2 The perimeter of a square is 96 inches. Find the length of a side.

You need to know the formula for the perimeter of a square. From the list inside the back cover, you can get the formula $P = 4s$, where s is the length of a side of a square.

The perimeter is given as 96 inches, so that $P = 96$. Substitute 96 for P in the formula $P = 4s$.

$$P = 4s$$
$$96 = 4s$$
$$\frac{1}{4}(96) = \frac{1}{4}(4s)$$
$$24 = s.$$

The side of the square is 24 inches long.

Example 3 The area of a triangle is 126 square inches. The base of the triangle is 21 inches. Find the height.

The formula for area of a triangle is $A = \frac{1}{2}bh$, where A is area, b is the base, and h is the height. Substitute 126 for A and 21 for b in the formula.

$$A = \frac{1}{2}bh, \quad 126 = \frac{1}{2}(21)h.$$

Simplify the problem by getting rid of the fraction $\frac{1}{2}$. Multiply both sides of the equation by 2.

$$2(126) = 2\left(\frac{1}{2}\right)(21)h$$

$$252 = 21h.$$

Now multiply both sides by 1/21.

$$\frac{1}{21}(252) = \frac{1}{21}(21h)$$

$$12 = h.$$

The height of the triangle is 12 inches.

Example 4 The length of a rectangle is 2 more than the width. The perimeter is 40. Find the width and the length.

Let x represent the width of the rectangle. Then $x + 2$ represents the length. The formula for the perimeter of a rectangle is $P = 2l + 2w$. Substitute x for w, $x + 2$ for l, and 40 for P.

$$\begin{aligned} P &= 2l + 2w \\ 40 &= 2(x + 2) + 2x \\ 40 &= 2x + 4 + 2x \\ 40 &= 4x + 4 \\ 36 &= 4x \\ 9 &= x. \end{aligned}$$

The width of the rectangle is 9, and the length is $x + 2 = 9 + 2 = 11$.

Example 5 How much simple interest will be earned on a deposit of \$600 for 4 years at an interest rate of 6% per year?

The formula for simple interest is $I = prt$, where I represents interest, p represents principal, r is rate, and t is time in years. Using the given information, we have

$$\begin{aligned} I &= prt \\ I &= (600)(.06)(4) \qquad 6\% = .06 \\ I &= 144 \end{aligned}$$

The deposit will earn interest of \$144.

Example 6 Two cars start from the same point at the same time and travel in the same direction at constant speeds of 34 and 45 miles per hour, respectively. In how many hours will they be 33 miles apart?

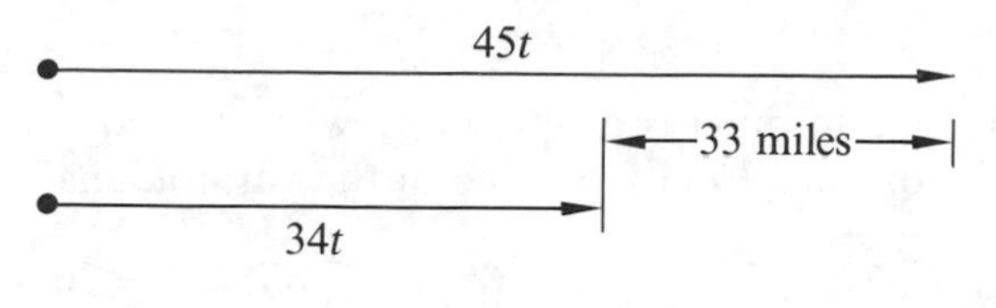

Figure 2-2

To work this problem, we need to know the formula for distance:

$$d = rt,$$

where d represents distance, r is rate (or speed), and t is time.

Let t represent the unknown number of hours. The distance traveled by the slower car is its rate multiplied by its time, or $34t$. The distance traveled by the faster car is $45t$. The numbers $34t$ and $45t$ represent different distances. From the information in the problem we know that these distances differ by 33 miles, or $45t - 34t = 33$. Now solve the equation to find t.

$$45t - 34t = 33$$
$$11t = 33$$
$$t = 3.$$

In 3 hours the two cars will be 33 miles apart.

Sometimes it is necessary to solve a large number of problems which involve the same formula. For example, we might need to solve several problems which involve the formula for the area of a rectangle, $A = lw$. Suppose that in each problem we are given the area and the length of a rectangle and we need to find the width. We can save time in the long run by rewriting the formula $A = lw$ so that the number we do not know, w, is alone on one side of the equals sign. To do this, multiply both sides of $A = lw$ by $1/l$:

$$A = lw$$
$$\frac{1}{l}A = \frac{1}{l}(lw)$$
$$\frac{A}{l} = w$$

For example, if $A = 50$ and $l = 10$, then w is found as follows:

$$w = \frac{A}{l} = \frac{50}{10} = 5$$

The width of the rectangle is 5. This process of isolating one letter from a formula is called *solving for a specified variable.*

EXERCISES

In Exercises 1–14, a formula is given, along with the value of some of the variables of the formula. Find the value of the variable that is not given.

1. $P = 4s$; $s = 32$

2. $P = 2l + 2w$; $l = 5, w = 3$

3. $A = \frac{1}{2}bh$; $b = 6, h = 12$

4. $A = \frac{1}{2}bh$; $b = 6, h = 24$

5. $A = \frac{1}{2}bh$; $A = 20, b = 5$

6. $A = \frac{1}{2}bh$; $A = 30, b = 6$

7. $P = 2l + 2w$; $P = 40, w = 6$

8. $V = \frac{1}{3}Bh$; $V = 80, B = 24$

9. $A = \frac{1}{2}(b + B)h$; $b = 6, B = 8, h = 3$

10. $A = \frac{1}{2}(b + B)h$; $b = 10, B = 12, h = 3$

11. $d = rt$; $d = 8, r = 2$

12. $d = rt$; $d = 100, t = 5$

13. $C = 2\pi r$; $C = 9.42, \pi = 3.14$*

14. $C = 2\pi r$; $C = 25.12, \pi = 3.14$*

In Exercises 15–28, solve the given formula for the specified variable.

15. $A = lw$; l

16. $d = rt$; r

17. $d = rt$; t

18. $V = lwh$; w

19. $V = lwh$; h

20. $I = prt$; p

21. $I = prt$; t

22. $C = 2\pi r$; r

23. $A = \frac{1}{2}bh$; b

24. $A = \frac{1}{2}bh$; h

25. $P = 2l + 2w$; w

26. $a + b + c = p$; b

27. $A = \frac{1}{2}(b + B)h$; b

28. $C = \frac{5}{9}(F - 32)$; F

In Exercises 29–51, write an equation and solve it. Check your solution in the original statement of the problem. Formulas are listed inside the back cover.

29. The area of a rectangle is 60 and the width is 6. Find the length.

30. The perimeter of a square is 80. Find the length of a side.

* Actually, π is not equal to 3.14, but is *approximately* equal to 3.14.

31. The radius of a circle is 6. Find the circumference. (Let π be approximated by 3.14.)

32. The length of a rectangle is 15, and the perimeter is 50. Find the width.

33. The perimeter of a triangle is 72. One side is 16, and another side is 32. Find the third side.

34. Harry drove 480 miles in 12 hours on a mountain road. Find his speed.

35. A train goes 55 miles per hour for 7 hours. Find the distance that it traveled.

36. The shorter base of a trapezoid is 16 and the longer base is 20. The height is 6. Find the area.

37. The perimeter of a square is seven times the length of a side, decreased by 12. Find the length of a side.

38. The circumference of a circle is five times the radius increased by 2.56 meters.* Find the radius of the circle. (Use 3.14 as an approximation for π.)

39. The perimeter of a rectangle is 16 times the width. The length is 12 centimeters more than the width. Find the dimensions of the rectangle.

40. The numerical value of the area of a triangle is five times the length of the base. The height is fifty times the reciprocal of the base. Find the length of the base.

41. The width of a rectangle is one less than the length. The perimeter is five times the length, decreased by 5. Find the length of the rectangle.

42. From a point on a straight road, John and Fred ride ten-speed bicycles in opposite directions. John rides 10 miles per hour and Fred rides 12 miles per hour. In how many hours will they be 55 miles apart?

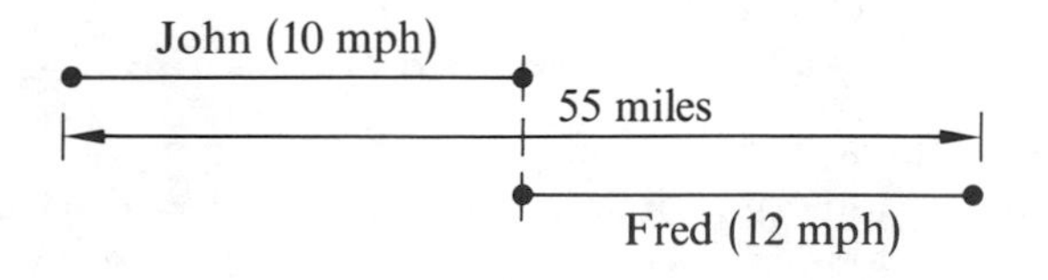

43. Two trains leave Los Angeles at the same time. One travels north at 60 miles per hour and the other south at 80 miles per hour. In how many hours will they be 280 miles apart?

* A *meter* is a unit of length in the metric system. An introduction to the metric system is an appendix at the back of this book.

44. Two cars are 400 miles apart. Both start at the same time and travel toward one another. They meet four hours later. If the average speed of one car is 20 miles per hour faster than the other, what is the average speed of each car?

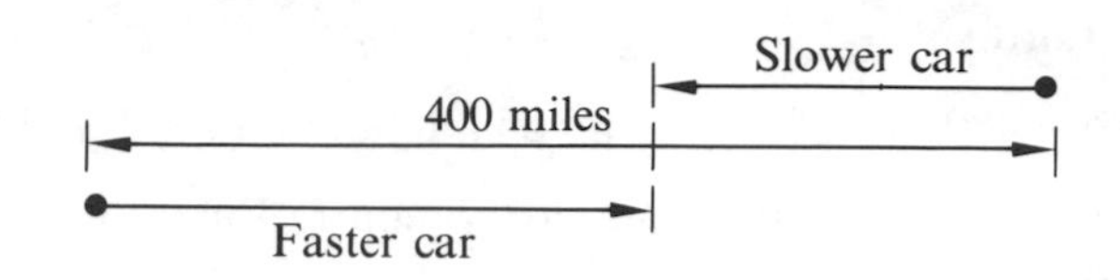

45. Ann has saved \$163 for a trip to Disneyland. Transportation will cost \$28, tickets for the park entrance and rides will cost \$15 per day, and lodging and meals will cost \$30 per day. How many days can she spend there?

46. Hamburgers cost 90 cents each, and a bag of french fries costs 40 cents. How many hamburgers and how many bags of french fries can Ted buy with \$8.80 if he wants twice as many hamburgers as bags of french fries?

47. Ms. Sullivan has \$10,000 to invest. She wants to invest part at $5\frac{1}{2}\%$ interest and part at $6\frac{1}{2}\%$. The total annual interest will be \$625. How much should be invested at each rate?

48. Mr. Jones received \$16,000 from his mother's estate. He invested part at 8% per year. He put \$2000 less than twice that amount in a safe 5% bond. His total annual income from interest is \$980. How much did he invest at each rate?

49. A boat travels upstream for three hours. The return trip requires two hours. If the speed of the current is 5 miles per hour, find the speed of the boat in still water.

50. In an automobile motorcross, a driver was 120 miles from the finish line after 5 hours. Another driver, who was in a later heat, traveled at the same speed as the first driver. After 3 hours, she was 250 miles from the finish. What was the average speed of each driver?

51. Tom Harrison earns \$8600 per year. He is given a 4% annual raise in salary. At the same time, inflation in the economy is 6% per year. By how many dollars has his purchasing power declined during the year?

2.7 THE ADDITION PROPERTY OF INEQUALITY

Inequalities are statements in which algebraic expressions are related by

	$<$ is less than		$>$ is greater than
or	$\leq$ is less than or equal to	or	$\geq$ is greater than or equal to.

We assume that the domain of any inequality we study is the set of all real numbers. So when you solve an inequality, you must find all real number solutions for it. For example, $x \leq 2$ represents all real numbers that are less than or equal to 2, and not just the integers less than or equal to 2.

To graph all real numbers satisfying $x \leq 2$, place a dot at 2 on a number line, and draw an arrow extending from the dot to the left (to represent the fact that all numbers less than 2 are also part of the graph). The graph is in Figure 2.3.

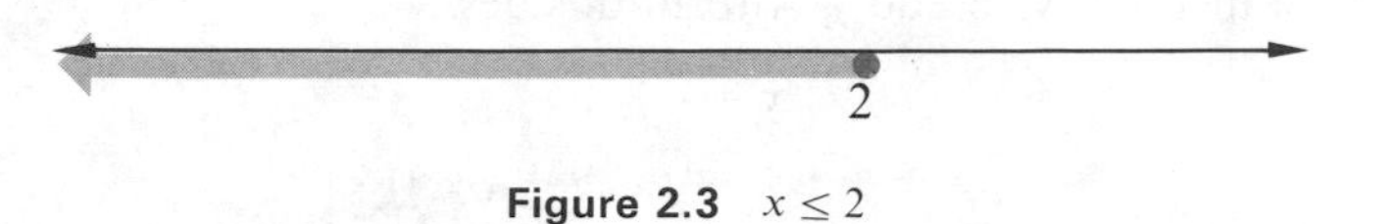

Figure 2.3 $x \leq 2$

Example 1 Graph $x > -5$.

The statement $x > -5$ says that x can represent numbers greater than -5, but x cannot equal -5 itself. To show this on a graph, place an open circle at -5, with an arrow to the right, as in Figure 2.4.

−5

Figure 2-4 $x > -5$

Example 2 Graph $-3 \leq x < 2$.

The statement $-3 \leq x < 2$ is read "-3 is less than or equal to x and x is less than 2." To graph this inequality, place a heavy dot at -3 (because of the $\leq$ symbol), and an open circle at 2 (because of the $<$ symbol). Then draw a line segment between the two circles, as in Figure 2.5.

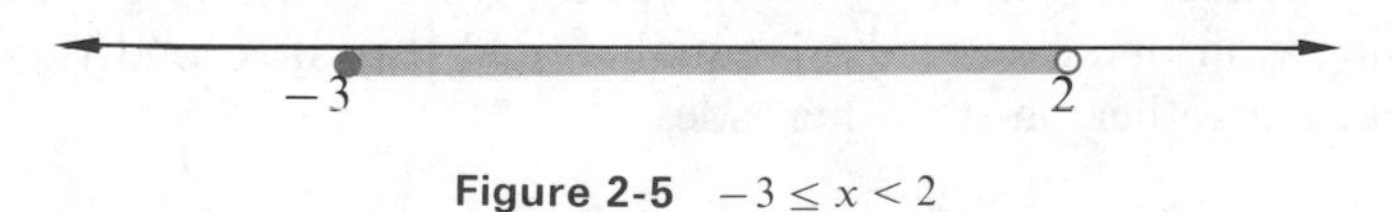

Figure 2-5 $-3 \leq x < 2$

We have mentioned these methods of graphing inequalities since such graphs give us a good way to show the solution of inequalities. We shall see that inequalities such as $x + 4 \leq 9$ can be solved in much the same way that we solved equations.

To solve an inequality such as $x + 4 \leq 9$, we must find all values for the variable that make the given inequality true. The inequality $x + 4 \leq 9$ tells us that $x + 4$ is less than or equal to 9. If the same term were added to both $x + 4$ and 9, the resulting sums would still be unequal. And they would be unequal in the *same order*. This fact is the **addition property of inequality,** which states that the same term can be added to both sides of an inequality.

For any expressions A, B, and C, the inequalities

$$A < B$$

and

$$A + C < B + C$$

have exactly the same solutions.

The addition property of inequality also works with $>$, $\leq$, or $\geq$.

Going back to $x + 4 \leq 9$, you can solve this by using the addition property of inequality to add -4 to both sides.

$$x + 4 \leq 9$$
$$x + 4 + (-4) \leq 9 + (-4)$$
$$x \leq 5.$$

Any number less than or equal to 5 satisfies the inequality $x + 4 \leq 9$. This solution can be written as a solution set, $\{x | x \leq 5\}$. In words, "The set of all numbers x such that x is less than or equal to 5." In this book, the solution of such inequalities is written $x \leq 5$. This is shorthand for "The solution is the set of all real numbers less than or equal to 5." The graph of the solution $x \leq 5$ is in Figure 2.6.

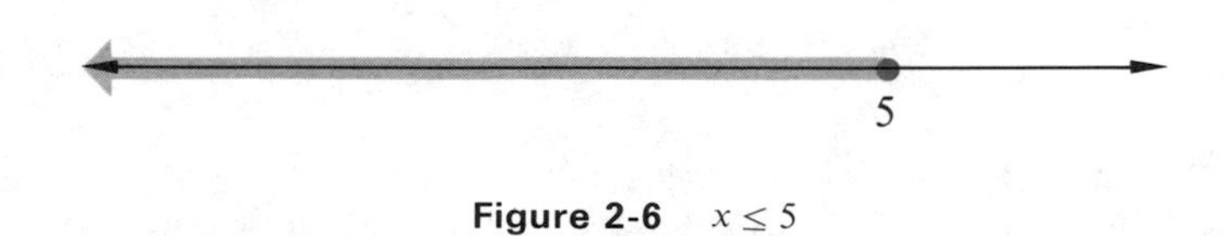

Figure 2-6 $x \leq 5$

Example 3 Solve the inequality $7 + 3k > 2k - 5$.

Use the addition property of inequality twice—once to get the terms containing k on one side of the inequality, and then a second time to get the integers together on the other side.

$$7 + 3k > 2k - 5$$
$$7 + 3k + (-2k) > 2k - 5 + (-2k)$$
$$7 + k > -5$$
$$7 + k + (-7) > -5 + (-7)$$
$$k > -12.$$

The graph of the solution $k > -12$ is in Figure 2.7.

-12

Figure 2-7 $k > -12$

Example 4 Solve $6 + 3y \geq 4y - 5$.

First add $-3y$ to both sides.

$$6 + 3y + (-3y) \geq 4y - 5 + (-3y)$$
$$6 \geq y - 5.$$

Then add 5 to both sides.

$$6 + 5 \geq y - 5 + 5$$
$$11 \geq y.$$

This solution is perfectly correct, but it is customary to write the solution to an inequality with the variable on the left. The statement $11 \geq y$ says that 11 is greater than or equal to y. We can say the same thing in another way by saying that y is *less* than or equal to 11, or

$$y \leq 11.$$

The graph of the solution $y \leq 11$ is in Figure 2.8.

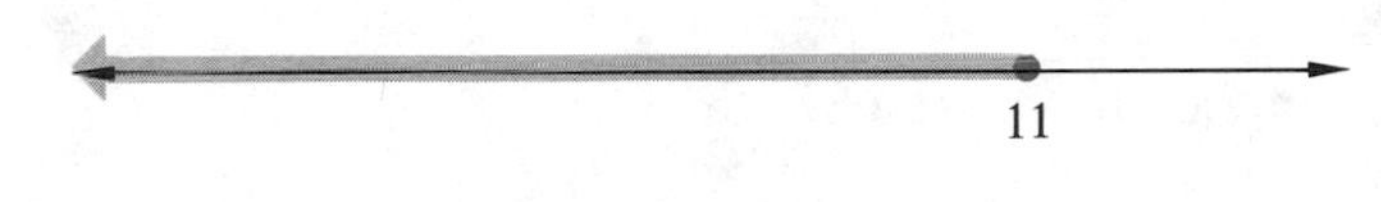

Figure 2.8 $y \leq 11$

If the inequality $6 + 3y \geq 4y - 5$ in Example 4 were solved by first adding the term $-4y$ to both sides, then

$$6 + 3y + (-4y) \geq 4y - 5 + (-4y)$$
$$6 - y \geq -5$$
$$6 - y + (-6) \geq -5 + (-6)$$
$$-y \geq -11.$$

With an answer like $-y \geq -11$, we cannot go any further with the methods discussed so far. To finish this solution we need the multiplication property of inequality, which we postpone until the next section. In the exercises that follow, if you get an answer like $-y \geq -11$, then start over—add terms to both sides in a different way.

Example 5 If 2 is added to five times a number, the result is greater than or equal to 5 more than four times the number. Find the number.

Let x represent the number you want to find. Then "2 is added to five times a number" is expressed as $5x + 2$. And "5 more than four times the number" is $4x + 5$. The two expressions are related by "greater than or equal to."

$$5x + 2 \geq 4x + 5.$$

To solve the inequality $5x + 2 \geq 4x + 5$, first add $-4x$ to both sides.

$$5x + 2 + (-4x) \geq 4x + 5 + (-4x)$$
$$x + 2 \geq 5.$$

Then add -2 to both sides.

$$x + 2 + (-2) \geq 5 + (-2)$$
$$x \geq 3.$$

The number you want is greater than or equal to 3.

EXERCISES

Graph each of the inequalities in Exercises 1–10 on a number line.

1. $x \leq 4$

2. $x \leq -3$

3. $x \geq -5$

4. $x \geq 6$

5. $x < 3$

6. $x > 4$

7. $-2 \leq x \leq 5$

8. $8 \leq x \leq 10$

9. $3 \leq x < 5$

10. $0 < x \leq 10$

Solve each of the inequalities of Exercises 11–24.

11. $x + 6 < 8$

12. $x - 4 < 2$

13. $x - 3 \geq -2$

14. $x + 2 \geq -6$

15. $x - 8 \leq 4$

16. $2 + x \geq 5$

17. $-3 + x \geq 2$

18. $-8 + x < -10$

19. $x + 6 \leq 6$

20. $x + 5 > 5$

21. $4x < 3x + 6$

22. $5x \leq 4x - 8$

23. $9x \leq 8x - 9$

24. $12x > 11x + 5$

Solve each of the inequalities of Exercises 25–34. Graph each solution.

25. $3x + 5 \leq 2x - 6$

26. $5x - 2 < 4x - 5$

27. $2x - 8 > x - 3$

28. $4x + 6 \leq 3x - 5$

29. $3(x - 5) + 2 < 2(x - 4)$

30. $4(x + 6) - 5 > 3(x + 1)$

31. $-6(x + 2) + 3 \geq -7(x - 5)$

32. $-3(x - 5) + 8 < -4(x + 2)$

33. $5(2x + 3) - 2(x - 8) > 3(2x + 4) + x - 2$

34. $2(3x - 5) + 4(x + 6) \geq 2(3x + 2) + 3(x - 5)$

In each of Exercises 35–36, write an inequality using the information given in the problem, and then solve it.

35. Betty made 82 on her first algebra test, 76 on the second, and 88 on the third. What is the lowest grade that she can get on the fourth test so that the average of the four tests is at least 80?

36. If the length of a rectangle is to be twice the width, and the difference between the two dimensions is to be less than or equal to 7, what is the largest possible value for the width?

2.8 THE MULTIPLICATION PROPERTY OF INEQUALITY

With the addition property of inequality alone, we cannot solve inequalities such as $-y \geq -11$. To solve this inequality, we need the multiplication property of inequality. To see how this property works, let's look at some examples.

First take the inequality $3 < 7$ and multiply both sides by the positive number 2.

$$\begin{aligned} 3 &< 7 \\ 2(3) &< 2(7) \\ 6 &< 14. \qquad \text{True} \end{aligned}$$

Multiply both sides of $3 < 7$ by the negative number -5.

$$\begin{aligned} 3 &< 7 \\ -5(3) &< -5(7) \\ -15 &< -35. \qquad \text{False} \end{aligned}$$

To get a true statement when we multiply both sides by -5, we would have to reverse the direction of the inequality symbol.

$$\begin{aligned} 3 &< 7 \\ -5(3) &> -5(7) \\ -15 &> -35. \qquad \text{True} \end{aligned}$$

Take the inequality $-6 < 2$ as another example. Multiply both sides by the positive number 4.

$$\begin{aligned} -6 &< 2 \\ 4(-6) &< 4(2) \\ -24 &< 8. \qquad \text{True} \end{aligned}$$

If we multiply both sides of $-6 < 2$ by -5, and *at the same time reverse the direction of the inequality symbol*, we get

$$\begin{aligned} -6 &< 2 \\ (-5)(-6) &> (-5)(2) \\ 30 &> -10. \qquad \text{True} \end{aligned}$$

In summary, we have the two parts of the **multiplication property of inequality.**

For any expressions A, B, and C:

(1) If C is positive, then the inequalities

$$A < B$$

and

$$AC < BC$$

have exactly the same solutions.

(2) If C is negative, then the inequalities

$$A < B$$

and

$$AC > BC$$

have exactly the same solutions.

The multiplication property of inequality also works with $>$, $\leq$, or $\geq$.

Important (1) When you multiply both sides of an inequality by a positive number, the direction of the inequality symbol does not change. Also, adding terms to both sides does not change the symbol. (2) But when you multiply both sides of an inequality by a negative number, then the direction of the symbol does change. *You reverse the symbol of inequality only when multiplying by a negative number, not otherwise.*

Example 1 Solve the inequality $3r < 18$.

To simplify this inequality, use the multiplication property of inequality, and multiply both sides by $\frac{1}{3}$. Since $\frac{1}{3}$ is a positive number, the direction of the inequality symbol does not change.

$$3r < 18$$

$$\frac{1}{3}(3r) < \frac{1}{3}(18)$$

$$r < 6.$$

The graph of this solution is shown in Figure 2.9.

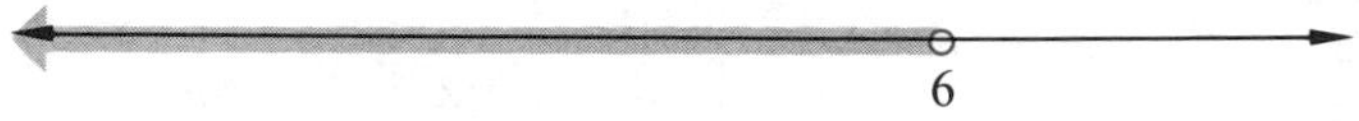

6

Figure 2-9 $r < 6$

Example 2 Solve the inequality $-4t \geq 8$.

Here we need to multiply both sides of the inequality by $-\frac{1}{4}$, a negative number. This does change the direction of the inequality symbol.

$$-4t \geq 8$$
$$\left(-\frac{1}{4}\right)(-4t) \leq \left(-\frac{1}{4}\right)(8)$$
$$t \leq -2.$$

The solution is graphed in Figure 2.10.

-2

Figure 2-10 $t \leq -2$

Example 3 Solve the inequality $-x \leq -11$.

Use the multiplication property of inequality and multiply both sides by -1, a negative number. Since -1 is negative, change the direction of the inequality symbol.

$$-x \leq -11$$
$$(-1)(-x) \geq (-1)(-11)$$
$$x \geq 11.$$

The solution is graphed in Figure 2.11.

11

Figure 2-11 $x \geq 11$

We can now summarize the steps involved in solving an inequality.

Step 1 Use the associative, commutative, and distributive properties to combine terms on both sides of the inequality.

Step 2 Use the addition property of inequality to simplify the inequality to one of the form $ax < b$, where a and b are real numbers.

Step 3 Use the multiplication property of inequality to simplify further to an inequality of the form $x < c$ or $x > c$, where c is a real number.

(In this summary, $<$ can be replaced with $>$, $\leq$, or $\geq$.)

Example 4 Solve the inequality $3z + 2 - 5 > -z + 7 + 2z$.

Step 1 Simplify and combine terms.

$$3z + 2 - 5 > -z + 7 + 2z$$
$$3z - 3 > z + 7.$$

Step 2 Use the addition property of inequality.

$$3z - 3 + 3 > z + 7 + 3$$
$$3z > z + 10$$
$$3z + (-z) > z + 10 + (-z)$$
$$2z > 10.$$

Step 3 Use the multiplication property of inequality.

$$\frac{1}{2}(2z) > \frac{1}{2}(10)$$
$$z > 5.$$

Since $\frac{1}{2}$ is positive, the direction of the inequality symbol was not changed in step 3. A graph of the solution is shown in Figure 2.12.

5

Figure 2-12 $z > 5$

Example 5 Solve $5(k - 3) - 7k \geq 4(k - 3) + 9$.

Step 1 Combine terms.

$$5(k - 3) - 7k \geq 4(k - 3) + 9$$
$$5k - 15 - 7k \geq 4k - 12 + 9$$
$$-2k - 15 \geq 4k - 3.$$

Step 2 Use the addition property.

$$-2k - 15 + (-4k) \geq 4k - 3 + (-4k)$$
$$-6k - 15 \geq -3$$
$$-6k - 15 + 15 \geq -3 + 15$$
$$-6k \geq 12.$$

Step 3 Multiply both sides by $-\frac{1}{6}$, a negative number. Change the direction of the inequality symbol.

$$\left(-\frac{1}{6}\right)(-6k) \leq \left(-\frac{1}{6}\right)(12)$$
$$k \leq -2.$$

A graph of the solution is shown in Figure 2.13.

−2

Figure 2-13 $k \leq -2$

EXERCISES

In Exercises 1–24, solve each inequality. Graph each solution.

1. $3x < 27$

2. $5h \geq 20$

3. $4r \geq -12$

4. $6a < -18$

5. $-2k \leq 12$

6. $-3v > 6$

7. $-8y > 72$

8. $-5z \leq 40$

9. $-5m > -35$

10. $-8x \leq -16$

11. $-6r < -16$

12. $-9a \geq -63$

13. $4k + 1 \geq 2k - 9$

14. $5y + 3 < 2y + 12$

15. $3 + 2r > 5r - 27$

16. $8 + 6t \leq 8t + 12$

17. $4q + 1 - 5 < 8q + 4$

18. $5x - 2 \leq 2x + 6 - x$

19. $10p + 20 - p > p + 3 - 23$

20. $-3v + 6 + 3 - 2 > -5v - 19$

21. $-k + 4 + 5k \leq -1 + 3k + 5$

22. $6y - 2y - 4 + 7y > 3y - 4 + 7y$

23. $2(x - 5) + 3x < 4(x - 6) + 3$

24. $5(t + 3) - 6t \leq 3(2t + 1) - 4t$

In Exercises 25–28, write an inequality using the information given in the problem. Then solve it.

25. A student has test grades of 75 and 82. What must he score on a third test to have an average of 80 or higher?

26. In Exercise 25, if 100 is the highest score possible on the third test, how high an average (to the nearest tenth) can the student make? What is the lowest average possible for the three tests?

27. Twice a number added to three times the sum of a number and 2 is more than 17. Find the numbers that satisfy this condition.

28. Fred earned \$200 at odd jobs during July, \$300 during August, and \$225 during September. If his average salary for the four months from July through October is to be at least \$250, how much must he earn during October?

2.9 COMPOUND STATEMENTS (OPTIONAL)

A statement made up of two simple statements, such as

$$x < 2 \text{ and } x > -1,$$

is called a **compound statement.** To find the solutions of a compound statement, we must find all numbers that make the entire statement true. Compound statements containing the word "and" are true only when both parts of the statement are true. This is the same as the common meaning of "and."

The solution set of a compound statement involving "and" is the **intersection,** or *overlap,* of the solution sets of the individual parts. To find the solution of the compound statement

$$x < 2 \text{ and } x > -1,$$

graph $x < 2$ and graph $x > -1$ separately. Then take the intersection or overlap of these two graphs to get the solution of the compound statement. Figure 2.14 shows the graphs of $x < 2$, $x > -1$, and the intersection. From these graphs, we see that the intersection is the set of all points between -1 and 2. Thus, the solution of the compound statement $x < 2$ and $x > -1$ can be written as $-1 < x < 2$.

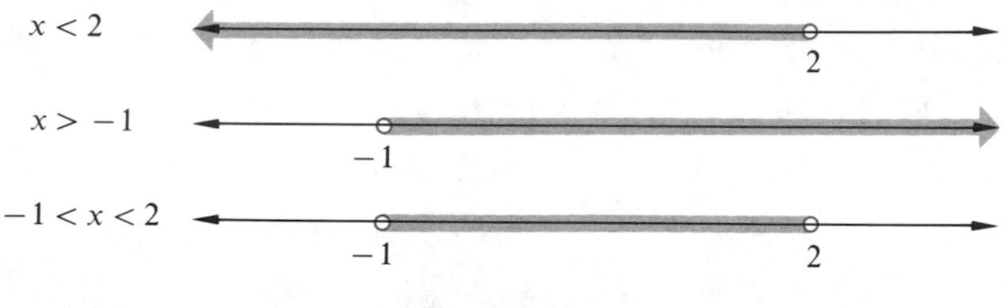

Figure 2-14

Example 1 Solve the compound statement $x < 4$ and $x < 2$.

Graph $x < 4$, and graph $x < 2$ separately, as shown in Figure 2.15. From the figure, we see that the intersection, and hence the solution, is $x < 2$.

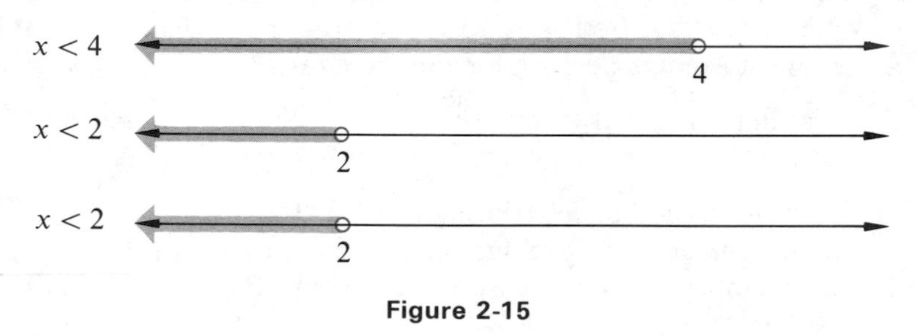

Figure 2-15

Example 2 Solve the compound statement $x \leq -4$ and $x \geq -1$.

As shown in Figure 2.16, there are no points common to both $x \leq -4$ and to $x \geq -1$, so there is no solution for the compound statement.

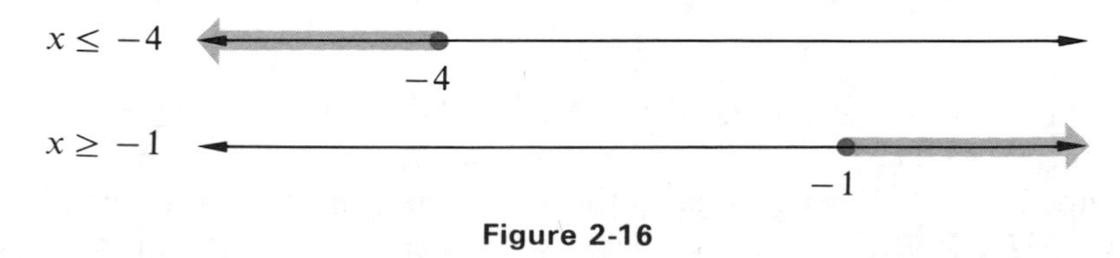

Figure 2-16

Compound statements can also be formed by using the word "or." A compound statement using the word "or" is true if *either one* of its parts is true or if *both parts* are true. To find the solution for the compound statement

$$x < -2 \text{ or } x \geq 3,$$

graph both $x < -2$ and $x \geq 3$ separately, as shown in Figure 2.17. Then take the **union** (put the two sets together) of these two graphs. The union is shown on the third number line of Figurc 2.17. The solution here is written $x < -2$ or $x \geq 3$. There is no compact way to write it.

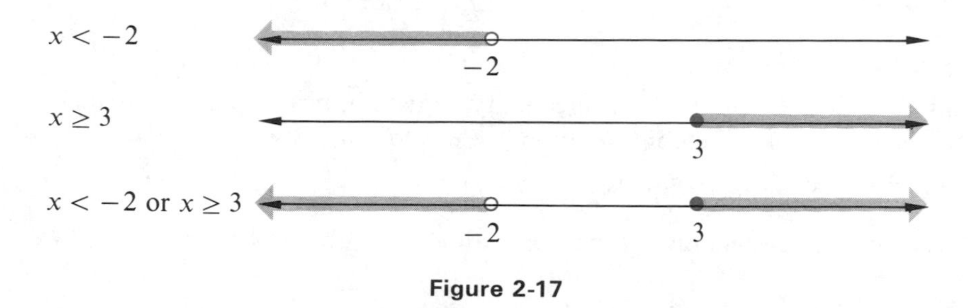

Figure 2-17

Example 3 Solve the compound statement $x > 2$ or $x > -1$.
As shown in Figure 2.18, the union, and thus the solution, is $x > -1$.

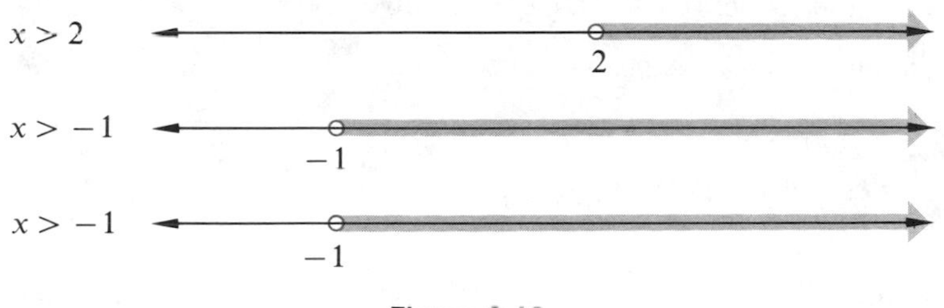

Figure 2-18

EXERCISES

In Exercises 1–12, write *true* or *false* for each compound statement.

1. $4 = 5 - 1$ and $3 \leq 5$

2. $3 + 2 > 3$ and $7 > 0$

3. $5 = 6 - 1$ and $-2 \geq 3$

4. $8 \geq 8$ and $3 \geq 2$

5. $5 + 4 < 9$ and $6 \neq 5 + 1$

6. $-8 < 4$ and $2 \neq 3$

7. $9 = 7 + 2$ or $2 = 3 - 1$

8. $9 \leq 10$ or $5 \leq 6$

9. $3 < 3$ or $4 = 5 - 1$

10. $8 - 1 = 7$ or $9 = 2 + 5$

11. $9 - 1 = 6$ or $3 \leq 2$

12. $5 < 5$ or $8 = 7 + 2$

In Exercises 13–28, graph the solution of each compound statement.

13. $x < 2$ and $x > 0$

14. $x > 3$ and $x < 5$

15. $x \geq 6$ and $x \leq 9$

16. $x \leq -5$ and $x \geq -10$

17. $x \leq 5$ and $x \leq 3$

18. $x \geq -2$ and $x \geq 0$

19. $x \leq 4$ and $x \geq 6$

20. $x \leq -3$ and $x \geq -1$

21. $x \leq 5$ or $x \geq 8$

22. $x \leq 2$ or $x \geq 5$

23. $x > -4$ or $x < -6$

24. $x > -3$ or $x < -10$

25. $x < 8$ or $x < 10$

26. $x \geq 3$ or $x \leq 1$

27. $x \geq 8$ or $x \leq 10$

28. $x \leq -4$ or $x \geq -6$

In Exercises 29–34, write an inequality, equation, or compound statement, as necessary, using the information given. *Do not solve.*

29. Twice a number is more than 8 and the number plus 6 is less than 14.

30. Bill is five years older than Norman, and the sum of their ages is less than 23.

31. Twice a number is three and the number is smaller than 2.

32. The sum of a number and two is less than fourteen, and the number minus 4 is greater than five.

33. A number is between 8 and 12, inclusive.

34. A number is less than 6, or more than 9.

CHAPTER 2 SUMMARY

Key Words

Equation	Term	Compound statement
Linear equation	Collecting terms	Intersection
Inequality	Combining terms	Union

Properties

For any expressions A, B, and C:

Addition Property of Equality

The equations $A = B$ and $A + C = B + C$ have the same solutions.

Multiplication Property of Equality

The equations $A = B$ and $AC = BC$ have the same solutions.

Addition Property of Inequality

The inequalities $A < B$ and $A + C < B + C$ have the same solutions.

Multiplication Property of Inequality

If $C > 0$, then $A < B$ and $AC < BC$ have the same solutions.
If $C < 0$, then $A < B$ and $AC > BC$ have the same solutions.

(All statements about inequalities are true for $>$, $\leq$, or $\geq$ also.)

Solving an Equation

Step 1 Combine terms as needed, using the commutative, associative, and distributive properties.

Step 2 Simplify the equation by using the addition property so that the variable is on one side of the equals sign, and the number term is on the other side.

Step 3 Use the multiplication property to further simplify an equation until you get the form $x =$ number.

Step 4 Check the solution by substitution into the original equation.

Solving an Inequality

Step 1 Use the commutative, associative, and distributive properties to combine terms on both sides of the inequality.

Step 2 Use the addition property to simplify the inequality until you get the form $ax < b$, $ax > b$, etc.

Step 3 Use the multiplication property to further simplify an inequality until you get the form $x < c$ or $x > c$, etc.

CHAPTER 2 TEST

Simplify the following expressions by combining terms.

1. $2x + 5 + 5x - 3x - 3$

2. $k - 3k + 5k - 6k + 4k$

3. $9r + 3r - 4r - r - 8r$

4. $3z - 6z + 8 - 9 + 4z - 9$

5. $4(2m + 1) - 3(m + 5) + 2m - 1$

6. $3(2z - 8) - 2(2 - 3z) + 6$

Solve each of the following equations.

7. $x + 7 = 10$

8. $2m - 5 = 3$

9. $6v + 3 = 8v - 7$

10. $3(a + 2) = 1 - 2(a - 5)$

11. $4(p + 3) - 5 = 3(p + 4) - 10$

12. $\frac{m}{5} = 2$

13. $\frac{2}{3}z = 18$

14. $-(r + 4) = 2 + r$

15. Solve the formula $I = prt$ for p.

16. Solve the formula $A = \frac{1}{2}(b + B)h$ for h.

Solve each of the following inequalities. Graph the solution.

17. $x + 4 \leq 8$

18. $5z \geq -10$

19. $-2m < -14$

20. $-3k < k - 8$

21. $5(k - 2) + 3 \leq 2(k - 3) + 2k$

22. $-4r + 2(r - 3) \geq 5r - (3 + 6r) + 1 - 8$

Write an equation for each of the following. Then solve it.

23. Dick's lunch cost \$1.20. Sandwiches are 35¢ each and milk is 15¢ per glass. He bought one glass of milk. How many sandwiches did he buy?

24. A rectangle has a perimeter which is two inches less than three times the length. The perimeter is 190 inches. Find the dimensions of the rectangle.

25. Joe bicycled from here to there, a distance of 21 miles, in five hours. During the last two hours, he became tired and slowed down by two miles per hour. What was his speed for the first three hours?

26. Ken is five years older than Ted. In five years Ken will be one-and-one-half times as old as Ted. What are their ages now?

27. One of the many regulations of the Post Office says that a package can be mailed only if the length plus the girth (distance around at the widest point) does not exceed 72 inches. That is, if l represents the length of the package, w the width, and h the height, then

$$l + 2w + 2h \leq 72.$$

Find the longest acceptable package with width 10 inches and height 6 inches.

Graph the solution of each of the following.

28. $x < 4$ and $x > 2$

29. $x < 6$ and $x > 8$

30. $x \leq 4$ or $x \geq 6$

31. $x \leq 6$ or $x \leq 8$

3 POLYNOMIALS

3.1 EXPONENTS

Expressions like xx and xxx occur so frequently in algebra that it is convenient to write them in shortened form. For example, write

xx	as x^2,	read	"x squared"
xxx	as x^3,	read	"x cubed"
$xxxx$	as x^4,	read	"x to the fourth power" or "x to the fourth"
$xxxxx$	as x^5,	read	"x to the fifth"

and so on. In the symbol x^3, for example, x is called the **base** and 3 is called the **exponent** or **power**. The symbol x^3 itself is called an **exponential**.

Example 1 Write $3 \cdot 3 \cdot 3 \cdot 3 \cdot 3$ in exponential form. Evaluate the exponential.

Since there are five factors of the number three, the base is 3, and the exponent is 5. The exponential is 3^5. The value is

$$3^5 = 3 \cdot 3 \cdot 3 \cdot 3 \cdot 3 = 243.$$

Example 2 Evaluate 5^4. Name the base and the exponent.

$$5^4 = 5 \cdot 5 \cdot 5 \cdot 5 = 625.$$

The base is 5; the exponent is 4.

We must be sure to identify the base carefully in any problem involving exponents. For example, let us agree that

$$-5^4 = -(5^4) = -(5 \cdot 5 \cdot 5 \cdot 5) = -625,$$

so that the base is 5. But

$$(-5)^4 = (-5)(-5)(-5)(-5) = 625,$$

where the base is -5. Note also that

$$-(-5)^3 = -(-5)(-5)(-5) = -(-125) = 125.$$

Here the base is also -5.

We can use the definition of exponents to decide how to multiply two exponentials. For example,

$$2^4 \cdot 2^3 = (2 \cdot 2 \cdot 2 \cdot 2)(2 \cdot 2 \cdot 2).$$

Altogether, 2 is written seven times in the product, so

$$2^4 \cdot 2^3 = (2 \cdot 2 \cdot 2 \cdot 2)(2 \cdot 2 \cdot 2) = 2 \cdot 2 \cdot 2 \cdot 2 \cdot 2 \cdot 2 \cdot 2$$
$$2^4 \cdot 2^3 = 2^7.$$

The exponent 7 in the result is the sum of the exponents 4 and 3.

Example 3 Find the product of a^2 and a^6.

$$a^2 \cdot a^6 = (a \cdot a)(a \cdot a \cdot a \cdot a \cdot a \cdot a)$$
$$a^2 \cdot a^6 = a^8.$$

Again, the exponent 8 is the sum of the exponents 2 and 6.

To get a general rule from these two examples, we can look at the product of the exponentials a^m and a^n, where m and n are both positive integers. By the definition of exponent, the exponential a^m means that the base a is written as a factor m times. Also, a^n means that the base a is written as a factor n times. The product of a^m and a^n would mean that the base a is written as a factor $m + n$ times.

$$a^m \cdot a^n = (\underbrace{a \cdot a \cdot a \cdots a}_{m \text{ times}})(\underbrace{a \cdot a \cdot a \cdots a}_{n \text{ times}})$$
$$= \underbrace{a \cdot a \cdot a \cdots a}_{m + n \text{ times}}$$
$$= a^{m+n}.$$

Hence, for positive integers m and n, we have the **product rule for exponentials**:

$$a^m \cdot a^n = a^{m+n}.$$

The base must be the same before the product rule for exponentials can be applied. Thus $6^3 \cdot 6^5 = 6^8$, but we cannot use the product rule to simplify $6^3 \cdot 4^5$ since the bases, 6 and 4, are different.

Example 4 Simplify $(-4)^5(-4)^3$.

Since the bases are the same, we use the product rule.

$$(-4)^5(-4)^3 = (-4)^{5+3} = (-4)^8.$$

Example 5 Multiply $2x^3$ by $3x^7$.

$$2x^3 \cdot 3x^7 = 2 \cdot 3 \cdot x^3 \cdot x^7 = 6x^{3+7} = 6x^{10}.$$

The rule for division of exponentials is similar to the product rule. We shall state the general rule after some examples.

$$\frac{6^5}{6^2} = \frac{6 \cdot 6 \cdot 6 \cdot 6 \cdot 6}{6 \cdot 6} = 6 \cdot 6 \cdot 6 = 6^3.$$

Here, the difference of the exponents, $5 - 2$, gives the new exponent 3.

$$\frac{5^8}{5^3} = \frac{5 \cdot 5 \cdot 5 \cdot 5 \cdot 5 \cdot 5 \cdot 5 \cdot 5}{5 \cdot 5 \cdot 5} = 5 \cdot 5 \cdot 5 \cdot 5 \cdot 5 = 5^5.$$

The difference $8 - 3$ gives the new exponent 5.

It would seem from these examples that as long as the base is the same (and is not 0), and the exponent in the numerator is larger than the exponent in the denominator, we can divide exponentials by the rule

$$\frac{a^m}{a^n} = a^{m-n} \qquad (m \text{ is a positive integer larger than } n).$$

We must have $a \neq 0$ since we cannot divide by zero.

On the other hand, suppose the exponent on the denominator is larger than the exponent on the numerator, as in the following example:

$$\frac{3^4}{3^6} = \frac{3 \cdot 3 \cdot 3 \cdot 3}{3 \cdot 3 \cdot 3 \cdot 3 \cdot 3 \cdot 3} = \frac{1}{3 \cdot 3} = \frac{1}{3^2}.$$

The difference $6 - 4$ gives the new exponent 2. Also,

$$\frac{8^7}{8^{11}} = \frac{1}{8^4}.$$

In this example, $11 - 7 = 4$.

In general, if the exponent in the denominator is larger than that in the numerator, then for $a \neq 0$,

$$\frac{a^m}{a^n} = \frac{1}{a^{n-m}} \qquad (n \text{ is a positive integer larger than } m).$$

There is only one other possibility: what happens if m and n are equal? For example,

$$\frac{6^5}{6^5} = \frac{6 \cdot 6 \cdot 6 \cdot 6 \cdot 6}{6 \cdot 6 \cdot 6 \cdot 6 \cdot 6} = 1.$$

However, by subtraction of exponents as above, we would have

$$\frac{6^5}{6^5} = 6^{5-5} = 6^0.$$

This means that $6^0 = 1$.

In general, for any number a, but $a \neq 0$, we define

$$a^0 = 1.$$

Example 6 **(a)** $60^0 = 1$

(b) $(-60)^0 = 1$

(c) $-60^0 = -(60^0) = -1$

Note the difference between Examples 6(b) and 6(c). In Example 6(b) the base is -60 and the exponent is 0. Any non-zero base raised to a zero exponent is 1. But in Example 6(c), the base is 60. Then $60^0 = 1$, so that $-60^0 = -1$.

Taking all three cases from above into account, the **quotient rule for exponentials** can be expressed as follows. For positive integers m and n, and $a \neq 0$,

$$\frac{a^m}{a^n} = \begin{cases} a^{m-n} & \text{If } m \text{ is larger than } n \\ a^0 = 1 & \text{If } m = n \\ \dfrac{1}{a^{n-m}} & \text{If } n \text{ is larger than } m \end{cases}$$

Example 7

$$\begin{aligned} \frac{3^2x^5}{3^4x^3} &= \frac{3^2}{3^4} \cdot \frac{x^5}{x^3} \\ &= \frac{1}{3^2} \cdot \frac{x^2}{1} \\ &= \frac{x^2}{3^2} \\ &= \frac{x^2}{9}. \end{aligned}$$

How can we simplify an expression such as $(8^3)^4$? The exponent 4 tells us that the base 8^3 is a factor four times.

$$(8^3)^4 = 8^3 \cdot 8^3 \cdot 8^3 \cdot 8^3.$$

We know that $8^3 = 8 \cdot 8 \cdot 8$, and so we can write

$$\begin{aligned} (8^3)^4 &= (8 \cdot 8 \cdot 8)(8 \cdot 8 \cdot 8)(8 \cdot 8 \cdot 8)(8 \cdot 8 \cdot 8) \\ &= 8^{12}. \end{aligned}$$

Looking at the exponents, we see $3 \cdot 4 = 12$.

In general, to evaluate $(a^m)^n$, where m and n are positive integers, we use the **power rule for exponentials**.

$$(a^m)^n = a^{mn}.$$

We can use the properties studied in Chapter 1 to develop two more rules for exponentials. By definition,

$$\begin{aligned}(4 \cdot 8)^3 &= (4 \cdot 8)(4 \cdot 8)(4 \cdot 8)\\ &= 4 \cdot 4 \cdot 4 \cdot 8 \cdot 8 \cdot 8 \qquad \text{(Commutative and associative)}\\ &= 4^3 \cdot 8^3.\end{aligned}$$

Based on this example, we have the following rule. For any positive integer m,

$$(ab)^m = a^m b^m.$$

Example 8 $(3xy)^2 = 3^2x^2y^2 = 9x^2y^2$.

Since a/b can be written as $a \cdot (1/b)$, the rule discussed above together with some of the properties of real numbers gives us the final rule for exponentials. For any positive integer m, and $b \neq 0$,

$$\left(\frac{a}{b}\right)^m = \frac{a^m}{b^m}.$$

Example 9 $\left(\frac{2}{3}\right)^5 = \frac{2^5}{3^5} = \frac{32}{243}$.

Example 10 Simplify $\frac{(4^2)^3}{4^5}$.

Use the power rule and then the first part of the quotient rule.

$$\frac{(4^2)^3}{4^5} = \frac{4^6}{4^5} = 4^1 = 4.$$

Example 11 Use the rules for exponents to simplify $(2x)^3(2x)^2$.

In this example, use the product rule first.

$$(2x)^3(2x)^2 = (2x)^5 = 2^5x^5 = 32x^5.$$

Example 12 Use the rules for exponents to simplify $\left(\frac{2x^3}{5}\right)^4$.

By the last two rules given above,

$$\left(\frac{2x^3}{5}\right)^4 = \frac{2^4x^{12}}{5^4} = \frac{16x^{12}}{625}.$$

Summary of Rules for Exponentials

If m and n are any positive integers, then

$a^m \cdot a^n = a^{m+n}$ — Product rule

$$\frac{a^m}{a^n} = \begin{cases} a^{m-n} & \text{if } m \text{ is larger than } n \\ a^0 = 1 & \text{if } m = n \\ \dfrac{1}{a^{n-m}} & \text{if } n \text{ is larger than } m \end{cases} \quad (a \neq 0)$$ — Quotient rule

$(a^m)^n = a^{mn}$ — Power rule

$(ab)^n = a^n b^n$

$\left(\dfrac{a}{b}\right)^m = \dfrac{a^m}{b^m} \quad (b \neq 0)$

EXERCISES

In Exercises 1–10, identify the base and the exponent.

1. 5^{12}
2. a^6
3. $(3m)^4$
4. -2^4
5. -125^3
6. $(-1)^8$
7. $(-24)^2$
8. $-(-3)^5$
9. $3m^2$
10. $5y^3$

In Exercises 11–20, write each expression using exponents.

11. $3 \cdot 3 \cdot 3 \cdot 3 \cdot 3$
12. $4 \cdot 4 \cdot 4$
13. $5 \cdot 5 \cdot 5 \cdot 5$
14. $3 \cdot 3 \cdot 3 \cdot 3 \cdot 3 \cdot 3 \cdot 3 \cdot 3 \cdot 3$
15. $(-2)(-2)(-2)(-2)(-2)$
16. $(-1)(-1)(-1)(-1)$
17. $\dfrac{1}{4 \cdot 4 \cdot 4 \cdot 4 \cdot 4}$
18. $\dfrac{1}{(-2)(-2)(-2)}$
19. $\dfrac{1}{3 \cdot 3 \cdot 3 \cdot 3}$
20. $\dfrac{1}{2 \cdot 2 \cdot 2 \cdot 2 \cdot 2}$

In Exercises 21–32, use the product rule to simplify. Write each answer in exponential form.

21. $4^2 \cdot 4^3$
22. $3^5 \cdot 3^4$
23. $9^5 \cdot 9^3$
24. $8^6 \cdot 8^4$
25. $3^4 \cdot 3^7$
26. $2^5 \cdot 2^{15}$
27. $(4^3)^2$
28. $(2^3)^4$
29. $(-3)^3(-3)^2$
30. $(-4)^5(-4)^3$
31. $(-2)^3(-2)^6$
32. $(-3)^4(-3)^6$

In Exercises 33–42, evaluate the expressions.

33. $3^2 + 3^4$

34. $2^8 - 2^6$

35. $4^2 + 4^3$

36. $3^3 + 3^4$

37. $2^2 + 2^5$

38. $4^2 + 4^1$

39. $4^0 + 5^0$

40. $3^0 + 8^0$

41. $(-9)^0 + 9^0$

42. $(-8^0) + (-8^0)$

In Exercises 43–50, use the quotient rule to simplify each expression.

43. $\dfrac{4^3}{4^2}$

44. $\dfrac{11^5}{11^6}$

45. $\dfrac{4^2}{4^4}$

46. $\dfrac{14^{11}}{14^{15}}$

47. $\dfrac{(-14)^6}{(-14)^5}$

48. $\dfrac{(-3)^7}{(-3)^8}$

49. $\dfrac{-19^0}{(-18)^0}$

50. $\dfrac{14^0}{-16^0}$

In Exercises 51–70, use the rules for exponentials as necessary to simplify. Write each answer in exponential form.

51. $x^4 \cdot x^5$

52. $m^2 \cdot m^7$

53. $r^3 \cdot r^8$

54. $p^4 \cdot p^{10}$

55. $\dfrac{(y^3)^3}{(y^2)^2}$

56. $\dfrac{(r^2)^4}{(r^3)^2}$

57. $\dfrac{a^{11}}{(a^2)^4}$

58. $\dfrac{s^{14}}{(s^5)^2}$

59. $\dfrac{(k^2)^9}{(k^6)^2}$ 18 12

60. $\dfrac{(w^4)^2}{(w^7)^3}$

61. $(5m)^3$

62. $(2xy)^4$

63. $\left(\dfrac{a}{5}\right)^3$

64. $\left(\dfrac{9}{x}\right)^2$

65. $\left(\dfrac{3mn}{2}\right)^5$

66. $\left(\dfrac{2x^3}{3y^2}\right)^4$

67. $\dfrac{x^7x^8(x^3)^2}{x^9x^7}$

68. $\dfrac{(m^3)^2(m^2)^4m^8}{(m^9)^3}$

69. $\dfrac{b^{11}(b^2)^4}{(b^3)^3(b^2)^6}$

70. $\dfrac{(8m^2)^3(8m^4)^2}{(8m^3)^4}$

3.2 POLYNOMIALS

In an expression such as

$$4x^3 + 6x^2 + 5x + \frac{1}{x},$$

the quantities $4x^3$, $6x^2$, $5x$, and $1/x$, which are to be added, are called **terms**. This meaning of "term" is more general than that given in Section 2.1, since it includes quotients (like $1/x$) as well as products. In the term $4x^3$, the number 4 is called the **numerical coefficient**, or simply the **coefficient**, of x^3. In the same way, 6 is the coefficient of x^2 in the term $6x^2$, and 5 is the coefficient of x in the term $5x$. In the expression

$$x - 6x^4,$$

the coefficient of x is 1 because $x = 1 \cdot x$. The coefficient of x^4 is -6 since $x - 6x^4$ can be written as the sum $x + (-6x^4)$.

Example 1 **(a)** The coefficient in the term $4x^3$ is 4.
(b) The coefficient in the term $-v^3$ is -1 since $-v^3 = -(1v^3) = -1v^3$.
(c) The number 5 is the coefficient in the term 5, since $5 = 5x^0$.

Like terms have exactly the same variable with the same exponent. Only the coefficients may be different. Examples include

$$19m^5 \quad \text{and} \quad 14m^5,$$
$$6y^9, \quad -37y^9, \quad \text{and} \quad y^9.$$

To add like terms, use the distributive property.

Example 2 **(a)** $-4x^3 + 6x^3 = (-4 + 6)x^3 = 2x^3$.
(b) $3x^4 + 5x^4 = (3 + 5)x^4 = 8x^4$.
(c) $9x^6 - 14x^6 + x^6 = (9 - 14 + 1)x^6 = -4x^6$.
(d) $12m^2 + 5m + 4m^2 = (12 + 4)m^2 + 5m = 16m^2 + 5m$.

Example 2(d) shows that it is not possible to add $16m^2$ and $5m$. These two terms are unlike because the exponents on the variables are different. Unlike terms have different variables or different exponents on the same variables.

Example 3 **(a)** $4m^2 - 5m^3$ cannot be combined or simplified any further.
(b) $8x^4 + 9y^4$ cannot be combined or simplified.
(c) $3z^2 + 3z + 3$ cannot be combined or simplified.

One of the basic ideas of algebra is that of a polynomial. A **polynomial** is defined as any finite sum of terms which are the product of a number and a variable raised to a power, such as

$$4y^3 + 3x^2 - 2m.$$

[Recall that $4y^3 + 3x^2 - 2m = 4y^3 + 3x^2 + (-2)m$.] On the other hand,

$$2x^3 - x^2 + \frac{4}{x^4}$$

is not a polynomial because the last term is the *quotient* (not the product) of a number and a variable raised to a power.

In general, we shall be concerned only with polynomials containing a single variable, such as x. A **polynomial in x** is a polynomial whose terms contain only variables which are powers of x (including the zero power of x). Thus

$$16x^8 - 7x^6 + 5x^5 + 5x^3 - 3x + 2$$

is a polynomial in x. (Note that $2 = 2x^0$.) This last polynomial is written in **descending powers** of the variable, since the exponents on x decrease from left to right.

The **degree** of a term with one variable is the exponent on the variable. Thus $3x^4$ has degree 4, $6x^{17}$ has degree 17, $5x$ has degree 1, and -7 has degree 0 (-7 can be written as $-7x^0$). The **degree of a polynomial** in one variable is the highest exponent found in any non-zero term of the polynomial. Thus $3x^4 - 5x^2 + 6$ is of degree 4, while $5x$ is of degree 1, and 3 $(3x^0)$ is of degree 0.

Three types of polynomials are very common, and thus are given special names. A polynomial with exactly three terms is called a **trinomial**. (*Tri* means "three," as in *triangle*.) Examples are:

$$9m^3 - 4m^2 + 6,$$
$$19y^2 + 8y^9 + 5,$$
$$-3m^5 - 9m^2 + 2.$$

A polynomial with exactly two terms is called a **binomial**. (*Bi* means "two," as in *bicycle*.) Examples are

$$-9x^4 + 9x^3,$$
$$8m^2 + 6m,$$
$$3m^5 - 9m^2.$$

A polynomial with only one term is called a **monomial**. (*Mo* means "one," as in *monaural*.) Examples are:

$$9m, \qquad -6y^5, \qquad a^2, \qquad 6.$$

A polynomial represents different numbers for different values of the variable. For example, the polynomial

$$3x^4 - 5x^3 - 4x - 4$$

has the following value when x is -2.

$$\begin{aligned} 3x^4 - 5x^3 - 4x - 4 &= 3(-2)^4 - 5(-2)^3 - 4(-2) - 4 \\ &= 3(16) - 5(-8) - 4(-2) - 4 \\ &= 48 + 40 + 8 - 4 \\ &= 92. \end{aligned}$$

Example 4 Find the value of $3x^4 - 5x^3 - 4x - 4$ when $x = 1$.
Substitute 1 for x.

$$\begin{aligned} 3x^4 - 5x^3 - 4x - 4 &= 3(1)^4 - 5(1)^3 - 4(1) - 4 \\ &= 3(1) - 5(1) - 4 - 4 \\ &= 3 - 5 - 4 - 4 \\ &= -10. \end{aligned}$$

We sometimes use a capital letter to represent a polynomial. For example, if we let $P(x)$ represent the polynomial $3x^4 - 5x^3 - 4x - 4$, then $P(x) = 3x^4 - 5x^3 - 4x - 4$ where $P(x)$ is read "P of x." We sometimes express the fact that $P(x) = 92$ when $x = -2$ by writing $P(-2) = 92$. (Read $P(-2)$ as "P of -2.")

Example 5 If $P(x) = 9x^3 - 8x + 6$, find **(a)** $P(-3)$; **(b)** $P(1)$.
(a) If we replace x by -3, we have

$$\begin{aligned} P(-3) &= 9(-3)^3 - 8(-3) + 6 \\ &= 9(-27) + 24 + 6 \\ &= -243 + 30 \\ P(-3) &= -213. \end{aligned}$$

(b) $P(1) = 9(1)^3 - 8(1) + 6 = 9 - 8 + 6 = 7$.

EXERCISES

In Exercises 1–10, let $P(x) = x^3 - 3x^2 + 2x - 3$ and $Q(x) = x^4 - 1$. Find the following values.

1. $P(-1)$
2. $P(0)$
3. $P(2)$
4. $Q(2)$
5. $P(-2)$
6. $Q(1)$
7. $Q(-2)$
8. $P(-2) + Q(-2)$
9. $P(-1) \cdot Q(-2)$
10. $P(0) \cdot Q(0)$

In Exercises 11–30, combine terms whenever possible. You may need to use the rules for exponents to simplify first.

11. $3m^5 + 5m^5$
12. $-4y^3 + 3y^3$
13. $2r^5 + (-3r^5)$
14. $-19y^2 + 9y^2$
15. $2m^5 - 5m^2$
16. $-9y + 9y^2$
17. $3x^5 + 2x^5 - 4x^5$
18. $6x^3 + 8x^7 - 9x^3$
19. $-4p^7 + 8p^7 - 5p^7$
20. $-3a^8 + 4a^8 - 3a^8 + 2a^8$
21. $(2y)^2 + 3y^2$
22. $(4r^2)^3 - 8r^6$

23. $(2ab^2)^3 + 4a^3b^2$

24. $(-6b^3)^2 + (7b^2)^3$

25. $(8m^5)^0 + (8m^5)^0$

26. $(3x^2)^2 + 4x^4$

27. $(5c^2)^2 - 2c^4$

28. $(k^5)^2 - 4(k^2)^5$

29. $(2m^3)^0 + (4m^2)^3$

30. $(3r^4)^3 + (2r^6)^2$

For each polynomial in Exercises 31–40, give the degree of the polynomial, and tell whether it is (a) a monomial, (b) a binomial, (c) a trinomial, (d) none of these.

31. $5x^4 - 8x$

32. $4y - 8y$

33. $23x^9 - \frac{1}{2}x^2 + x$

34. $2m^7 - 3m^6 + 2m^5 + m$

35. $x^8 + 3x^7 - 5x^4$

36. $2x - 2x^2$

37. $\frac{3}{5}x^5 + \frac{2}{5}x^5$

38. $\frac{9}{11}x^2$

39. -8

40. $2m^8 - 5m^9 + 2m^{10}$

In Exercises 41–49, find the value of the polynomial when $x = 2$; and when $x = -1$.

41. $2x^2 - 4x$

42. $8x + 5x^2 + 2$

43. $2x^5 - 4x^4 + 5x^3 - x^2$

44. $9x + 1$

45. $2x^2 + 5x + 1$

46. $-3x^2 + 4x + 2(5x - 1)$

47. $2x^2 + 3$

48. $-3x^2 + 4x + 5$

49. $x^2 - x^3$

In Exercises 50–55, write *always, sometimes,* or *never* for each of the statements.

50. A binomial is a polynomial.

51. A polynomial is a trinomial.

52. A trinomial is a binomial.

53. A monomial has no coefficient.

54. A binomial is a trinomial.

55. A polynomial of degree 4 has 4 terms.

3.3 ADDITION AND SUBTRACTION OF POLYNOMIALS

In the last section we used the distributive property to add two like terms such as

$$-6x^3 + 4x^3 = (-6 + 4)x^3 = -2x^3.$$

We can now use the same property to add two polynomials.

Example 1 Add the polynomials $6x^3 - 4x^2 + 3$ and $-2x^3 + 7x^2 - 5$. We wish to find the sum

$$(6x^3 - 4x^2 + 3) + (-2x^3 + 7x^2 - 5).$$

Regroup to collect like terms, and change subtractions to additions of inverses.

$$[6x^3 + (-2)x^3] + [-4x^2 + 7x^2] + [3 + (-5)].$$

Now each group of like terms can be combined using the distributive property. The result is the trinomial

$$4x^3 + 3x^2 - 2.$$

We can find the sum of two polynomials in another way by placing one directly over the other, with like terms lined up vertically in columns.

Example 2 Add $6x^3 - 4x^2 + 3$ and $-2x^3 + 7x^2 - 5$.

We write the like terms in columns.

$$\begin{array}{r} 6x^3 - 4x^2 + 3 \\ -2x^3 + 7x^2 - 5 \\ \hline \end{array}$$

Then we add, column by column.

$$\begin{array}{r} 6x^3 \\ -2x^3 \\ \hline 4x^3 \end{array} \qquad \begin{array}{r} -4x^2 \\ 7x^2 \\ \hline 3x^2 \end{array} \qquad \begin{array}{r} 3 \\ -5 \\ \hline -2 \end{array}$$

Adding the three sums, we get the same answer as we found in Example 1.

$$4x^3 + 3x^2 + (-2) = 4x^3 + 3x^2 - 2.$$

In Section 1.5 the difference $a - b$ was defined as $a + (-b)$. For example,

$$7 - 2 = 7 + (-2) = 5,$$

and

$$-8 - (-2) = -8 + 2 = -6.$$

Since polynomials are expressions that represent numbers, we can use the same method to subtract two polynomials.

Example 3 Subtract the polynomial $6x^3 - 4x^2 + 2$ from the polynomial $11x^3 + 2x^2 - 8$.

By the definition of subtraction,

$$(11x^3 + 2x^2 - 8) - (6x^3 - 4x^2 + 2)$$
$$= (11x^3 + 2x^2 - 8) + [-(6x^3 - 4x^2 + 2)].$$

To simplify $-(6x^3 - 4x^2 + 2)$ recall the fact that

$$-(a + b) = -a + (-b).$$

Thus

$$\begin{aligned} -(6x^3 - 4x^2 + 2) &= -6x^3 - (-4x^2) - 2 \\ &= -6x^3 + 4x^2 - 2. \end{aligned}$$

All the signs inside the parentheses, including the understood + on $6x^3$, have been changed. We can now complete the problem.

$$\begin{aligned} &(11x^3 + 2x^2 - 8) - (6x^3 - 4x^2 + 2) \\ &\quad = (11x^3 + 2x^2 - 8) + [-(6x^3 - 4x^2 + 2)] \\ &\quad = (11x^3 + 2x^2 - 8) + (-6x^3 + 4x^2 - 2) \\ &\quad = [11x^3 + (-6x^3)] + (2x^2 + 4x^2) + [-8 + (-2)] \\ &\quad = 5x^3 + 6x^2 - 10. \end{aligned}$$

To check a subtraction problem such as this, use the fact that if $a - b = c$, then $a = b + c$. For example, $6 - 2 = 4$. To check this, write $6 = 2 + 4$. For the polynomials, to check subtraction, add $6x^3 - 4x^2 + 2$ and $5x^3 + 6x^2 - 10$. Since the sum is $11x^3 + 2x^2 - 8$, the subtraction was performed correctly.

Example 4 $(14y^3 - 6y^2 + 2y - 5) - (2y^3 - 7y^2 - 4y + 6)$

$$\begin{aligned} &= (14y^3 - 6y^2 + 2y - 5) + [-(2y^3 - 7y^2 - 4y + 6)] \\ &- (14y^3 - 6y^2 + 2y - 5) + (-2y^3 + 7y^2 + 4y - 6) \\ &= 14y^3 - 6y^2 + 2y - 5 - 2y^3 + 7y^2 + 4y - 6 \\ &= 14y^3 - 2y^3 - 6y^2 + 7y^2 + 2y + 4y - 5 - 6 \\ &= 12y^3 + y^2 + 6y - 11. \end{aligned}$$

To check, add:

$$(2y^3 - 7y^2 - 4y + 6) + (12y^3 + y^2 + 6y - 11) = 14y^3 - 6y^2 + 2y - 5.$$

Subtractions can also be done in columns. For example, subtract $6x^3 - 4x^2 + 2$ from $11x^3 + 2x^2 - 8$.

Step 1 Write the problem with like terms arranged in columns:

$$\begin{array}{r} 11x^3 + 2x^2 - 8 \\ 6x^3 - 4x^2 + 2 \\ \hline \end{array}$$

Step 2 Take the inverse of each term in the second polynomial:

$$\begin{array}{r} 11x^3 + 2x^2 - 8 \\ -6x^3 + 4x^2 - 2 \\ \hline \end{array}$$

Step 3 Add column by column:

$$\begin{array}{r} 11x^3 + 2x^2 - 8 \\ -6x^3 + 4x^2 - 2 \\ \hline 5x^3 + 6x^2 - 10 \end{array}$$

Example 5 Use the method of subtracting by columns to find $(14y^3 - 6y^2 + 2y - 5) - (2y^3 - 7y^2 - 4y + 6)$.

Step 1 Arrange like terms in columns

$$\begin{array}{r} 14y^3 - 6y^2 + 2y - 5 \\ 2y^3 - 7y^2 - 4y + 6 \\ \hline \end{array}$$

Step 2 Change all signs in the second row, and then add.

$$\begin{array}{r} 14y^3 - 6y^2 + 2y - \ 5 \\ -2y^3 + 7y^2 + 4y - \ 6 \\ \hline 12y^3 + \ y^2 + 6y - 11 \end{array}$$

Either the horizontal or the vertical method may be used for adding and subtracting polynomials. The choice is a matter of personal preference.

Example 6 $(6a^3 + 2a^2 - 5a) + (-3a^3 + 3a) - (5a^3 - a^2 + a)$

$$= (6a^3 + 2a^2 - 5a) + (-3a^3 + 3a) + (-5a^3 + a^2 - a)$$
$$= (6a^3 - 3a^3 - 5a^3) + (2a^2 + a^2) + (-5a + 3a - a)$$
$$= -2a^3 + 3a^2 - 3a.$$

EXERCISES

In Exercises 1–12, add or subtract as indicated.

1. Add:

$$\begin{array}{r} 3m^2 + 5m \\ 2m^2 - 2m \\ \hline \end{array}$$

2. Add:

$$\begin{array}{r} 4a^3 - 4a^2 \\ 6a^3 + 5a^2 \\ \hline \end{array}$$

3. Subtract:

$$\begin{array}{r} 12x^4 - \ x^2 \\ 8x^4 + 3x^2 \\ \hline \end{array}$$

4. Subtract:

$$\begin{array}{r} 2a + 5d \\ 3a - 6d \\ \hline \end{array}$$

1A 11D

5. Subtract:

$$\begin{array}{r} 2n^5 - 5n^3 + 6 \\ 3n^5 + 7n^3 + 8 \\ \hline \end{array}$$

$1N^5$ $-12N^3$ $+2$

6. Subtract:

$$\begin{array}{r} 3r^2 - 4rs + 2r^2s \\ 7r^2 + 2rs - 3r^2s \\ \hline \end{array}$$

$-4R^2$ $-6RS$ $5R^2S$

7. Add:

$$\begin{array}{r} 9m^3 - 5m^2 + 4m - 8 \\ 3m^3 + 6m^2 + 8m - 6 \\ \hline \end{array}$$

8. Add:

$$\begin{array}{r} 12r^5 + 11r^4 - 7r^3 - 2r^2 - 5r - 3 \\ -8r^5 - 10r^4 + 3r^3 + 2r^2 - 5r + 7 \\ \hline \end{array}$$

9. Add:

$$\begin{array}{r} 12m^2 - 8m + 6 \\ 3m^2 + 5m - 2 \\ \hline \end{array}$$

10. Subtract:

$$\begin{array}{r} 5a^4 - 3a^3 + 2a^2 \qquad\quad \\ a^3 - a^2 + a - 1 \\ \hline \end{array}$$

11. Add:

$$\begin{array}{r} 5b^2 + 6b + 2 \\ 3b^2 - 4b + 5 \\ -4b^2 + 5b + 1 \\ \hline \end{array}$$

12. Add:

$$\begin{array}{r} 3w^2 - 5w + 2 \\ 4w^2 + 6w - 5 \\ 8w^2 + 7w - 2 \\ \hline \end{array}$$

In Exercises 13–28, perform the indicated operations.

13. $(2r^2 + 3r) - (3r^2 + 5r)$

14. $(3r^2 + 5r - 6) + (2r - 5r^2)$

15. $(8m^2 - 7m) - (3m^2 + 7m)$

16. $(x^2 + x) - (3x^2 + 2x - 1)$

17. $8 - (6s^2 - 5s + 7)$

18. $2 - [3 - (4 + s)]$

19. $(8s - 3s^2) + (-4s + 5s^2)$

20. $(3x^2 + 2x + 5) + (8x^2 - 5x - 4)$

21. $(16x^3 - x^2 + 3x) + (-12x^3 + 3x^2 + 2x)$

22. $(-2b^6 + 3b^4 - b^2) - (b^6 + 2b^4 + 2b^2)$

23. $(7y^4 + 3y^2 + 2y) - (18y^4 - 5y^2 - y)$

24. $(3x^2 + 2x + 5) + (-7x^2 - 8x + 2) + (3x^2 - 4x + 7)$

25. $(9a^4 - 3a^2 + 2) + (4a^4 - 4a^2 + 2) + (-12a^4 + 6a^2 - 3)$

26. $(4m^2 - 3m + 2) + (5m^2 + 13m - 4) - (16m^2 + 4m - 3)$

27. $[(8m^2 + 4m - 7) - (2m^2 - 5m + 2)] - (m^2 + m + 1)$

28. $(9b^3 - 4b^2 + 3b + 2) + (-2b^3 - 3b^2 + b) - (8b^3 + 6b + 4)$

In Exercises 29–32, write each of the statements as an equation or an inequality. Do not try to solve.

29. $4 + x^2$ is larger than 8.

30. The difference between $5 + 2x + x^2$ and $6 + 3x$ is larger than $8x + x^2$.

31. The sum of $5 + x^2$ and $3 - 2x$ is not equal to 5.

32. The sum of $3 - 2x + x^2$ and $8 - 9x + 3x^2$ is negative.

Solve each of the following word problems.

33. The sum of a certain polynomial and $6x^2 - 4x + 2$ is $3x^2 - 6x + 1$. Find the polynomial.

34. The difference between a certain polynomial and $-9x^2 + 3x - 4$ is $4x^2 - 3x + 2$. Find the polynomial.

3.4 MULTIPLYING POLYNOMIALS

We can evaluate products of monomials by using the rules for exponents and the commutative and associative properties. For example,

$$(6x^3)(4x^4) = 6 \cdot 4 \cdot x^3 \cdot x^4 = 24x^7.$$

Also, $$(-8m^6)(-9n^6) = (-8)(-9)m^6n^6 = 72m^6n^6.$$

To extend this method to include finding the product of a monomial and another polynomial with more than one term, use the distributive property, as shown in the following examples.

Example 1 $$4x^2(3x + 5) = (4x^2)(3x) + (4x^2)(5)$$
$$= 12x^3 + 20x^2.$$

Example 2 $$-8m^3(4m^3 + 3m^2 + 2m - 1)$$
$$= (-8m^3)(4m^3) + (-8m^3)(3m^2) + (-8m^3)(2m) + (-8m^3)(-1)$$
$$= -32m^6 - 24m^5 - 16m^4 + 8m^3$$

We also use the distributive property to find the product of any two polynomials. Suppose we want to find the product of the polynomials $x + 1$ and $x - 4$. If we work with $x + 1$ as a single quantity, we can use the distributive property to write

$$(x + 1)(x - 4) = (x + 1)x + (x + 1)(-4).$$

Now use the distributive property to multiply $(x + 1)x$ and $(x + 1)(-4)$.

$$(x + 1)x + (x + 1)(-4) = x(x) + 1(x) + x(-4) + 1(-4)$$
$$= x^2 + x + (-4x) + (-4)$$
$$= x^2 - 3x - 4.$$

Example 3 Multiply $(2x + 1)(3x + 5)$.

$$(2x + 1)(3x + 5) = (2x + 1)(3x) + (2x + 1)(5)$$
$$= (2x)(3x) + (1)(3x) + (2x)(5) + (1)(5)$$
$$= 6x^2 + 3x + 10x + 5$$
$$= 6x^2 + 13x + 5.$$

The work involved in multiplication can often be simplified by writing one polynomial over the other.

$$\begin{array}{r} 2x + 1 \\ 3x + 5 \\ \hline \end{array}$$

We need not worry about lining up the like terms in columns, since we are multiplying, and any terms may be multiplied. To begin, multiply each of the terms in the top row by 5.

$$\begin{array}{r} 2x + 1 \\ 3x + 5 \\ \hline 10x + 5 \end{array}$$

This is similar to ordinary multiplication. Then multiply $3x$ times each term in the top row. Be careful to place the like terms in columns, since the final step will involve addition, as in multiplying two numbers.

$$\begin{array}{rrr} & 2x & + 1 \\ & 3x & + 5 \\ \hline & 10x & + 5 \\ 6x^2 & + 3x & \\ \hline 6x^2 & + 13x & + 5 \end{array}$$

Thus $(2x + 1)(3x + 5) = 6x^2 + 13x + 5$.

Example 4

$$\begin{array}{rrr} & 3p & - 5 \\ & 2p & + 6 \\ \hline & 18p & - 30 \\ 6p^2 & - 10p & \\ \hline 6p^2 & + 8p & - 30 \end{array}$$

Example 5

$$\begin{array}{rrrrr} & & 4m^3 & - 2m^2 & + 4m \\ & & & m^2 & + 5 \\ \hline & & 20m^3 & - 10m^2 & + 20m \\ 4m^5 & - 2m^4 & + 4m^3 & & \\ \hline 4m^5 & - 2m^4 & + 24m^3 & - 10m^2 & + 20m \end{array}$$

Example 6

$$
\begin{array}{rrrrrr}
 & & 3y^3 & & +\ 4y & -\ 5 \\
 & & & y^2 & +\ 2y & +\ 3 \\
\hline
 & & 9y^3 & & +\ 12y & -\ 15 \\
 & 6y^4 & & +\ 8y^2 & -\ 10y & \\
3y^5 & & +\ 4y^3 & -\ 5y^2 & & \\
\hline
3y^5 & +\ 6y^4 & +\ 13y^3 & +\ 3y^2 & +\ 2y & -\ 15
\end{array}
$$

Example 7

$$
\begin{array}{rrrrrr}
 & x^4 & +\ x^3 & +\ x^2 & +\ x & +\ 1 \\
 & & & & x & -\ 1 \\
\hline
 & -\ x^4 & -\ x^3 & -\ x^2 & -\ x & -\ 1 \\
x^5 & +\ x^4 & +\ x^3 & +\ x^2 & +\ x & \\
\hline
x^5 & & & & & -\ 1
\end{array}
$$

The product is $x^5 - 1$.

Example 8 Find $(n + 2)^2$.

$$
\begin{aligned}
(n + 2)^2 &= (n + 2)(n + 2) \\
&= (n + 2)n + (n + 2)2 \\
&= n^2 + 2n + 2n + 4 \\
&= n^2 + 4n + 4.
\end{aligned}
$$

EXERCISES

Find each product.

1. $2m(3m + 2)$
2. $-5p(6 - 3p)$
3. $3p(-2p^3 + 4p^2)$
4. $4x(3 + 2x + 5x^3)$
5. $-8z(2z + 3z^2 + 3z^3)$
6. $7y(3 - 4y + 5y^2 - 2y^3)$
7. $2y(3 + 2y + 5y^4)$
8. $-2m^4(3m^2 + 5m + 6)$
9. $-3m^3(2m^2 - 3m^3)$
10. $2r^2(3r - 4r^2 + 2r^3)$
11. $10b^4(b^2 + 2b - 1)$
12. $12a(a^5 - 3a^4 + 2a^3 + a^2 + 2)$
13. $-4x^3(x^4 + 2x^3 - 4x^2 - 3x + 5)$
14. $-3y^4(3 - y + 2y^2 - 4y^3 + 8y^4)$
15. $(m + 7)(m + 5)$
16. $(n - 1)(n + 4)$
17. $(x + 5)(x - 5)$
18. $(y + 8)(y - 8)$
19. $(t - 4)(t + 4)$
20. $(x - 4)(x + 2)$
21. $(6p + 5)(p - 1)$
22. $(2x + 3)(6x - 4)$

23. $(4m - 3)(4m + 3)$

24. $(3x - 2)(x + 5)$

25. $(b + 8)(6b - 2)$

26. $(5a + 1)(2a + 7)$

27. $(8 - 2a)(3 + a)$

28. $(6 - 4m)(2 + 3m)$

29. $(-4 + k)(2 - k)$

30. $(6 - 3x)(4 + 2x)$

31. $(6x + 1)(2x^2 + 4x + 1)$

32. $(9y - 2)(8y^2 - 6y + 1)$

33. $(9a + 2)(9a^2 + a + 1)$

34. $(2r - 1)(3r^2 + 4r - 4)$

35. $(4m + 3)(5m^3 - 4m^2 + m - 5)$

36. $(y + 4)(3y^4 - 2y^2 + 1)$

37. $(2x - 1)(3x^5 - 2x^3 + x^2 - 2x + 3)$

38. $(2a + 3)(a^4 - a^3 + a^2 - a + 1)$

39. $(5x^2 + 2x + 1)(x^2 - 3x + 5)$

40. $(2m^2 + m - 3)(m^2 - 4m + 5)$

41. $(3a^3 + a + 5)(2a^3 + a^2 + 4)$

42. $(2y^2 + y - 2)(-y^2 + 2y + 1)$

43. $(2m^2 + m - 3)(m^2 - 4m + 5)$

44. $(3k^2 + 2k - 1)(k^2 + 4k - 2)$

45. $(3x - 1)(x + 1)(2x - 5)$

46. $(m + 2)(m - 2)(m - 2)$

47. $(3y + 4)(2y - 5)(y + 1)$

48. $(2m - 1)(m + 3)(m - 2)$

49. $(m - 5)^3$

50. $(x - 3)(x - 4)(x - 5)(x + 3)$

51. $(x + 3)(x + 1)(x - 2)(x + 3)$

3.5 PRODUCTS OF BINOMIALS

The procedures described in Section 3.4 can be used to find the product of any two polynomials, and are really the only practical methods for polynomials with three or more terms. We can, of course, use the same methods to multiply two binomials. However, in practice, a large proportion of the polynomials that must be multiplied are binomials, so for binomials we need a method that can be used quickly without writing out all the steps. To find such a shortcut, let us look carefully at the process of multiplying two binomials. If we multiply $(x + 3)(x + 5)$ using the distributive property, we have

$$\begin{aligned}(x + 3)(x + 5) &= (x + 3)x + (x + 3)5 \\ &= (x)(x) + (3)(x) + (x)(5) + (3)(5) \\ &= x^2 + 3x + 5x + 15 \\ &= x^2 + 8x + 15.\end{aligned}$$

Let us find where each term in the second step comes from. The first term, $(x)(x)$, is found by multiplying the two first terms:

$$(x + 3)(x + 5)$$

Multiply the first terms: $(x)(x)$

The second term, $(3)(x)$, is the product of the two middle terms, sometimes called the *inner product.*

$$(x + 3)(x + 5)$$

Multiply the inner terms: $(3)(x)$

The third term, $(x)(5)$, is the product of the first term of the first binomial and the last term of the second binomial. This is called the *outer product.*

$$(x + 3)(x + 5)$$

Multiply the outer terms: $(x)(5)$

Finally, $(3)(5)$ is the product of the two last terms.

$$(x + 3)(x + 5)$$

Multiply the last terms: $(3)(5)$

In the third step of the multiplication, we add the inner product and the outer product. This step should be performed mentally, so that the three terms of the answer can be written down without any extra steps.

In summary, these steps should be followed:

1. Multiply the two first terms to get the first term of the answer.
2. Find the inner product and the outer product and add them to get the middle term of the answer.
3. Multiply the two last terms to get the last term of the answer.

Example 1 Multiply $(x + 8)(x - 6)$ by the shortcut method.

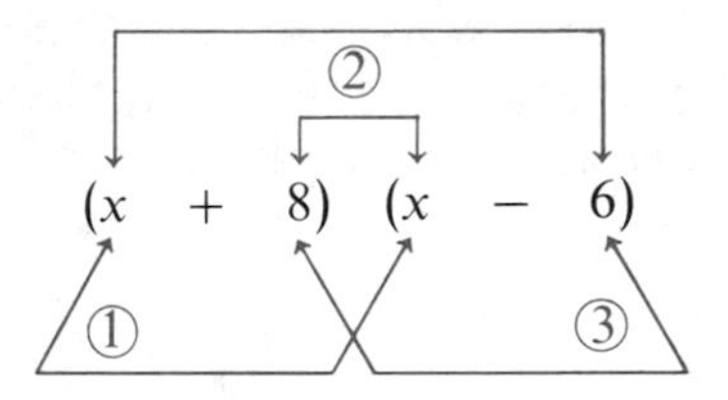

Step 1 Multiply $(x)(x)$ to get x^2.
Step 2 Multiply $(8)(x)$ and multiply $(-6)(x)$. Add to get $2x$.
Step 3 Multiply $(8)(-6)$ to get -48.
The result is $x^2 + 2x - 48$.

Example 2 Multiply $9x - 2$ and $3x + 1$ by the shortcut method.

Step 1 $(9x - 2)(3x + 1)$ $27x^2$

Step 2 $(9x - 2)(3x + 1)$ $9x - 6x = 3x$

Step 3 $(9x - 2)(3x + 1)$ -2

The result is $27x^2 + 3x - 2$.

Special types of binomial products occur so often in practice that the form of the answers should be memorized. For example, we shall frequently need to find the *square of a binomial.*

Example 3 Find $(2m + 3)^2$.

Squaring $2m + 3$ by the shortcut method gives

$$(2m + 3)(2m + 3) = 4m^2 + 12m + 9.$$

In the result, note that we have the square of both the first and the last terms of the binomial, $(2m)^2 = 4m^2$ and $3^2 = 9$. We also have twice the product of the two terms of the binomial, that is, $2(2m)(3) = 12m$.

In general, then, the square of a binomial is a trinomial composed of the *square* of the first term, plus *twice the product* of the two terms, plus the *square* of the last term of the binomial, or

$$(x + y)^2 = x^2 + 2xy + y^2.$$

Example 4 **(a)** $(5z - 1)^2 = (5z)^2 + 2(5z)(-1) + (-1)^2$
$= 25z^2 - 10z + 1.$

Recall that $(5z)^2 = 5^2z^2 = 25z^2$.

(b) $(3b + 5r)^2 = (3b)^2 + 2(3b)(5r) + (5r)^2$
$= 9b^2 + 30br + 25r^2.$

(c) $(2a - 9x)^2 = 4a^2 - 36ax + 81x^2$.

The fact that $(x + y)^2 = x^2 + 2xy + y^2$ can sometimes be used to find the square of a number without using tables.

Example 5

$$\begin{aligned} 21^2 = (20 + 1)^2 &= 20^2 + 2(20)(1) + 1^2 \\ &= 400 + 40 + 1 \\ &= 441. \end{aligned}$$

Example 6

$$\begin{aligned} 29^2 = (30 - 1)^2 &= 30^2 + 2(30)(-1) + (-1)^2 \\ &= 900 - 60 + 1 \\ &= 841. \end{aligned}$$

Binomial products of the form $(x + y)(x - y)$ also occur frequently. In these products, one binomial is the sum of two terms, while the other is the difference of the same two terms. For example, let us multiply $(a + 2)(a - 2)$ by the shortcut method:

$$(a + 2)(a - 2) = a^2 - 2a + 2a - 4$$
$$= a^2 - 4.$$

In general,

$$(x + y)(x - y) = x^2 - y^2,$$

where the result, $x^2 - y^2$, is called the **difference of two squares**.

Example 7 Multiply $(5m + 3)(5m - 3)$.
Use the formula which gives the difference of two squares.

$$(5m + 3)(5m - 3) = (5m)^2 - 3^2$$
$$= 25m^2 - 9.$$

Example 8 $(4x + y)(4x - y) = (4x)^2 - y^2 = 16x^2 - y^2$.

The fact that $(x + y)(x - y) = x^2 - y^2$ can also be used to help with numerical calculations.

Example 9 $19(21) = (20 - 1)(20 + 1) = 20^2 - 1^2 = 400 - 1 = 399$.

Example 10 $23(27) = (25 - 2)(25 + 2) = 25^2 - 2^2 = 625 - 4 = 621$.

EXERCISES

In Exercises 1–42, find each product.

1. $(r - 1)(r + 3)$
2. $(x + 2)(x - 5)$
3. $(x - 7)(x - 3)$
4. $(r + 3)(r + 6)$
5. $(2x - 1)(3x + 2)$
6. $(4y - 5)(2y + 1)$
7. $(6z + 5)(z - 3)$
8. $(8a + 3)(6a + 1)$
9. $(a + 4)(2a + 1)$
10. $(3x - 1)(2x + 3)$
11. $(2r - 1)(4r + 3)$
12. $(5m + 2)(3m - 4)$
13. $(2a + 4)(3a - 2)$
14. $(11m - 10)(10m + 11)$
15. $(5x + 4)(4x - 5)$
16. $(8x + 3)(x - 2)$
17. $(-3 + 2r)(4 + r)$
18. $(-5 + 6z)(2 - z)$
19. $(-a + 3)(-2a - 5)$
20. $(-3y + 6)(-4y + 1)$
21. $(m + n)^2$
22. $(x + y)^2$
23. $(5 + x)^2$
24. $(y - 2)^2$
25. $(x + 2y)^2$
26. $(3m - n)^2$
27. $(2z - 5x)^2$
28. $(6a - b)^2$
29. $(5p + 2q)^2$
30. $(8a - 3b)^2$

31. $(4a + 5b)(4a + 5b)$
32. $(9y + z)(9y + z)$
33. $(m - n)(m + n)$
34. $(p + q)(p - q)$
35. $(r + z)(r - z)$
36. $(a + b)(a - b)$
37. $(6a - p)(6a + p)$
38. $(5y + 3x)(5y - 3x)$
39. $(2m - 5)(2m + 5)$
40. $(3a - 5)(3a + 5)$
41. $(7y + 10)(7y - 10)$
42. $(6x + 3)(6x - 3)$

In Exercises 43–48, use $(x + y)^2 = x^2 + 2xy + y^2$ to find the products. See Examples 5 and 6.

43. 18^2
44. 19^2
45. 31^2
46. 42^2
47. 101^2
48. 1001^2

In Exercises 49–54, use $(x + y)(x - y) = x^2 - y^2$ to find the products. See Examples 9 and 10.

49. 39(41)
50. 13(17)
51. 99(101)
52. 18(22)
53. 999(1001)
54. 47(53)

In Exercises 55–58, write each statement as an equation or an inequality using x to represent the unknown number.

55. The square of 3 more than a number is 5.

56. The square of the sum of a number and 6 is less than 3.

57. When 3 plus a number is multiplied by the number less 4, the result is greater than 7.

58. Twice a number plus 4, multiplied by 6 times the number, less 5, gives 8.

3.6 DIVIDING A POLYNOMIAL BY A MONOMIAL

To divide a polynomial by a monomial use the fact that

$$\frac{a}{b} = a \cdot \frac{1}{b}.$$

To divide $5m^5 - 10m^3$ by $5m^2$, we multiply $5m^5 - 10m^3$ by $1/5m^2$.

$$\frac{5m^5 - 10m^3}{5m^2} = (5m^5 - 10m^3)\frac{1}{5m^2}.$$

Now apply the distributive property to the product on the right, and use the quotient rule for exponents, as follows.

$$\begin{aligned}(5m^5 - 10m^3)\frac{1}{5m^2} &= (5m^5)\frac{1}{5m^2} - (10m^3)\frac{1}{5m^2} \\ &= \frac{5m^5}{5m^2} - \frac{10m^3}{5m^2} \\ &= m^3 - 2m.\end{aligned}$$

Therefore,

$$(5m^5 - 10m^3)\frac{1}{5m^2} = m^3 - 2m.$$

To check, multiply: $5m^2(m^3 - 2m) = 5m^5 - 10m^3$. Since we cannot divide by 0, the quotient $(5m^5 - 10m^3)/5m^3$ has no value if $m = 0$. In the rest of this chapter, we assume that all denominators are not 0.

Example 1 Divide: $\dfrac{16a^5 - 12a^4 + 8a^2}{4a^3}$.

$$\begin{aligned}\frac{16a^5 - 12a^4 + 8a^2}{4a^3} &= (16a^5 - 12a^4 + 8a^2)\frac{1}{4a^3}\\ &= (16a^5)\frac{1}{4a^3} - (12a^4)\frac{1}{4a^3} + (8a^2)\frac{1}{4a^3}\\ &= \frac{16a^5}{4a^3} - \frac{12a^4}{4a^3} + \frac{8a^2}{4a^3}\\ &= 4a^2 - 3a + \frac{2}{a}.\end{aligned}$$

The result is *not* a polynomial because of the expression $2/a$. Thus, we see that although the sum, difference, and product of two polynomials always result in a polynomial, the quotient of two polynomials may not.

In Chapter 1, the set of real numbers was said to be closed under the operations of addition, multiplication, subtraction, and division (except for division by 0). Whenever these operations are performed on real numbers, the result is a real number. The set of all polynomials with coefficients that are real numbers is thus closed under addition, subtraction, and multiplication—but not under division.

Example 2 Divide the polynomial

$$180y^{10} - 150y^8 + 120y^6 - 90y^4 + 120y^2$$

by the monomial $30y^2$.

Using the methods of this section, we have

$$\begin{aligned}&\frac{180y^{10} - 150y^8 + 120y^6 - 90y^4 + 120y^2}{30y^2}\\ &= \frac{180y^{10}}{30y^2} - \frac{150y^8}{30y^2} + \frac{120y^6}{30y^2} - \frac{90y^4}{30y^2} + \frac{120y^2}{30y^2}\\ &= 6y^8 - 5y^6 + 4y^4 - 3y^2 + 4.\end{aligned}$$

Example 3 Divide: $\dfrac{12x^4 - 7x^3 + x - 4}{4x}$.

$$\frac{12x^4 - 7x^3 + x - 4}{4x} = \frac{12x^4}{4x} - \frac{7x^3}{4x} + \frac{x}{4x} - \frac{4}{4x}$$

$$= 3x^3 - \frac{7x^2}{4} + \frac{1}{4} - \frac{1}{x}$$

$$= 3x^3 - \frac{7}{4}x^2 + \frac{1}{4} - \frac{1}{x}.$$

EXERCISES

In Exercises 1–10, divide each polynomial by the monomial $2m$.

1. $60m^4 - 20m^2$
2. $16m^3 - 8m^2$
3. $120m^6 - 60m^3 + 80m^2$
4. $10m^5 - 16m^2 + 8m^3$
5. $6m^5 - 4m^3 + 2m^2$
6. $8m^5 - 4m^3 + 4m^2$
7. $8m^3 - 4m^2 + 6m$
8. $2m^5 - 4m^2 + 8m$
9. $m^2 + m + 1$
10. $2m^2 - 2m + 5$

In Exercises 11–20, divide each polynomial by $3x$.

11. $3x^4 + 9x^3 + 3x^2 + 6x$
12. $15x^2 - 9x$
13. $12x^4 - 3x^3 + 3x$
14. $45x^3 + 15x^2 - 9x$
15. $27x^3 - 9x^4 + 18x^5$
16. $-12x^6 + 6x^5 + 3x^4 - 9x^3 + 3x^2$
17. $36x + 24x^2 + 3x^3$
18. $6x^5 - 3x^4 + 18x^3 + 9x^2 + 27$
19. $x^3 + 6x^2 - x$
20. $4x^4 - 3x^3 + 2x$

In Exercises 21–30, perform the division.

21. $\dfrac{120y^6 - 80y^5}{40y^4}$
22. $\dfrac{15x^3 - 2x^2}{6x^2}$
23. $\dfrac{100p^5 - 50p^4 + 30p^3 - 30p}{10p^2}$
24. $\dfrac{25 + 10p + p^2}{p}$
25. $\dfrac{36m^5 - 24m^4 + 16m^3 - 8m^2}{4m^3}$

26. $\dfrac{8x + 16x^2 + 10x^3}{4x^4}$

27. $(16y^5 - 8y^2 + 12y) \div 4y^2$

28. $(20a^4 - 15a^5 + 25a^3) \div 15a^4$

29. $(120x^{11} - 60x^{10} + 140x^9 - 100x^8) \div 10x^{12}$

30. $(5 + x + 6x^2 + 8x^3) \div 3x^4$

Solve each of the following.

31. What polynomial, when divided by $3x^2$, yields $4x^3 + 3x^2 - 4x + 2$ as a quotient?

32. What polynomial, when divided by $4m^3$, yields $-6m^2 + 4m$ as a quotient?

33. The quotient of a certain polynomial and $-7y^2$ is $9y^2 + 3y + 5 - 2/y$. Find the polynomial.

34. The quotient of a certain polynomial and $1/a$ yields $2a^2 + 3a + 5a^3$. Find the certain polynomial.

3.7 THE QUOTIENT OF TWO POLYNOMIALS

To divide a polynomial by any other polynomial, we use a method of "long divison." This is similar to long division of two whole numbers.

Step 1
Divide 27 into 6696.

$$27\overline{)6696}$$

Step 1
Divide $2x + 3$ into $8x^3 - 4x^2 - 14x + 15$.

$$2x + 3\overline{)8x^3 - 4x^2 - 14x + 15}$$

Step 2
27 divides into 66
2 times; $2 \cdot 27 = 54$.

$$\begin{array}{r} 2 \\ 27\overline{)6696} \\ 54 \\ \hline \end{array}$$

Step 2
$2x$ divides into $8x^3$
$4x^2$ times;
$4x^2(2x + 3) = 8x^3 + 12x^2$.

$$\begin{array}{l} \ \ 4x^2 \\ 2x + 3\overline{)8x^3 - 4x^2 - 14x + 15} \\ \ 8x^3 + 12x^2 \\ \hline \end{array}$$

Step 3 Subtract and bring down the next term.

$$\begin{array}{r} 2 \\ 27\overline{)6696} \\ 54 \\ \hline 129 \end{array}$$

$$\begin{array}{l} \ \ 4x^2 \\ 2x + 3\overline{)8x^3 - 4x^2 - 14x + 15} \\ \ 8x^3 + 12x^2 \\ \hline \ - 16x^2 - 14x \end{array}$$

(To subtract two polynomials, change the sign of the second, and add.)

Step 4

27 divides into 129
4 times; $4 \cdot 27 = 108$.

$$\begin{array}{r} 24 \\ 27\overline{)6696} \\ 54 \\ \hline 129 \\ 108 \\ \hline \end{array}$$

Step 4

$2x$ divides into $-16x^2$
$-8x$ times;
$-8x(2x+3) = -16x^2 - 24x$.

$$\begin{array}{r} 4x^2 - 8x \qquad\quad \\ 2x+3\overline{)8x^3 - 4x^2 - 14x + 15} \\ 8x^3 + 12x^2 \qquad\qquad\quad \\ \hline -16x^2 - 14x \qquad \\ -16x^2 - 24x \qquad \\ \hline \end{array}$$

Step 5 Subtract and bring down the next term.

$$\begin{array}{r} 24 \\ 27\overline{)6696} \\ 54 \\ \hline 129 \\ 108 \\ \hline 216 \end{array}$$

$$\begin{array}{r} 4x^2 - 8x \qquad\quad \\ 2x+3\overline{)8x^3 - 4x^2 - 14x + 15} \\ 8x^3 + 12x^2 \qquad\qquad\quad \\ \hline -16x^2 - 14x \qquad \\ -16x^2 - 24x \qquad \\ \hline 10x + 15 \end{array}$$

Step 6

27 divides into 216
8 times; $8 \cdot 27 = 216$.

$$\begin{array}{r} 248 \\ 27\overline{)6696} \\ 54 \\ \hline 129 \\ 108 \\ \hline 216 \\ 216 \\ \hline \end{array}$$

Step 6

$2x$ divides into $10x$
5 times; $5(2x+3) = 10x + 15$.

$$\begin{array}{r} 4x^2 - 8x + 5 \qquad \\ 2x+3\overline{)8x^3 - 4x^2 - 14x + 15} \\ 8x^3 + 12x^2 \qquad\qquad\quad \\ \hline -16x^2 - 14x \qquad \\ -16x^2 - 24x \qquad \\ \hline 10x + 15 \\ 10x + 15 \\ \hline \end{array}$$

From the last step we get the quotients. Neither division led to a remainder.

6696 divided by 27 is 248

$8x^3 - 4x^2 - 14x + 15$ divided by $2x + 3$ is $4x^2 - 8x + 5$.

Step 7 Check by multiplication.

$27 \cdot 248 = 6696$

$(2x + 3)(4x^2 - 8x + 5) = 8x^3 - 4x^2 - 14x + 15$.

Example 1 Divide $4x^3 - 4x^2 + 5x - 8$ by $2x - 1$.

$$
\begin{array}{r}
2x^2 - x + 2 \\
2x - 1 \overline{)\,4x^3 - 4x^2 + 5x - 8} \\
\underline{4x^3 - 2x^2} \\
-2x^2 + 5x \\
\underline{-2x^2 + x} \\
4x - 8 \\
\underline{4x - 2} \\
-6
\end{array}
$$

1. $2x$ divides into $4x^3$, $2x^2$ times; $2x^2(2x - 1) = 4x^3 - 2x^2$.
2. Subtract; bring down the next term.
3. $2x$ divides into $-2x^2$, $-x$ times; $-x(2x - 1) = -2x^2 + x$.
4. Subtract; bring down the next term.
5. $2x$ divides into $4x$, 2 times; $2(2x - 1) = 4x - 2$.
6. Subtract. The remainder is -6.

Thus $2x - 1$ divides into $4x^3 - 4x^2 + 5x - 8$ with a quotient of $2x^2 - x + 2$ and a remainder of -6. The result is not a polynomial because of the remainder.

$$\frac{4x^3 - 4x^2 + 5x - 8}{2x - 1} = 2x^2 - x + 2 + \frac{-6}{2x - 1}.$$

To check: $(2x - 1)\left(2x^2 - x + 2 + \dfrac{-6}{2x - 1}\right) = 4x^3 - 4x^2 + 5x - 8.$

Example 2 Divide $x^3 - 1$ by $x - 1$.

Here the polynomial $x^3 - 1$ is missing the x^2 term and the x term. When this is the case, the polynomial should be filled in with 0 coefficients for the missing terms.

$$x^3 - 1 = x^3 + 0x^2 + 0x - 1.$$

We can write the problem as follows.

$$
\begin{array}{r}
x^2 + x + 1 \\
x - 1 \overline{)\,x^3 + 0x^2 + 0x - 1} \\
\underline{x^3 - x^2} \\
x^2 + 0x \\
\underline{x^2 - x} \\
x - 1 \\
\underline{x - 1}
\end{array}
$$

There is no remainder. The quotient is $x^2 + x + 1$. Check by multiplication:

$$(x^2 + x + 1)(x - 1) = x^3 - 1.$$

Example 3 Divide $x^4 + 2x^3 + 2x^2 - x - 1$ by $x^2 + 1$.

The denominator $x^2 + 1$ has a missing x term which we fill in with $0x$, since $x^2 + 1 = x^2 + 0x + 1$. We then proceed as usual through the division process.

$$\begin{array}{r}
x^2 + 2x + 1 \\
x^2 + 0x + 1 \overline{)x^4 + 2x^3 + 2x^2 - x - 1} \\
\underline{x^4 + 0x^3 + x^2} \qquad\qquad \\
2x^3 + x^2 - x \quad \\
\underline{2x^3 + 0x^2 + 2x} \quad \\
x^2 - 3x - 1 \\
\underline{x^2 + 0x + 1} \\
-3x - 2
\end{array}$$

When the result of subtracting $(-3x - 2)$ is a polynomial of smaller degree than the divisor $(x^2 + 0x + 1)$, that polynomial is the remainder. We write the quotient as

$$x^2 + 2x + 1 + \frac{-3x - 2}{x^2 + 1}.$$

EXERCISES

Perform each of the divisions.

1. $\dfrac{x^2 - x - 6}{x - 3}$
2. $\dfrac{m^2 - 2m - 24}{m + 4}$
3. $\dfrac{2y^2 + 9y - 35}{y + 7}$
4. $\dfrac{y^2 + 2y + 1}{y + 1}$
5. $\dfrac{p^2 + 2p - 24}{p + 6}$
6. $\dfrac{x^2 + 11x + 24}{x + 8}$
7. $\dfrac{r^2 - 8r + 15}{r - 3}$
8. $\dfrac{t^2 - 3t - 10}{t - 5}$
9. $\dfrac{12m^2 - 20m + 3}{2m - 3}$
10. $\dfrac{2y^2 - 5y - 3}{2y + 1}$
11. $\dfrac{2a^2 - 11a - 21}{2a + 3}$
12. $\dfrac{9w^2 + 6w - 8}{3w - 2}$
13. $\dfrac{2x^2 + 5x + 3}{2x + 1}$
14. $\dfrac{4m^2 - 4m + 2}{2m - 1}$
15. $\dfrac{2a^2 - 3a + 4}{2a + 1}$
16. $\dfrac{4p^2 - 4p + 1}{2p - 1}$

17. $\dfrac{2d^2 - 2d + 5}{2d + 4}$

18. $\dfrac{6m^4 + 8m^3 - 10m^2 - 3}{m + 3}$

19. $\dfrac{2x^3 - x^2 + 3x + 2}{2x + 1}$

20. $\dfrac{12t^3 - 11t^2 + 9t + 18}{4t + 3}$

21. $\dfrac{8k^4 - 12k^3 - 2k^2 + 7k - 6}{2k - 3}$

22. $\dfrac{27r^4 - 36r^3 - 6r^2 + 26r - 24}{3r - 4}$

23. $\dfrac{3y^3 + y^2 + 3y + 1}{y^2 + 1}$

24. $\dfrac{2x^5 + 6x^4 - x^3 + 3x^2 - x}{2x^2 + 1}$

25. $\dfrac{x^4 - x^2 - 6x}{x^2 - 2}$

26. $\dfrac{x^4 - 2x^2 + 5}{x^3 - 1}$

27. $\dfrac{x^3 + 1}{x + 1}$

28. $\dfrac{x^4 - 1}{x^2 - 1}$

29. $\dfrac{x^4 - 1}{x^2 + 1}$

30. $\dfrac{x^5 - 1}{x^2 - 1}$

3.8 NEGATIVE INTEGER EXPONENTS

We want to define negative integer exponents so that all the rules for exponents are still valid. (See the list of rules on page 92.) For example, by the quotient rule we subtract exponents.

$$\frac{3^8}{3^6} = 3^{8-6} = 3^2.$$

However, consider the quotient $3^0/3^2$. Since $3^0 = 1$, we have

$$\frac{3^0}{3^2} = \frac{1}{3^2}.$$

Suppose we subtract exponents in this case.

$$\frac{3^0}{3^2} = 3^{0-2} = 3^{-2}.$$

The exponential 3^{-2} is read "Three to the negative two power." In order to have negative exponents satisfy the same properties as positive exponents, we must have

$$3^{-2} = \frac{1}{3^2}.$$

Thus, in general, for $a \neq 0$ and n any integer, we define

$$a^{-n} = \frac{1}{a^n}.$$

Example 1 **(a)** $3^{-2} = \dfrac{1}{3^2} = \dfrac{1}{9}$ **(c)** $\left(\dfrac{1}{2}\right)^{-3} = \dfrac{1}{\left(\dfrac{1}{2}\right)^3} = \dfrac{1}{\dfrac{1}{8}} = 1 \cdot \dfrac{8}{1} = 8$

(b) $5^{-3} = \dfrac{1}{5^3} = \dfrac{1}{125}$

By defining a^{-n} as we have, it can be shown that all the rules for exponentials are still valid. This is true of the product rule, $a^m \cdot a^n = a^{m+n}$.

$$5^3 \cdot 5^{-5} = 5^{3+(-5)} = 5^{-2} = \frac{1}{5^2} = \frac{1}{25};$$

$$(-8)^{-5} \cdot (-8)^7 = (-8)^{-5+7} = (-8)^2 = 64.$$

The power rule, $(a^m)^n = a^{mn}$, is still valid.

$$(3^{-4})^2 = 3^{(-4)(2)} = 3^{-8} = \frac{1}{3^8}; \quad (3m^{-2})^3 = 3^3m^{-6} = 3^3 \cdot \frac{1}{m^6} = \frac{27}{m^6}.$$

Example 2 $\left(\dfrac{6}{5}\right)^{-2} = \dfrac{6^{-2}}{5^{-2}} = \dfrac{\dfrac{1}{6^2}}{\dfrac{1}{5^2}} = \dfrac{1}{6^2} \cdot \dfrac{5^2}{1} = \dfrac{5^2}{6^2} = \dfrac{25}{36}.$

Example 3 $(2^3 \cdot 2^{-4})^2 = (2^{3+(-4)})^2 = (2^{-1})^2 = 2^{-2} = \dfrac{1}{2^2} = \dfrac{1}{4}.$

Furthermore, using the definition of negative exponent presented in this section, the quotient rule can now be simplified.

For all integers m and n, and $a \neq 0$,

$$\frac{a^m}{a^n} = a^{m-n}. \qquad \text{Quotient rule}$$

Example 4 $\dfrac{5^{-3}}{5^{-7}} = 5^{-3-(-7)} = 5^4.$

Example 5

$$\left(\frac{3x^{-2}}{4^{-1}y^3}\right)^{-3} = \frac{3^{-3}x^6}{4^3y^{-9}} = \frac{\dfrac{1}{3^3} \cdot x^6}{4^3 \cdot \dfrac{1}{y^9}} = \frac{\dfrac{x^6}{3^3}}{\dfrac{4^3}{y^9}} = \frac{x^6}{3^3} \cdot \frac{y^9}{4^3} = \frac{x^6y^9}{27 \cdot 64} = \frac{x^6y^9}{1728}.$$

EXERCISES

Evaluate each expression in Exercises 1–20.

1. 3^{-3}
2. 4^{-2}
3. 5^{-2}
4. 2^{-5}
5. 9^{-1}
6. $(-12)^{-1}$
7. $(-6)^{-2}$
8. 8^{-3}
9. 7^{-1}
10. 12^{-2}
11. $\left(\frac{1}{2}\right)^{-5}$
12. $\left(\frac{1}{5}\right)^{-2}$
13. $\left(\frac{1}{2}\right)^{-1}$
14. $\left(\frac{3}{4}\right)^{-1}$
15. $\left(\frac{2}{3}\right)^{-3}$
16. $\left(\frac{5}{4}\right)^{-2}$
17. $2^{-1} + 3^{-1}$
18. $3^{-1} - 4^{-1}$
19. $5^{-1} + 4^{-1}$
20. $3^{-1} + 6^{-1}$

In Exercises 21–52, perform the indicated operations. Write the answers without negative exponents, and with each variable occurring only once. Any variable represents a positive real number.

21. $3^4 \cdot 3^{-5}$
22. $5^{-6} \cdot 5^7$
23. $9^8 \cdot 9^{-7}$
24. $4^3 \cdot 4^{-5}$
25. $9^{-4} \cdot 9^{-2} \cdot 9^5$
26. $6^{-2} \cdot 6^3 \cdot 6^{-4}$
27. $\frac{8^9 \cdot 8^{-11}}{8^{-5}}$
28. $\frac{7^4 \cdot 7^{-3}}{7^5}$
29. $\frac{4^3 \cdot 4^{-5}}{4^7}$
30. $\frac{2^5 \cdot 2^{-4}}{2^{-1}}$
31. $\frac{5^{-3} \cdot 5^{-2}}{5^4}$
32. $\frac{8^{-2} \cdot 8^5}{8^6}$
33. $\frac{m^4 \cdot m^{-5}}{m^{-6}}$
34. $\frac{p^3 \cdot p^{-5}}{p^5}$
35. $\frac{m^{11} \cdot m^{-7}}{m^5}$
36. $\frac{z^3 \cdot z^{-5}}{z^{10}}$
37. $\frac{r^5 \cdot r^{-8}}{r^{-6} \cdot r^4}$
38. $\frac{x^3 \cdot x^{-1}}{x^8 \cdot x^{-2}}$
39. $\frac{a^6 \cdot a^{-3}}{a^{-5} \cdot a}$
40. $\frac{b^{10} \cdot b^{-2}}{b^{-8} \cdot b^6}$
41. $(3x^{-5})^2$
42. $(-5p^{-4})^{-2}$
43. $(9^{-1}y^5)^{-2}$
44. $(4^{-2}m^{-3})^{-2}$
45. $\left(\frac{a}{b}\right)^{-1}$
46. $\left(\frac{2a}{3}\right)^{-2}$
47. $\left(\frac{5m^{-2}}{m^{-1}}\right)^2$
48. $\left(\frac{4x^3}{3^{-1}}\right)^{-1}$
49. $\frac{(3x^2)^{-2}(5x^{-1})^3}{3x^{-5}}$
50. $\frac{(2y^{-1}z)^2(3y^{-2}z^{-1})^3}{(y^2z^3)^{-1}}$
51. $\frac{(4a^2b^3)^{-2}(2ab^{-1})^3}{(a^2b)^{-4}}$
52. $\frac{(m^6n)^{-2}(m^2n^{-2})^3}{m^{-1}n^{-2}}$

3.9 AN APPLICATION OF EXPONENTIALS: SCIENTIFIC NOTATION

One example of the use of exponentials comes from the field of science. The numbers occurring in science are often extremely large (such as the distance from the earth to the nearest star, the sun, which is 93,000,000 miles) or else extremely small (the wavelength of yellow-green light is approximately .0000006 meters). Because of the difficulty of working with many zeros, scientists often express such numbers as exponentials. Each number is written as the product of a number a (where $1 \leq a < 10$) and some power of 10. This form is called **scientific notation.** There is always one nonzero integer before the decimal point. For example, 35 is written 3.5×10^1, or 3.5×10; while 56,200 is written 5.62×10^4, since

$$56{,}200 = 5.62 \times 10{,}000 = 5.62 \times 10^4.$$

Example 1 **(a)** $93{,}000{,}000 = 9.3 \times 10^7$
(b) $463{,}000{,}000{,}000{,}000 = 4.63 \times 10^{14}$
(c) $63{,}200{,}000{,}000 = 6.32 \times 10^{10}$
(d) $302{,}100 = 3.021 \times 10^5$

Example 2 **(a)** Express 6.2×10^3 without exponents.
The number 6.2×10^3 is in scientific notation. We write

$$6.2 \times 10^3 = 6.2 \times 1000 = 6200.$$

(b) $4.283 \times 10^5 = 4.283 \times 100{,}000 = 428{,}300.$

To use scientific notation for very small numbers, note that $.1 = 1/10 = 10^{-1}$; $.01 = 1/100 = 10^{-2}$, and so on. Thus, we can use negative exponents to write

$$.004 = 4 \times 10^{-3},$$
$$.0000762 = 7.62 \times 10^{-5},$$
$$.000000000834 = 8.34 \times 10^{-10}.$$

EXERCISES

Express the numbers in Exercises 1–18 in scientific notation.

1. 6,835,000,000
2. 321,000,000,000,000
3. 8,360,000,000,000
4. 6850
5. 215
6. 683
7. 25,000
8. 110,000,000
9. .035
10. .005
11. .0101
12. .0000006
13. .000012
14. .000000982
15. .00027
16. .00045
17. .000000000114
18. .00000000037

In Exercises 19–38, write each of the numbers without exponents.

19. 8.1×10^9

20. 3.5×10^2

21. 9.132×10^6

22. 2.14×10^0

23. 3.24×10^8

24. 4.35×10^4

25. 3.2×10^{-4}

26. 5.76×10^{-5}

27. 4.1×10^{-2}

28. 1.79×10^{-3}

29. 7.8×10^{-6}

30. 7.698×10^{-7}

31. $(2 \times 10^8) \times (4 \times 10^{-3})$

32. $(5 \times 10^4) \times (3 \times 10^{-2})$

33. $(4 \times 10^{-1}) \times (1 \times 10^{-5})$

34. $(6 \times 10^{-5}) \times (2 \times 10^4)$

35. $\dfrac{9 \times 10^5}{3 \times 10^{-1}}$

36. $\dfrac{12 \times 10^{-4}}{4 \times 10^4}$

37. $\dfrac{4 \times 10^{-3}}{8 \times 10^{-2}}$

38. $\dfrac{5 \times 10^{-1}}{25 \times 10^{-5}}$

Write the numbers in the following applications in scientific notation.

39. Light visible to the human eye has a wavelength between .0004 mm. and .0008 mm.

40. In the ocean, the amount of oxygen per cubic mile of water is 4,037,000,000 tons, while the amount of radium is .0003 tons.

41. Each tide in the Bay of Fundy carries more than 3,680,000,000,000,000 cubic feet of water into the bay.

42. The mean (average) diameter of the sun is about 865,000 miles.

Write the numbers in the following applications without exponents.

43. 1×10^3 cubic millimeters equals 6.102×10^{-2} cubic inch.

44. In the food chain which links the largest sea creature, the whale, to the smallest, the diatom, 4×10^{14} diatoms sustain a medium-sized whale for only a few hours.

45. Surprisingly, ocean trenches are nearly all the same depth, 3.5×10^4 feet.

46. The average lifespan of a human in seconds is about 1×10^9.

47. The disassociation constant for aqueous ammonia is 1.86×10^{-5}.

CHAPTER 3 SUMMARY

Key Words

Exponential
Exponential form
Exponent (Power)
Base
Scientific notation

Term
Coefficient
Like terms
Unlike terms

Polynomial
Monomial
Binomial
Trinomial

Polynomial in x
Descending powers
Degree

Rules for Exponentials

$a^m \cdot a^n = a^{m+n}$ Product Rule

$\frac{a^m}{a^n} = a^{m-n}, a \neq 0$ Quotient Rule

$(a^m)^n = a^{mn}$ Power Rule

$(ab)^m = a^m b^m$; $\left(\frac{a}{b}\right)^m = \frac{a^m}{b^m}, b \neq 0$; $a^{-n} = \frac{1}{a^n}$; $a^0 = 1$

Operations with Polynomials

To add any two polynomials, add like terms.

To subtract any two polynomials, add the inverse of the second polynomial to the first polynomial.

To divide a polynomial by a monomial, multiply each term of the polynomial by the reciprocal of the monomial. Use the Quotient Rule to simplify.

Multiplying Two Binomials

Step 1 Multiply the two first terms to get the first term of the answer.

Step 2 Find the inner product and the outer product, and add them to get the middle term of the answer.

Step 3 Multiply the two end terms to get the last term of the answer.

Special Products of Binomials

$(x + a)^2 = x^2 + 2xa + a^2$
$(x - a)^2 = x^2 - 2xa + a^2$ Square of a binomial

$(x + a)(x - a) = x^2 - a^2$ Difference of two squares

CHAPTER 3 TEST

If $P(x) = x^4 + 2x^2 - 7x + 2$, find the following values.

1. $P(2)$

2. $P(-1)$

Write in exponential form, and then evaluate.

3. $4 \cdot 4 \cdot 4 \cdot 4 \cdot 4$

4. $\frac{6^4}{6^6}$

5. $(-2)^2(-2)^3$

6. $2^{-3} \cdot 2^7$

7. $(2^3)^2$

8. $(2/3)^2$

9. 8^{-2}

10. $\frac{8^0 \cdot 8^{13}}{8^{12}}$

11. $\frac{(3)^{10}(3)^4}{(3^3)^5}$

12. Use the rules for exponentials to simplify $(2x^2y^3)^{-3}$, and write with only positive exponents.

For each polynomial, combine terms, then give the degree of the polynomial. Finally, select the most specific description from:

(a) trinomial (b) binomial (c) monomial (d) none of these.

13. $3x^2 + 6x - 4x^2$

14. $11m^3 - m^2 + m^4$

15. $3x^3 - 4x^2 + 2x - 1$

16. 7

17. $8m^5 - 3m^2 + 4m^9$

18. $-7x^2 + 8x^2 - 2x^2$

Perform the indicated operation.

19. $(3x^3 + 2x^2 - 5x + 3) + (7x^3 - 4x^2 - 3x - 3)$

20. $(2x^5 + 3x^3 - 4x + 7) - (x^5 - 3x^3 + x^2 - 2x - 5)$

21. $(y^2 - 5y - 3) + (3y^2 + 2y) - (y^2 - y - 1)$

22. $(10x^3x^4)(-4x^2x)$

23. $6m^2(m^3 + 2m^2 - 3m + 7)$

24. $(r - 5)(r + 2)$

25. $(3t + 4)(2t - 3)$

26. $(2p + 5)^2$

27. $(x - 8)(x + 8)$

28. $(k^3 + 2k^2 - 5k)(k^2 + 2k + 3)$

29. $(15r^4 - 10r^3 + 25r^2 - 15r) \div 5r$

30. $(3x^5 + 12x^4 - 9x^3 + 6x^2 - 60x - 120) \div (x - 2)$

Write in scientific notation.

31. 6,000,000

32. 245,000,000

33. .000025

34. .0104

Write without exponents.

35. 4.8×10^4

36. 2.91×10^8

37. 6.45×10^{-2}

38. 1.03×10^{-5}

39. $\dfrac{2 \times 10^5 \times 5 \times 10^{-3}}{5 \times 10^4}$

40. $\dfrac{1 \times 10^{-2} \times 4 \times 10^4}{2 \times 10^{-1} \times 1 \times 10^3}$

41. The period of a radio wave is .000001 seconds.

4 FACTORING

4.1 FACTORS

Since the product of 6 and 2 is 12, we say that 6 and 2 are factors of 12, or that $6 \cdot 2$ is a **factored form** of 12. Other factored forms of 12 are $(-6)(-2)$, $3 \cdot 4$, $(-3)(-4)$, $(12)(1)$, and $(-12)(-1)$. Note that $(1/2)(24)$ and $(2/3)(18)$ might also be considered factored forms of 12. However, factors are usually limited to integers. So then the factors of 12 are

$$1, \quad -1, \quad 2, \quad -2, \quad 3, \quad -3, \quad 4, \quad -4, \quad 6, \quad -6, \quad 12, \quad -12.$$

In general, an integer a is a **factor** of an integer b if b/a is an integer.

Example 1 The factors of 36 are 1, -1, 2, -2, 3, -3, 4, -4, 6, -6, 9, -9, 12, -12, 18, -18, 36, -36.

Example 2 The factors of 11 are 11, -11, 1, and -1.

An integer which is a factor of two or more integers is called a **common factor** of those integers. For example, 6 is a common factor of 18 and 24, since 6 is a factor of both 18 and 24. Other common factors of 18 and 24 are -6, -3, -2, -1, 1, 2, and 3. The **greatest common factor** of a set of integers is the largest common factor of the set. Thus, 6 is the greatest common factor of 18 and 24 because 6 is the largest common factor of these numbers.

Example 3 **(a)** The greatest common factor of 7, 11, and 14 is 1, since 1 is the largest number that divides all three numbers.
(b) The greatest common factor of 2, 4, 6, and 8 is 2.
(c) The number 3 is the greatest common factor of 9, 27, and 6.
(d) The number 4 is the greatest common factor of the numbers 8, 12, 16, and 32.

You can also find the greatest common factor of a collection of terms. For example, the terms x^4, x^5, x^6, and x^7 have x^4 as the greatest common factor, because x^4 is the highest power of x that is a factor of each of the terms x^4, x^5, x^6, and x^7. To see this, write the terms in factored form.

$$x^4 = x^4 \cdot 1$$
$$x^5 = x^4 \cdot x$$
$$x^6 = x^4 \cdot x^2$$
$$x^7 = x^4 \cdot x^3.$$

Example 4 Find the greatest common factor of the terms y^2, y^5, y^7, and y^{15}.

Here, y^2 is the greatest common factor, since it is the highest power of y that is a factor of each term.

Example 5 Find the greatest common factor of the terms $21m^7$, $-18m^6$, $45m^8$, $-24m^5$.

First, note that 3 is the greatest common factor of the coefficients 21, -18, 45, and -24. Also, m^5 is the greatest common factor of the variable expressions m^7, m^6, m^8, and m^5. Hence $3m^5$ is the greatest common factor of the terms $21m^7$, $-18m^6$, $45m^8$, and $-24m^5$.

Example 6 Find the greatest common factor of the terms x^4y^2, x^7y^5, x^3y^7, x^6y^{15}.

In Example 4 we found y^2 to be the common factor of y^2, y^5, y^7, and y^{15}. In the same way, x^3 is the greatest common factor of x^4, x^7, x^3, and x^6. Therefore, x^3y^2 is the greatest common factor of the given terms.

Example 7 Find the greatest common factor of the terms $48y^{12}$, $-36y^{10}$, $24y^7$.

The greatest common factor of the coefficients 48, -36, and 24 is 12. The greatest common factor of y^{12}, y^{10}, and y^7 is y^7. Thus, the greatest common factor of the given terms is $12y^7$.

Suppose the terms in Example 7 were in a polynomial.

$$48y^{12} - 36y^{10} + 24y^7.$$

The greatest common factor of the terms $48y^{12}$, $-36y^{10}$, and $24y^7$ is $12y^7$. Factor each term into a product so that $12y^7$ is one factor.

$$48y^{12} - 36y^{10} + 24y^7 = 12y^7(4y^5) + 12y^7(-3y^3) + 12y^7(2).$$

Then, by the distributive property,

$$12y^7(4y^5) + 12y^7(-3y^3) + 12y^7(2) = 12y^7(4y^5 - 3y^3 + 2).$$

The factored form of the polynomial is $(12y^7)(4y^5 - 3y^3 + 2)$. The process we have described here is called **factoring out the greatest common factor.**

Example 8 Factor out the greatest common factor from $x^5 - x^3$.
The greatest common factor is x^3.

$$x^5 - x^3 = x^3(x^2) - x^3(1) = x^3(x^2 - 1).$$

The factored form is $x^3(x^2 - 1)$. Note that it was necessary to express x^3 as $x^3(1)$ in order to factor the second term so that one factor would be x^3.

Example 9 Factor out the greatest common factor from $6x^3 - 3x^2 + 9x$.
The greatest common factor is $3x$.

$$\begin{aligned} 6x^3 - 3x^2 + 9x &= 3x(2x^2) - 3x(x) + 3x(3) \\ &= 3x(2x^2 - x + 3). \end{aligned}$$

Example 10 Factor out the greatest common factor from $-20m^7p^2 - 36m^3p^4 + 48m^4p^3$.
The greatest common factor is $4m^3p^2$.

$$\begin{aligned} &-20m^7p^2 - 36m^3p^4 + 48m^4p^3 \\ &\quad = (4m^3p^2)(-5m^4) + (4m^3p^2)(-9p^2) + (4m^3p^2)(12mp) \\ &\quad = (4m^3p^2)(-5m^4 - 9p^2 + 12mp). \end{aligned}$$

EXERCISES

Complete the factoring in Exercises 1–14.

1. $12 = 6(\quad)$

2. $18 = 9(\quad)$

3. $3x^2 = 3x(\quad)$

4. $8x^3 = 8x(\quad)$

5. $9m^4 = 3m^2(\quad)$

6. $12p^5 = 6p^3(\quad)$

7. $-8z^9 = -4z^5(\quad)$

8. $-15k^{11} = -5k^8(\quad)$

9. $x^2y^3 = xy(\quad)$

10. $a^3b^2 = a^2b(\quad)$

11. $6x^2y^3 = 6xy(\quad)$

12. $27a^3b^2 = 9a^2b(\quad)$

13. $14x^4y^3 = 2xy(\quad)$

14. $-16m^3n^3 = 4mn^2(\quad)$

In Exercises 15–44, factor out the greatest common factor.

15. $12x + 24$

16. $18m - 9$

17. $3 + 36d$

18. $15 + 25r$

19. $9a^2 - 18a$

20. $21m^5 - m^4$

21. $65y^9 - 35y^5$

22. $100a^4 - 16a^2$

23. $121p^5 - 33p^4$

24. $8p^2 - 4p^4$

25. $11z^2 - 121$

26. $12z^2 - 11y^4$

27. $9m^2 + 90m^3$

28. $16r^2s + 64rs^2$

29. $19y^3p^2 + 38y^2p^3$

30. $-12m^2n + 4mn^2$

31. $18x^2y^3 - 24x^4y$

32. $100m^5 - 50m^3 + 100$

33. $13y^6 + 26y^5 - 39y^3$

34. $5x^4 + 25x^3 - 20x^2$

35. $16a^3 + 8a^2 + 24a$

36. $6a^2 + 8c^2 - 4b^2$

37. $45q^4p^5 - 36qp^6 + 81q^2p^3$

38. $a^5 + 2a^5b + 3a^5b^2 - 4a^5b^3$

39. $a^3b^5 - a^2b^7 + ab^3$

40. $m^6n^5 - 2m^5 + 5m^3n^5$

41. $125z^5a^3 - 60z^4a^5 + 85z^3a^4$

42. $-30a^2m^2 + 60a^3m + 180a^3m^2$

43. $33y^8 - 44y^{12} + 77y^3 + 11y^4$

44. $26g^6h^4 + 13g^3h^4 - 39g^4h^3$

4.2 FACTORING TRINOMIALS

The product of two binomials is usually a trinomial. For example,

$$(x + 4)(x - 6) = x^2 - 2x - 24;$$
$$(y + 5)(y - 3) = y^2 + 2y - 15.$$

We said that the product of two binomials is *usually* a trinomial. Exceptions include $(x + 2)(x - 2) = x^2 - 4$, which is not a trinomial.

In this section our goal is to factor a trinomial as the product of two binomial factors. We limit ourselves to trinomials like $x^2 - 2x - 24$ or $y^2 + 2y - 15$, where the coefficient of the squared term is 1.

Let's try to factor $x^2 + 5x + 6$. We want to find integers a and b such that

$$x^2 + 5x + 6 = (x + a)(x + b).$$

To find these integers a and b, first multiply the right-hand side of the equation.

$$(x + a)(x + b) = x^2 + ax + bx + ab.$$

By the distributive property,

$$x^2 + ax + bx + ab = x^2 + (a + b)x + ab.$$

Thus, we want integers a and b such that

$$x^2 + 5x + 6 = x^2 + (a + b)x + ab.$$

The integers a and b must satisfy the conditions

$$a + b = 5$$
$$ab = 6.$$

Can we find two integers whose sum is 5 $(a + b = 5)$ and whose product is 6 $(ab = 6)$? Since many pairs of integers can be found which have a sum of 5, it is perhaps best to list first those pairs of integers which have a product of 6.

$$\begin{array}{rr} 1 \cdot 6 = 6 & 1 + 6 = 7 \\ 2 \cdot 3 = 6 & 2 + 3 = 5 \\ (-1)(-6) = 6 & -1 + (-6) = -7 \\ (-2)(-3) = 6 & -2 + (-3) = -5 \end{array}$$

All four pairs have a product of 6, but only the pair with 2 and 3 has a sum of 5. Then 2 and 3 are the integers we are looking for.

$$x^2 + 5x + 6 = (x + 2)(x + 3).$$

The trinomial $x^2 + 5x + 6$ has been *factored* into the product of the binomials $x + 2$ and $x + 3$.

To check, multiply the binomials.

$$(x + 2)(x + 3) = x^2 + 5x + 6.$$

Note that the method we used can only be used on trinomials where the coefficient of the squared term is 1. Methods for other trinomials will be presented later in the chapter.

Example 1 To factor $m^2 + 9m + 14$, look for two integers whose product is 14, and whose sum is 9. List the pairs of integers whose products are 14. Then examine the sums.

$$\begin{array}{rrr} 14, & 1 & 14 + 1 = 15 \\ 7, & 2 & 7 + 2 = 9 \\ -14, & -1 & -14 + (-1) = -15 \\ -7, & -2 & -7 + (-2) = -9 \end{array}$$

From the list, 7 and 2 are the integers we need, since $7 \cdot 2 = 14$ and $7 + 2 = 9$. Thus the binomial factors of $m^2 + 9m + 14$ are $(m + 2)(m + 7)$.

Example 2 To factor $p^2 - 2p - 15$, find two integers whose product is -15 and whose sum is -2. If these numbers do not come to mind right away, we can always find them (if they exist) by listing all the pairs of integers whose product is -15.

$$\begin{array}{rrr} 15, & -1 & 15 + (-1) = 14 \\ 5, & -3 & 5 + (-3) = 2 \\ -15, & 1 & -15 + 1 = -14 \\ -5, & 3 & -5 + 3 = -2 \end{array}$$

The integers we need are -5 and 3. The factored trinomial is

$$p^2 - 2p - 15 = (p - 5)(p + 3).$$

Example 3 To factor $x^2 + 5x + 12$, we list all pairs of integers whose product is 12.

12,	1	$12 + 1 = 13$
6,	2	$6 + 2 = 8$
3,	4	$3 + 4 = 7$
-12,	-1	$-12 + (-1) = -13$
-6,	-2	$-6 + (-2) = -8$
-3,	-4	$-3 + (-4) = -7$

None of the pairs of integers has a sum of 5. Thus the trinomial cannot be factored. This means that we cannot find *integers* a and b such that

$$x^2 + 5x + 12 = (x + a)(x + b).$$

Later in this book we will see that such numbers a and b do exist—but they are not integers. To express this, we say that trinomials like $x^2 + 5x + 12$ cannot be factored *over the integers*, that is, factored so that the factors have integer coefficients. On the other hand, the trinomial $x^2 + 5x + 6$ *can* be factored over the integers, as we found.

Example 4 Since there are no integers whose product is 11 and whose sum is -8, the trinomial $k^2 - 8k + 11$ cannot be factored over the integers.

Example 5 Factor $x^2 + 4x - 5$.

The only factors of -5 are -5 and 1 or -1 and 5. We need a sum of 4, so we factor $x^2 + 4x - 5$ as

$$x^2 + 4x - 5 = (x - 1)(x + 5).$$

Example 6 To factor the trinomial $4x^5 - 28x^4 + 40x^3$, note that it has a common factor of $4x^3$. Factor it out first.

$$4x^5 - 28x^4 + 40x^3 = 4x^3(x^2 - 7x + 10).$$

Then factor the trinomial $x^2 - 7x + 10$. The integers -5 and -2 have a product of 10 and a sum of -7. The complete factored form is

$$4x^5 - 28x^4 + 40x^3 = 4x^3(x - 5)(x - 2).$$

Example 7. To factor $z^2 - 2bz - 3b^2$, look for two numbers whose product is $-3b^2$ and whose sum is $-2b$. The numbers we need are $-3b$ and b. Thus

$$z^2 - 2bz - 3b^2 = (z - 3b)(z + b).$$

EXERCISES

Complete the factoring in Exercises 1–14.

1. $x^2 + 10x + 21 = (x + 7)(\quad)$

2. $p^2 + 11p + 30 = (p + 5)(\quad)$

3. $r^2 + 15r + 56 = (r + 7)(\quad)$

4. $x^2 + 15x + 44 = (x + 4)(\quad)$

5. $t^2 - 14t + 24 = (t - 2)(\quad)$

6. $x^2 - 9x + 8 = (x - 1)(\quad)$

7. $x^2 - 12x + 32 = (x - 4)(\quad)$

8. $y^2 - 2y - 15 = (y + 3)(\quad)$

9. $m^2 + 2m - 24 = (m - 4)(\quad)$

10. $x^2 + 9x - 22 = (x - 2)(\quad)$

11. $p^2 + 7p - 8 = (p + 8)(\quad)$

12. $y^2 - 7y - 18 = (y + 2)(\quad)$

13. $x^2 - 7x - 30 = (x - 10)(\quad)$

14. $k^2 - 3k - 28 = (k - 7)(\quad)$

In Exercises 15–54, factor as completely as possible. If a polynomial cannot be factored, write "cannot be factored."

15. $x^2 + 6x + 5$

16. $y^2 + y - 72$

17. $a^2 + 9a + 20$

18. $b^2 + 8b + 15$

19. $x^2 - 8x + 7$

20. $m^2 + m - 20$

21. $p^2 + 4p + 5$

22. $n^2 - 4n - 12$

23. $y^2 - 6y + 8$

24. $r^2 - 11r + 30$

25. $s^2 + 2s - 35$

26. $h^2 + 11h + 12$

27. $n^2 - 12n - 35$

28. $a^2 - 2a - 99$

29. $b^2 - 11b + 24$

30. $x^2 - 9x + 20$

31. $y^2 - 4y - 21$

32. $z^2 - 14z + 49$

33. $y^2 - 12y + 8$

34. $r^2 + r - 42$

35. $z^2 - 3z - 40$

36. $p^2 + 5p - 66$

37. $3m^3 + 12m^2 + 9m$

38. $3y^5 - 18y^4 + 15y^3$

39. $6a^2 - 48a - 120$

40. $h^7 - 5h^6 - 14h^5$

41. $3j^3 - 30j^2 + 72j$
42. $2x^6 - 8x^5 - 42x^4$
43. $3x^4 - 3x^3 - 90x^2$
44. $2y^3 - 8y^2 - 10y$
45. $x^2 + 4ax + 3a^2$
46. $x^2 - mx - 6m^2$
47. $y^2 - by - 30b^2$
48. $z^2 + 2zx - 15x^2$
49. $x^2 + xy - 30y^2$
50. $a^2 - ay - 56y^2$
51. $r^2 - 2rs + s^2$
52. $m^2 - 2mn - 3n^2$
53. $p^2 - 3pq - 10q^2$
54. $c^2 - 5cd + 4d^2$

4.3 MORE ABOUT FACTORING TRINOMIALS

Now we discuss methods of factoring trinomials where the coefficient of the squared term is *not* 1, such as

$$2x^2 + 7x + 6.$$

When we begin to factor this trinomial, we have no assurance that it is even factorable. But we assume it is and proceed from that assumption.

If we factor $2x^2 + 7x + 6$ into the product of two binomials, we will have four integers, a, b, c, and d, to consider. That is, we want to find a, b, c, and d, so that

$$2x^2 + 7x + 6 = (ax + b)(cx + d).$$

As we did in Section 4.2, we multiply the binomials to see how a, b, c, and d relate to the coefficients in the trinomial.

$$\begin{aligned}(ax + b)(cx + d) &= acx^2 + adx + bcx + bd \\ &= acx^2 + (ad + bc)x + bd.\end{aligned}$$

We see that $2 = ac$. Either

$$\begin{matrix} a = 2, & \text{or} & a = 1, \\ c = 1 & & c = 2. \end{matrix}$$

This is because the only positive factors of 2 are 2 and 1. (We usually consider only positive factors of the coefficient of x^2.)

$$2x^2 + 7x + 6 = (2x + b)(x + d)$$

or

$$2x^2 + 7x + 6 = (x + b)(2x + d).$$

To complete the factoring of $2x^2 + 7x + 6$, we must find values of b and d that are factors of 6, and also result in a middle term of $7x$. Theory is of little help here. Trial and error is the key to success in factoring problems of this type. Insight gained by experience in factoring also helps.

Let us begin by trying $b = 1$ and $d = 6$. Then the factors are $(2x + 1)(x + 6)$. Check by multiplying.

$$(2x + 1)(x + 6) = 2x^2 + 13x + 6 \neq 2x^2 + 7x + 6. \qquad \text{Wrong}$$

Exchange the 6 and 1.

$$(2x + 6)(x + 1) = 2x^2 + 8x + 6 \neq 2x^2 + 7x + 6. \qquad \text{Wrong}$$

Try 2 and 3, which are also factors of 6.

$$(2x + 2)(x + 3) = 2x^2 + 8x + 6 \neq 2x^2 + 7x + 6. \qquad \text{Wrong}$$

Here and in the previous trial we tried factors, $2x + 6$ and $2x + 2$, that themselves had common factors. But since the original polynomial had no common factors, none of its factors can either. Remembering this helps in eliminating possible factors of the polynomial.

Now let us try exchanging the 2 and 3.

$$(2x + 3)(x + 2) = 2x^2 + 7x + 6. \qquad \text{Correct}$$

The polynomial $2x^2 + 7x + 6$ can be factored as $(2x + 3)(x + 2)$. Trial and error is required for the solution of this problem. There is no way to avoid it.

Example 1 Factor $8p^2 + 14p + 5$.

The number 8 has many possible pairs of factors, while the number 5 has only 1 and 5 and -1 and -5. For this reason, we begin by considering the factors of 5. We can ignore any negative factors because all the terms of the trinomial have positive coefficients. Thus, since 5 has only 1 and 5 as suitable factors, if $8p^2 + 14p + 5$ can be factored, it will have to be factored as

$$(ap + 5)(cp + 1) \qquad \text{or} \qquad (ap + 1)(cp + 5).$$

We must now find correct integers for a and c. The number 8 has two pairs of suitable factors; 4 and 2, and 8 and 1. We must now resort to trial and error.

$$(8p + 5)(p + 1) = 8p^2 + 13p + 5 \qquad \text{Wrong}$$
$$(p + 5)(8p + 1) = 8p^2 + 41p + 5 \qquad \text{Wrong}$$
$$(2p + 5)(4p + 1) = 8p^2 + 22p + 5 \qquad \text{Wrong}$$
$$(4p + 5)(2p + 1) = 8p^2 + 14p + 5 \qquad \text{Correct}$$

Hence, through trial and error we have found that $8p^2 + 14p + 5$ can be factored over the integers as

$$(4p + 5)(2p + 1).$$

Example 2 Factor $6x^2 - 11x + 3$.

The integer 6 has several possible pairs of factors, while 3 has only 3 and 1, or -3 and -1. Thus we begin by factoring 3. Since the middle term here has a negative coefficient, we must consider negative factors.

Because of the negative coefficient on the middle term, let's try -3 and -1 as factors of 3. If $6x^2 - 11x + 3$ is factorable, we will be able to find integers a and c such that

$$6x^2 - 11x + 3 = (ax - 1)(cx - 3)$$

or

$$6x^2 - 11x + 3 = (ax - 3)(cx - 1).$$

We can factor 6 as $2 \cdot 3$ or $1 \cdot 6$. We can then begin trial and error.

$$(2x - 3)(3x - 1) = 6x^2 - 11x + 3 \qquad \text{Correct}$$

This time we got the answer on the first attempt.

Example 3 Factor $8x^2 + 6x - 9$.

The integer 8 has several possible factors, as does -9. Since the coefficient of the middle term is small, it probably would be wise to avoid large factors such as 8 or 9. Let us begin by trying 4 and 2 as factors of 8, and 3 and -3 as factors of -9.

$$(4x + 3)(2x - 3) = 8x^2 - 6x - 9 \qquad \text{Wrong}$$
$$(4x - 3)(2x + 3) = 8x^2 + 6x - 9 \qquad \text{Correct}$$

Example 4 Factor $12a^2 - ab - 20b^2$.

There are several possible factors of $12a^2$, including $12a$ and a, $6a$ and $2a$, and $3a$ and $4a$, just as there are many possible factors of $-20b^2$, including $-20b$ and b, $10b$ and $-2b$, $-10b$ and $2b$, $4b$ and $-5b$, and $-4b$ and $5b$. Once again, since our desired middle term is small we avoid the larger factors. Let us try as factors $6a$ and $2a$ and $4b$ and $-5b$:

$$(6a + 4b)(2a - 5b).$$

This cannot be correct, as we mentioned before, since $6a + 4b$ has a common factor, while the trinomial itself has none. For the same reason, $(6a - 5b)(2a + 4b)$ cannot be correct.

Let's try $3a$ and $4a$ with $4b$ and $-5b$.

$$(3a + 4b)(4a - 5b) = 12a^2 + ab - 20b^2 \qquad \text{Wrong}$$

Here the middle term has the wrong sign. What about $3a$ and $4a$ with $-4b$ and $5b$?

$$(3a - 4b)(4a + 5b) = 12a^2 - ab - 20b^2. \qquad \text{Correct}$$

Success at last.

Example 5 Factor $28x^5 - 58x^4 - 30x^3$.

First factor out $2x^3$, the greatest common factor.

$$28x^5 - 58x^4 - 30x^3 = 2x^3(14x^2 - 29x - 15).$$

We must use trial and error to factor $14x^2 - 29x - 15$. Let's try $7x$ and $2x$ as factors of $14x^2$, and -3 and 5 as factors of -15.

$$(7x - 3)(2x + 5) = 14x^2 + 29x - 15. \qquad \text{Wrong}$$

The middle term differs only in sign, and so we try

$$(7x + 3)(2x - 5) = 14x^2 - 29x - 15. \qquad \text{Correct}$$

The factored form of $28x^5 - 58x^4 - 30x^3$ is $2x^3(7x + 3)(2x - 5)$.

EXERCISES

Complete the factoring in Exercises 1–10.

1. $2x^2 - x - 1 = (2x + 1)(\quad)$

2. $3a^2 + 5a + 2 = (3a + 2)(\quad)$

3. $5b^2 - 16b + 3 = (5b - 1)(\quad)$

4. $2x^2 + 11x + 12 = (2x + 3)(\quad)$

5. $4y^2 + 17y - 15 = (y + 5)(\quad)$

6. $7z^2 + 10z - 8 = (z + 2)(\quad)$

7. $15x^2 + 7x - 4 = (3x - 1)(\quad)$

8. $12c^2 - 7c - 12 = (4c + 3)(\quad)$

9. $2m^2 + 19m - 10 = (2m - 1)(\quad)$

10. $6x^2 + x - 12 = (2x + 3)(\quad)$

In each of Exercises 11–56, factor as completely as possible.

11. $2x^2 - 5x - 3$

12. $3x^2 - x - 2$

13. $3x^2 + 10x + 7$

14. $7a^2 + 90a - 13$

15. $4r^2 + r - 3$

16. $3p^2 + 2p - 8$

17. $15m^2 + m - 2$

18. $6x^2 + x - 1$

19. $8m^2 - 10m - 3$

20. $2a^2 - 17a + 30$

21. $5a^2 - 7a - 6$

22. $12s^2 + 11s - 5$

23. $3r^2 + r - 10$

24. $20x^2 - 28x - 3$

25. $4y^2 + 69y + 17$

26. $21m^2 + 13m + 2$

27. $38x^2 + 23x + 2$

28. $20y^2 + 39y - 11$

29. $10x^2 + 11x - 6$

30. $6b^2 + 7b + 2$

31. $6w^2 + 19w + 10$

32. $20q^2 - 41q + 20$

33. $6q^2 + 23q + 21$

34. $8x^2 + 47x - 6$

35. $10m^2 - 23m + 12$

36. $4t^2 - 5t - 6$

37. $8k^2 + 2k - 15$

38. $15p^2 - p - 6$

39. $10m^2 - m - 24$

40. $16a^2 + 30a + 9$

41. $3 - 14x + 8x^2$

42. $-5 - 37x + 24x^2$

43. $6 - m - 40m^2$

44. $8 + 22a + 15a^2$

45. $40m - 2m^2 - 2m^3$

46. $18n^2 - 39n^3 + 15n^4$

47. $24a^4 + 10a^3 - 4a^2$

48. $18x^5 + 15x^4 - 75x^3$

49. $32z^5 - 20z^4 - 12z^3$

50. $15x^2y^2 - 7xy^2 - 4y^2$

51. $12p^2 + 7pq - 12q^2$

52. $6m^2 - 5mn - 6n^2$

53. $25a^2 + 25ab + 6b^2$

54. $6x^2 - 5xy - y^2$

55. $6a^2 - 7ab - 5b^2$

56. $25g^2 - 5gh - 2h^2$

4.4 TWO SPECIAL FACTORIZATIONS

In this section we present methods of factoring two special types of polynomials. First we factor a binomial that is the **difference of two squares**,

$$x^2 - 25 = x^2 - 5^2.$$

To factor $x^2 - 25$, rewrite it as

$$x^2 + 0x - 25.$$

This trinomial can be factored into two binomials. We need two integers whose product is -25 and whose sum is 0. The integers are 5 and -5, since $5 + (-5) = 0$ and $5(-5) = -25$. Thus the binomial $x^2 - 25$ is factored as $(x + 5)(x - 5)$.

Example 1 **(a)** $x^2 - 49 = (x + 7)(x - 7)$
(b) $y^2 - m^2 = (y + m)(y - m)$
(c) $z^2 - 4 = (z + 2)(z - 2)$

Example 2 Factor $25m^2 - 16$.
This is the difference of two squares, $(5m)^2 - 4^2$. Thus

$$25m^2 - 16 = (5m + 4)(5m - 4).$$

Example 3 **(a)** $49z^2 - 64 = (7z)^2 - 8^2 = (7z + 8)(7z - 8)$
(b) $9a^2 - 4b^2 = (3a)^2 - (2b)^2 = (3a + 2b)(3a - 2b)$
(c) $81y^2 - 36 = (9y)^2 - 6^2 = (9y + 6)(9y - 6)$
$= 3(3y + 2) \cdot 3(3y - 2)$
$= 9(3y + 2)(3y - 2)$

It would have been more convenient to factor $81y^2 - 36$ by first factoring out the greatest common factor, 9.

$$81y^2 - 36 = 9(9y^2 - 4) = 9[(3y)^2 - 2^2] = 9(3y + 2)(3y - 2).$$

It is never possible to factor a *sum* of two squares. For example, $x^2 + 1$ does not factor. To see why, write $x^2 + 1$ as $x^2 + 0x + 1$. We need two integers whose sum is 0 and whose product is 1. No such integers exist.

A quantity is a *perfect square* if it can be factored as the square of another quantity. Thus, 144, $4x^2$, and $81m^6$ are all perfect squares, since

$$144 = 12^2, \qquad 4x^2 = (2x)^2, \qquad 81m^6 = (9m^3)^2.$$

A **perfect square trinomial** is a trinomial that is the square of a binomial. For example, $x^2 + 8x + 16 = (x + 4)^2$.

For a trinomial to be a perfect square, it is necessary that two of its terms be perfect squares. Thus $16x^2 + 4x + 8$ is not a perfect square trinomial since only the term $16x^2$ is a perfect square, $(4x)^2$.

On the other hand, just because two of the terms are perfect squares, we cannot be sure that the trinomial is a perfect square trinomial. For example, $x^2 + 6x + 36$ has two perfect square terms, but it is not a perfect square trinomial. (Try to find a binomial that can be squared to give $x^2 + 6x + 36$.)

In general, the square of a binomial is of the form

$$(a + b)^2 = a^2 + 2ab + b^2.$$

The middle term of a perfect square trinomial is always twice the product of the two terms in the squared binomial. We can use this fact to check any attempt to factor a trinomial that appears to be a perfect square.

Factor $x^2 + 10x + 25$. The term x^2 is a perfect square, and so is 25. We can try to factor the trinomial as

$$x^2 + 10x + 25 = (x + 5)^2.$$

To check, take twice the product of the two terms in the squared binomial.

$$2 \cdot x \cdot 5 = 10x.$$

Since $10x$ is the middle term of the trinomial, the trinomial *is* a perfect square, and can be factored as $(x + 5)^2$.

Example 4 **(a)** $x^2 - 22x + 121$ is a perfect square trinomial since it can be factored as $(x - 11)^2$.
(b) $9m^2 - 24m + 16 = (3m - 4)^2$.
(c) $25y^2 + 20y + 16$ is not a perfect square trinomial even though it has two terms that are perfect squares. If we try to factor it as $(5y + 4)^2$, we get a middle term of $40y$, which is not the middle term of the trinomial.

In summary, factor the difference of two squares as

$$x^2 - y^2 = (x + y)(x - y).$$

A perfect square trinomial can be factored as

$$x^2 + 2xy + y^2 = (x + y)^2.$$

EXERCISES

In Exercises 1–26, factor completely. The table of squares and square roots inside the front cover of this book may be helpful..

1. $x^2 - 16$
2. $m^2 - 25$
3. $p^2 - 4$
4. $r^2 - 9$
5. $m^2 - n^2$
6. $p^2 - q^2$
7. $a^2 - b^2$
8. $r^2 - t^2$
9. $9m^2 - 1$
10. $16y^2 - 9$
11. $25m^2 - 16$
12. $144y^2 - 25$
13. $4a^2 - 81$
14. $36x^2 - 49$
15. $121r^2 - 144$
16. $9m^2 - 16$
17. $64z^2 - 16$
18. $8z^2 - 32$
19. $36t^2 - 16$
20. $9 - 36a^2$
21. $25a^2 - 16r^2$
22. $100k^2 - 49m^2$
23. $a^4 - 1$
24. $x^4 - 16$
25. $m^4 - 81$
26. $p^4 - 256$

In Exercises 27–44, factor any expressions that are perfect square trinomials.

27. $a^2 + 4a + 4$

28. $p^2 + 6p + 9$

29. $x^2 + 14x + 49$

30. $y^2 + 8y + 16$

31. $4y^2 - 20y + 25$

32. $9m^2 - 12m + 4$

33. $9y^2 + 14y + 25$

34. $16m^2 + 42m + 49$

35. $16a^2 - 40ab + 25b^2$

36. $36y^2 - 60yp + 25p^2$

37. $100m^2 + 100m + 25$

38. $100a^2 - 140ab + 49b^2$

39. $49x^2 + 28xy + 4y^2$

40. $64y^2 - 48ya + 9a^2$

41. $4c^2 + 12cd + 9d^2$

42. $16t^2 - 40tr + 25r^2$

43. $25h^2 - 20hy + 4y^2$

44. $9x^2 + 24xy + 16y^2$

4.5 SOLVING QUADRATIC EQUATIONS

We have already discussed *linear equations*, which are equations involving first degree polynomials. Now we are ready to solve **quadratic equations**, equations involving second-degree polynomials, such as

$$x^2 - 5x = -6, \qquad 2a^2 - 5a = 3, \qquad \text{and} \qquad y^2 = 4.$$

Some quadratic equations can be solved by factoring. A general method for solving quadratic equations is presented in Chapter 9.

An important tool in solving quadratic equations by factoring is the **zero-factor property.**

If a and b represent real numbers, and if $ab = 0$, then $a = 0$ or $b = 0$.

In mathematics, the word "or" always means "either one or the other or both." Thus, the zero-factor property tells us that if the product of two factors is zero, either the first factor is zero, the second factor is zero, or both are zero. For example, if we know that $3c = 0$, then we can say that either $3 = 0$ or that $c = 0$ (or that both equal zero). Since we know that $3 \neq 0$, we must have $c = 0$.

To solve a quadratic equation such as

$$x^2 - 5x = -6,$$

we can use the zero-factor property. In order to apply the zero-factor property, we need a product equal to zero. To obtain such a product here, we first add 6 to both sides, which gives

$$x^2 - 5x + 6 = 0.$$

Next, we factor the trinomial $x^2 - 5x + 6$.

$$x^2 - 5x + 6 = (x - 3)(x - 2) = 0.$$

The product of the binomials $x - 3$ and $x - 2$ is zero.

By the zero-factor property, the product of two factors can be zero only if at least one of the factors is zero. Thus either $x - 3$ is zero or $x - 2$ is zero.

$$x - 3 = 0 \qquad \text{or} \qquad x - 2 = 0$$

Solving each equation for x, we have

$$x = 3 \qquad \text{or} \qquad x = 2.$$

To check, we let $x^2 - 5x + 6 = P(x)$, and calculate $P(3)$ and $P(2)$.

$$P(3) = (3)^2 - 5(3) + 6 = 9 - 15 + 6 = 0.$$
$$P(2) = (2)^2 - 5(2) + 6 = 4 - 10 + 6 = 0.$$

Since both 2 and 3 satisfy the equation $x^2 - 5x + 6 = 0$, the solutions are 2 and 3. Quadratic (or second-degree) equations commonly have two solutions. A quadratic equation may have less than two solutions, but never more than two.

Example 1 Solve the quadratic equation $x^2 - x - 20 = 0$.

First factor the polynomial $x^2 - x - 20$:

$$(x - 5)(x + 4) = 0.$$

By the zero-factor property this product can be zero only if

$$x - 5 = 0 \qquad \text{or} \qquad x + 4 = 0$$
$$x = 5 \qquad\qquad x = -4,$$

which gives solutions 5 and -4. Verify that for $P(x) = x^2 - x - 20$, both $P(5)$ and $P(-4)$ are 0.

Example 2 Solve the equation $2x^2 - 13x + 20 = 0$.

First factor the left side using the method of trial and error.

$$2x^2 - 13x + 20 = (2x - 5)(x - 4) = 0.$$

Since $(2x - 5)(x - 4) = 0$, we use the zero-factor property, and note that this product can be zero only if at least one factor is zero.

$$2x - 5 = 0 \qquad \text{or} \qquad x - 4 = 0$$
$$2x = 5 \qquad\qquad x = 4$$
$$x = \frac{5}{2}$$

Verify by checking that both $\frac{5}{2}$ and 4 are solutions.

The zero-factor property can be extended to more than two factors whose product is zero. We can solve any equation in which a factorable polynomial equals 0, such as $6x^3 - 6x = 0$.

Example 3 Solve $6x^3 - 6x = 0$.

First, we completely factor $6x^3 - 6x$.

$$\begin{aligned} 6x^3 - 6x &= 6x(x^2 - 1) \\ &= 6x(x + 1)(x - 1) \end{aligned}$$

Thus the equation can be written

$$6x(x + 1)(x - 1) = 0.$$

The product of these three factors can be zero only if at least one of the factors is zero, which means that either

$$\begin{array}{ccccc} 6x = 0 & \text{or} & x + 1 = 0 & \text{or} & x - 1 = 0 \\ x = 0 & & x = -1 & & x = 1. \end{array}$$

Verify by checking in the given equation that the solutions are 0, -1, and 1.

The equation $6x^3 - 6x = 0$ has *three* solutions. This does not contradict what we said before about quadratic equations, since $6x^3 - 6x = 0$ is *not* a quadratic equation—the highest exponent is 3, not 2.

EXERCISES

Solve the following equations.

1. $(x - 2)(x + 4) = 0$

2. $(y - 3)(2y + 5) = 0$

3. $(3x + 5)(2x - 1) = 0$

4. $(2m + 5)(3m - 2) = 0$

5. $3x(x + 1)(x - 1) = 0$

6. $2x(x + 2)(x + 3) = 0$

7. $5x(x - 1)(x + 3)(x - 2) = 0$

8. $x(x + 1)(x - 2)(x - 5) = 0$

9. $x^2 + 5x + 6 = 0$

10. $y^2 - 3y + 2 = 0$

11. $r^2 - 5r - 6 = 0$

12. $y^2 - y - 12 = 0$

13. $m^2 + 3m - 28 = 0$

14. $p^2 - p - 6 = 0$

15. $a^2 = 24 - 5a$

16. $r^2 = 2r + 15$

17. $x^2 = 3 + 2x$

18. $3m + 4 = m^2$

19. $z^2 = -2 - 3z$

20. $p^2 = 2p + 3$

21. $m^2 + 8m + 16 = 0$

22. $2a^2 + 3a - 20 = 0$

23. $25a^2 + 20a + 4 = 0$

24. $6r^2 - r - 2 = 0$

25. $2k^2 - k - 10 = 0$

26. $6x^2 = 7x + 5$

27. $6x^2 = 4 - 5x$

28. $6x^2 - 5x = 4$

29. $6a^2 = 5 - 13a$

30. $9s^2 + 12s = -4$

31. $m(m - 7) = -10$

32. $z(2z + 7) = 4$

33. $2(x^2 - 66) = -13x$

34. $3(m^2 + 4) = 20m$

35. $3r(r + 1) = (2r + 3)(r + 1)$

36. $(3k + 1)(k + 1) = 2k(k + 3)$

37. $y^2 - 4y + 4 = -y^2 + 4y - 4$

38. $12k(k - 4) = 3(k - 4)$

39. $16r^2 - 25 = 0$

40. $4k^2 - 9 = 0$

41. $9m^2 - 36 = 0$

42. $16x^2 - 64 = 0$

43. $(2r - 5)(3r^2 - 16r + 5) = 0$

44. $(3m - 4)(6m^2 + m - 2) = 0$

45. $(2x + 7)(x^2 - 2x - 3) = 0$

46. $(x - 1)(6x^2 + x - 12) = 0$

47. $(2m - 1)(m^2 + 4m + 3) = 0$

48. $x^3 - x = 0$

49. $x^3 = x^2$

50. $a^4 = a^2$

51. $m^3 = 4m$

52. $r^3 - 2r^2 - 8r = 0$

53. $y^3 - 6y^2 + 8y = 0$

54. $x^3 - x^2 - 6x = 0$

55. $a^3 + a^2 - 20a = 0$

56. $4b^3 - b = 0$

4.6 APPLICATIONS OF QUADRATIC EQUATIONS

It is important to learn how to solve quadratic equations because they occur so often in practical applications of mathematics. Many problems concerning area and volume require the solution of a quadratic equation.

Example 1 The width of a rectangle is 4 inches less than the length. The area is 96 square inches. Find the length and width.

We can use x to represent the length of the rectangle. Then, according to the statement of the problem, we can write the width as $x - 4$ (the width is less than the length). The area of a rectangle is given by the formula

$$A = (\text{length})(\text{width}).$$

In our problem, the area is 96, the length is x, and the width is $x - 4$. Substituting these values into the formula gives

$$96 = x(x - 4),$$
$$96 = x^2 - 4x.$$

This is the quadratic equation

$$x^2 - 4x - 96 = 0.$$

Factor:

$$(x - 12)(x + 8) = 0.$$

The solutions of the equation are $x = 12$ or $x = -8$.

However, the equation is only an algebraic representation of the area of a rectangle in the physical world. We must always be careful to check solutions against known physical facts. Since a rectangle cannot have a negative length, we discard the solution -8. Then 12 inches is the length of the rectangle, and $12 - 4 = 8$ inches is the width.

Example 2 Harry has a garden plot 60 feet by 100 feet. He wants to plant a strip of grass around the garden. (See Figure 4.1.) He has enough grass seed for 996 square feet. How wide a strip of grass can he plant?

We can use x to represent the width of the strip. We already know that the area of the grass strip is to be 996 square feet. To obtain an expression for the area of the grass strip, we can take the area of the large rectangle (garden plus grass) and subtract the area of the central garden plot. From the figure, we see that the width of the large rectangle is $60 + 2x$, while its length is $100 + 2x$.

$$\underbrace{(60 + 2x)(100 + 2x)}_{\text{area of large rectangle}} - \underbrace{60 \cdot 100}_{\text{area of central garden}} = \underbrace{996}_{\text{area of grass strip}}$$

If we multiply and combine terms, we obtain the equation

$$4x^2 + 320x - 996 = 0.$$

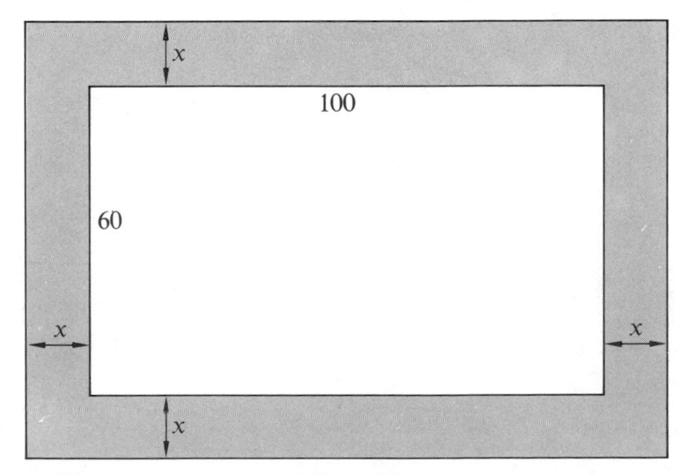

Figure 4-1

In the equation $4x^2 + 320x - 996$, we can factor out 4 to get

$$4(x^2 + 80x - 249) = 0.$$

Then we factor $x^2 + 80x - 249$, giving

$$4(x + 83)(x - 3) = 0.$$

Thus $4 = 0$ (impossible), or $x + 83 = 0$ or $x - 3 = 0$. The number -83 can't be the width of a grass strip, and so the only number which can be the answer to our problem is 3, the solution of $x - 3 = 0$. Thus the width of the strip should be 3 feet.

Example 3 Suppose we wish to use 130π square inches of tin to make a can whose height is three inches more than the radius of the base. What measure should we use for the radius of the base? (We shall ignore the need for material for seams or overlap.)

The area of a circle of radius r is given by the formula

$$A = \pi r^2.$$

Also, the formula for the circumference of (distance around) a circle of radius r is

$$C = 2\pi r.$$

To make a tin can, we need two round pieces of tin, one for the top and one for the bottom. If the radius of the top and the bottom is r, we will need πr^2 square inches of tin for each of the top and bottom pieces, or a total of $2\pi r^2$ square inches of tin. (See Figure 4.2.)

To form the side of the can, we need a rectangular piece of tin which can be rolled into a cylinder. The width of this rectangular piece of tin will be the height of the can, and the length will be the circumference of the top (Figure 4.2). Thus, if h is the height of the can and r is the radius of the top of the can, we can write the area of the rectangular piece of tin as

$$h(2\pi r).$$

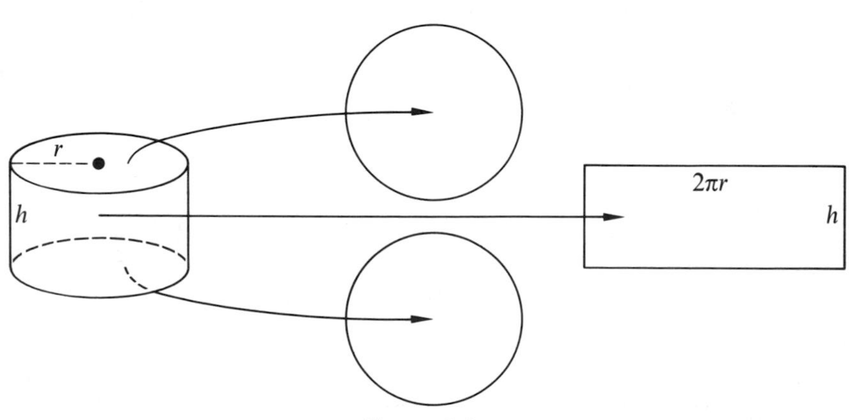

Figure 4-2

Hence the total amount of tin needed for our can is

$$\underbrace{2\pi r^2}_{\substack{\uparrow \\ \text{tin for} \\ \text{top and bottom}}} + \underbrace{h(2\pi r)}_{\substack{\uparrow \\ \text{tin for} \\ \text{side}}}.$$

We are using r to represent the radius of the base. From the statement of the problem the height of the can is $r + 3$ and the amount of tin available is 130π square inches. Using this information, our equation becomes

$$130\pi = 2\pi r^2 + (r + 3)(2\pi r).$$

We can simplify the equation and express it in factored form.

$$\begin{aligned} 130\pi &= 2\pi r^2 + 2\pi r^2 + 6\pi r \\ 130\pi &= 4\pi r^2 + 6\pi r \\ 0 &= 4\pi r^2 + 6\pi r - 130\pi \\ 0 &= 2\pi(2r^2 + 3r - 65) \\ 0 &= 2\pi(2r + 13)(r - 5). \end{aligned}$$

By the zero factor property,

$$\begin{array}{lclcl} 2\pi = 0 & \text{or} & 2r + 13 = 0 & \text{or} & r - 5 = 0 \\ \text{impossible} & & r = -13/2 & & r = 5. \end{array}$$

From these equations we see that the radius of the can is either $-\frac{13}{2}$ or 5. Since a tin can cannot have a negative radius, the only correct solution is $r = 5$.

Example 4 The product of two consecutive odd integers is one less than 5 times their sum. Find the integers.

Let us use s to represent the smaller of the two integers. Since the problem mentions consecutive *odd* integers, we use $s + 2$ to represent the larger of the two integers. Since the product is 1 less than 5 times the sum, we can write

$$\underbrace{s(s + 2)}_{\text{the product}} = \underbrace{5(s + s + 2)}_{\text{5 times sum}} \quad \underbrace{-1.}_{\text{one less than}}$$

The equation can be simplified.

$$\begin{aligned} &s^2 + 2s = 5(2s + 2) - 1 = 10s + 10 - 1, \\ &s^2 - 8s - 9 = 0. \end{aligned}$$

This last equation can be written by factoring as

$$(s - 9)(s + 1) = 0,$$

so that $s = 9$ or $s = -1$. If $s = 9$, then $s + 2 = 11$. Check: $9 \cdot 11 = 99$,

$5(9 + 11) - 1 = 99$. If $s = -1$, then $s + 2 = 1$. Check: $(1)(-1) = -1$, $5(1 + (-1)) - 1 = -1$. Thus 9 and 11 or -1 and 1 are the two pairs of consecutive odd integers that satisfy the conditions of the problem.

EXERCISES

1. The product of two consecutive integers is two more than twice their sum. Find the integers.
2. The product of two consecutive even integers is 60 more than twice the larger. Find the integers.
3. Find three consecutive even integers such that twice the sum of all three is half the product of the smaller two.
4. If the square of the sum of two consecutive integers is diminished by three times their product, the result is 31. Find the integers.
5. If the square of the larger of two numbers is diminished by six times the smaller, the result is five times the larger. The smaller is half the larger. Find the numbers.
6. Find two numbers whose sum is 8, if the sum of their squares is 40.
7. One number is four more than another. The square of the smaller increased by three times the larger is 66. Find the numbers.
8. The width of a rectangle is three less than its length. The area of the rectangle is 70. Find the dimensions of the rectangle.
9. The length of a rectangle is twice its width. If the width were increased by 2 inches, while the length remained the same, the resulting rectangle would have an area of 48 square inches. Find the dimensions of the original rectangle.
10. Bobby wishes to build a box to hold his boats. The box is to be 4 feet high, and the width of the box is to be one less than the length. The volume of the box will be 120 cubic feet. Find the dimensions of the box. (Hint: The formula for the volume of a box is given inside the back cover.)

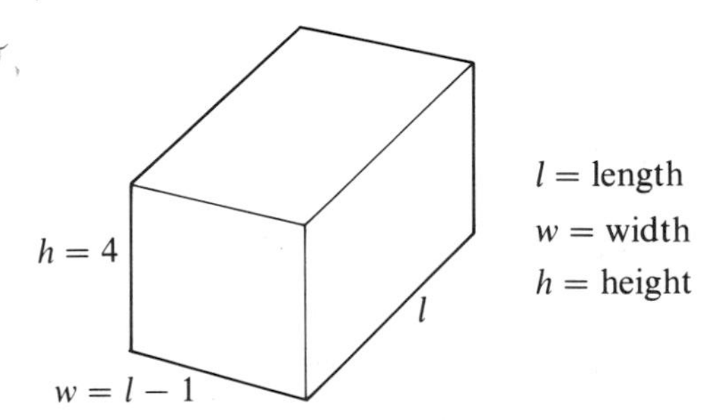

11. Find the radius of the base of a tin can if the radius equals the height, and 36π square inches of tin were used to make the can.

12. One square has sides one foot less than the length of the sides of a second square. If the difference of the areas of the two squares is 37 square feet, find the lengths of the sides of the two squares.

13. The volume of a right pyramid is 32 cubic inches. Suppose the numerical value of the height is 10 less than the numerical value of the area of the base. Find the height and the area of the base. (See formula for right pyramid inside the back cover.)

14. Suppose a right pyramid (see Exercise 13) has a rectangular base whose width is three inches less than the length. If the height is 8 and the volume is 144, find the dimensions of the base.

15. A rectangular waste filtration pond has an area of 108 square feet, with the width one-third of the length. A grass strip of area 112 square feet is planted around the pond. Find the width of the grass strip.

16. Tom has a pool 30 feet by 50 feet. He wants to add a concrete border of uniform width around the pool. How wide should the border be if he has enough concrete for 336 square feet?

17. Theresa wants to make a border of square Mexican tiles around her barbecue pit which is 3 feet by 5 feet. If she has enough tiles to lay 33 square feet, how wide can she make the border?

18. Jackson is cutting a mat to frame an 8″ × 10″ picture. He wants the border around the picture to be the same width on all four sides. He has a piece of mat-board the right shape with an area of 120 square inches. How wide will the border be?

19. Mary has a garden which borders her square lawn on three sides, as shown below. The garden is three feet wide on two sides and 5 feet wide on the third side. If the total area of the lawn and garden is 420 square feet, find the dimensions of the lawn.

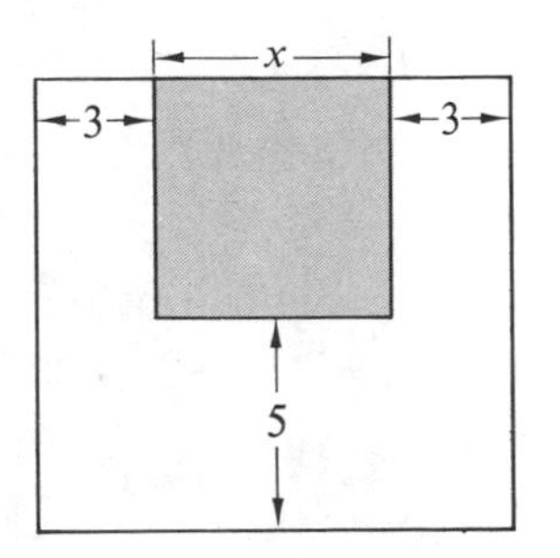

20. Jim wants to build a two-foot wide path around a tool shed which is twice as long as it is wide. He has determined that the area of the path will be twice the area of the tool shed. What are the dimensions of the shed?

4.7 SOLVING QUADRATIC INEQUALITIES (OPTIONAL)

A **quadratic inequality** is an inequality that involves a second-degree polynomial. Examples of quadratic inequalities include

$$2x^2 + 3x - 5 < 0, \qquad x^2 \leq 4, \qquad x^2 + 5x + 6 > 0.$$

To find the solution of a quadratic inequality such as $x^2 + 5x + 6 > 0$, we begin by finding the solution of the corresponding equation, which is $x^2 + 5x + 6 = 0$. The polynomial can be factored as

$$(x + 2)(x + 3)$$

so the solutions of the equation are $x = -2$ and $x = -3$. Since these are the *only* values of x which satisfy $x^2 + 5x + 6 = 0$, then all other values of x will make $x^2 + 5x + 6$ either less than zero or more than zero. The values -2 and -3 are boundaries of three regions on the number line. See Figure 4.3. Region A includes all numbers less than -3. Region B includes the numbers between -3 and -2. Region C includes all numbers greater than -2.

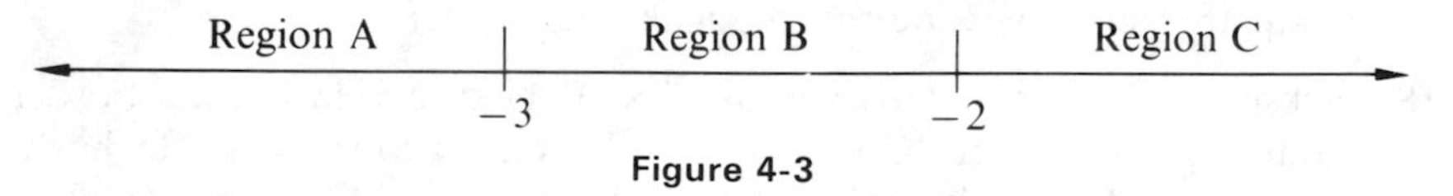

Figure 4-3

It can be shown that all values of x in a given region will make $x^2 + 5x + 6$ positive, or all values of x in a given region will make it negative. We need to test only one value of x from each region to see which regions satisfy $x^2 + 5x + 6 > 0$.

(a) Are the points of region A part of the solution? As a trial value, choose any number less than -3, say -4.

$$(-4)^2 + 5(-4) + 6 > 0$$
$$16 - 20 + 6 > 0$$
$$2 > 0. \qquad \text{True}$$

Since $2 > 0$ is true, the points of region A belong to the solution.

(b) For region B, choose a value between -3 and -2, say $-2\frac{1}{2}$, or $-\frac{5}{2}$.

$$\left(-\frac{5}{2}\right)^2 + 5\left(-\frac{5}{2}\right) + 6 > 0$$
$$\frac{25}{4} + \frac{-25}{2} + 6 > 0$$
$$-\frac{1}{4} > 0. \qquad \text{False}$$

Since $-\frac{1}{4} > 0$ is false, no point in region B belongs to the solution.
(c) For region C, try the number 0.

$$0^2 + 5(0) + 6 > 0$$
$$6 > 0. \qquad \text{True}$$

Since $6 > 0$ is true, the points of region C belong to the solution.

The solution is shown in Figure 4.4. The graph of the quadratic inequality $x^2 + 5x + 6 > 0$ includes all values of x less than -3, and all values of x greater than -2. We write the solution as

$$x < -3 \text{ or } x > -2.$$

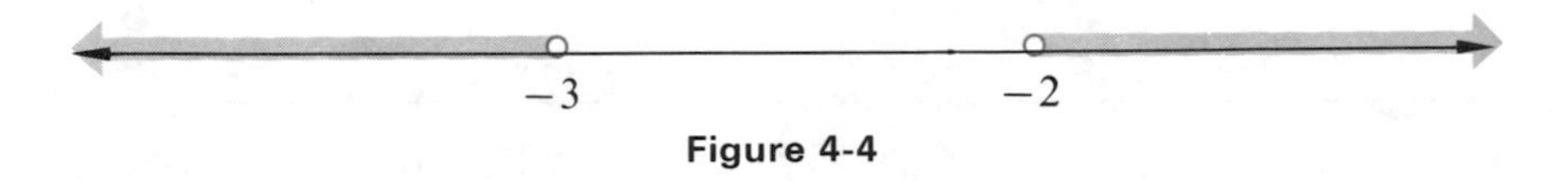

Figure 4-4

Example 1 Find the solution of the quadratic inequality $x^2 - 3x - 10 \leq 0$.

To begin, factor the corresponding equality $x^2 - 3x - 10 = 0$. The solutions of the equality are $x = 5$ and $x = -2$. These points are the boundaries of three regions on the number line. See Figure 4.5. This time, these boundary points will also belong to the solution, because we want all values of x which make $x^2 - 3x - 10 < 0$ or which make $x^2 - 3x - 10 = 0$.

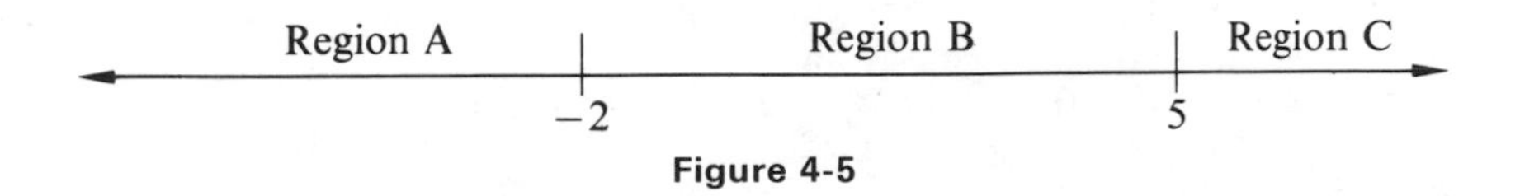

Figure 4-5

(a) Do the points of region A belong to the solution? To find out, select any point of region A, such as $x = -6$. Does $x = -6$ satisfy the original inequality?

$$(-6)^2 - 3(-6) - 10 < 0$$
$$36 + 18 - 10 < 0.$$
$$44 < 0. \qquad \text{False}$$

Since $44 < 0$ is false, the points of region A do not belong to the solution.
(b) What about region B? Let us try the value $x = 0$. We have

$$0^2 - 3(0) - 10 < 0$$
$$-10 < 0. \qquad \text{True}$$

Since $-10 < 0$ is true, the points of region B do belong to the solution.

(c) Let us try $x = 6$ to check region C.

$$6^2 - 3(6) - 10 < 0$$
$$36 - 18 - 10 < 0$$
$$8 < 0. \qquad \text{False}$$

Since $8 < 0$ is false, the points of region C do not belong to the solution.

The points of region B are the only ones that satisfy $x^2 - 3x - 10 < 0$. As shown in Figure 4.6, the solution includes the points of region B together with the endpoints $x = -2$ and $x = 5$. The solution is written

$$-2 \leq x \leq 5.$$

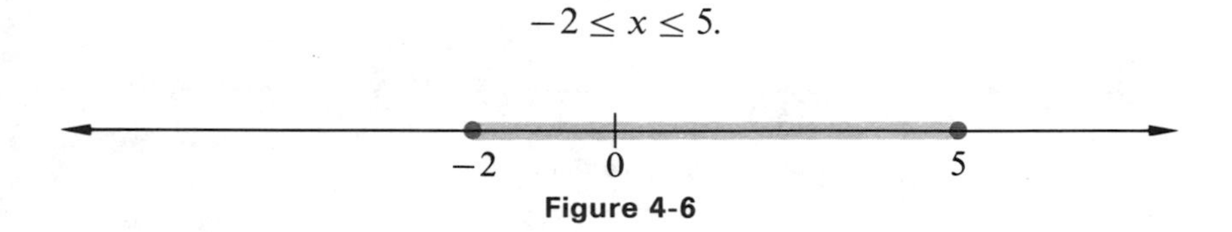

Figure 4-6

EXERCISES

Graph the solution of each quadratic inequality.

1. $(m + 2)(m - 5) < 0$
2. $(k - 1)(k + 3) > 0$
3. $(t + 6)(t + 5) \geq 0$
4. $(g - 2)(g - 4) \leq 0$
5. $(a + 3)(a - 3) < 0$
6. $(b - 2)(b + 2) > 0$
7. $(a + 6)(a - 7) \geq 0$
8. $(z - 5)(z - 4) \leq 0$
9. $m^2 + 5m + 6 > 0$
10. $y^2 - 3y + 2 < 0$
11. $z^2 - 4z - 5 \leq 0$
12. $x^2 + 2x + 1 \leq 0$
13. $x^2 - 2x + 1 < 0$
14. $2k^2 + 7k - 4 > 0$
15. $6r^2 - 5r - 4 < 0$
16. $p^2 > 0$
17. $a^2 - 16 < 0$
18. $9m^2 - 36 > 0$
19. $r^2 - 100 \geq 0$
20. $q^2 - 7q + 6 < 0$
21. $2k^2 - 7k - 15 \leq 0$
22. $6m^2 + m - 1 > 0$
23. $30r^2 + 3r - 6 \leq 0$
24. $x^3 + 2x^2 + x < 0$
25. $y^3 - 5y^2 + 6y > 0$

CHAPTER 4 SUMMARY

Key Words

Factor	Perfect square	Quadratic equation (second-degree equation)
Factored form	Perfect square trinomial	Quadratic inequality
Common factors	Difference of two squares	
Greatest common factor		

Zero-Factor Property

If a and b are real numbers, and if $ab = 0$, then $a = 0$ or $b = 0$.

Special Factored Forms

$x^2 - y^2 = (x + y)(x - y)$	Difference of two squares
$x^2 + 2ax + a^2 = (x + a)^2$	Perfect square trinomial

To Solve a Quadratic Equation

1. Express the equation as a polynomial equal to 0.
2. Factor the polynomial.
3. Set each factor equal to 0.
4. Solve each of the equations from step 3.

To Solve a Quadratic Inequality

1. Write the inequality as a polynomial greater than 0 or less than 0.
2. Set the polynomial from step 1 equal to 0.
3. Factor the polynomial.
4. Set each factor equal to 0.
5. Solve each equation from step 4.
6. The solutions from step 5 divide the number line into regions.
7. Select any number from each region, and test in the original inequality. If a test number satisfied the inequality, all numbers in that region belong to the solution.

CHAPTER 4 TEST

Factor completely.

1. $8m^3 - m^2$
2. $6mn + 12m^2$
3. $16m^2 - 4n^2$
4. $-3p^2q + 9pq + 6pq^2$
5. $x^2 + 11x + 30$
6. $x^2 + 6x - 7$
7. $2y^2 - 7y - 15$
8. $4x^2 + 12x + 9$
9. $x^2 - 25$
10. $4x^3t - 12x^2t^2 + 8xt^3$

Solve the following.

11. $y^2 + 3y + 2 = 0$
12. $x^2 - 4 = 0$
13. $3x^2 + 5x = 2$
14. $(p + 1)(p - 2)(p - 3) = 0$
15. $z^3 = 16z$
16. $x(x - 3) = 4(x + 2)$
17. Graph the solution of $m^2 + 2m - 24 > 0$.

For each problem, write an equation, and then solve it.

18. One number is nine larger than another. Their product is eleven more than five times their sum. Find the numbers.
19. The length of a certain rectangle is one less than twice the width. The area is 15 square inches. Find the dimensions.
20. A square has sides of 8 inches. If a certain number is added to one dimension, and subtracted from the other dimension, the new figure has an area 4 square inches less than the area of the original square. Find the number.

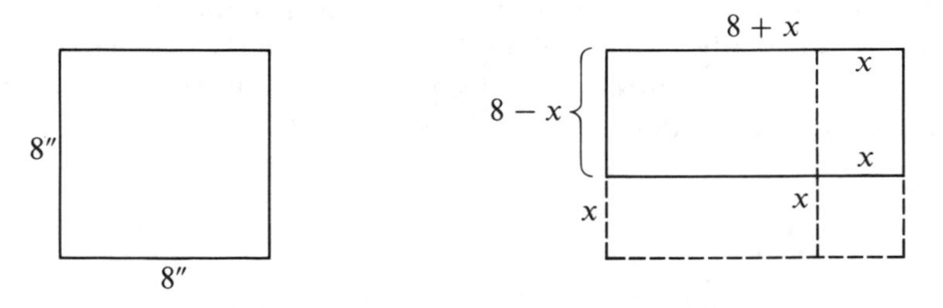

21. Jaime is building a border around a rectangular lawn fountain that is 9 feet by 10 feet. He has enough material for 20 square feet. How wide can he make the border?

5 RATIONAL NUMBERS AND RATIONAL EXPRESSIONS

5.1 MULTIPLICATION AND DIVISION OF RATIONAL NUMBERS

A **rational number** is the quotient of two integers a and b, where $b \neq 0$. We make the restriction $b \neq 0$ since division by zero is not meaningful in our number system. Using this definition, $5/6$, $-8/9$, $7/-2$, $0/6 = 0$, and $3/1 = 3$ are all rational numbers, while $\pi/8$ (π is not an integer) and $6/0$ are not. The words "fraction" and "rational number" are often used interchangeably. Actually, a fraction is the ratio of *any* two numbers, even irrational numbers, while a rational number is the ratio of two *integers*.

Since the easiest operation on rational numbers is multiplication, we begin with it. To multiply two rational numbers, multiply their numerators and multiply their denominators.

If a/b and c/d are any rational numbers, then their product is

$$\frac{a}{b} \cdot \frac{c}{d} = \frac{ac}{bd}.$$

Example 1 **(a)** $\dfrac{5}{11} \cdot \dfrac{-3}{7} = \dfrac{5(-3)}{11(7)} = \dfrac{-15}{77}$.

(b) $6 \cdot \dfrac{4}{5} = \dfrac{6}{1} \cdot \dfrac{4}{5} = \dfrac{6(4)}{1(5)} = \dfrac{24}{5}$.

Note that 6 must be written as a quotient, 6/1, before the rule for multiplication can be used.

(c) $\left(-\dfrac{1}{9}\right)\left(-\dfrac{5}{2}\right) = \left(\dfrac{-1}{9}\right)\left(\dfrac{-5}{2}\right) = \dfrac{(-1)(-5)}{(9)(2)} = \dfrac{5}{18}$.

A rational number is said to be **reduced to lowest terms** when the numerator and the denominator have no common integer factor other than 1 (or -1). For example, 7/21 can be reduced to lowest terms as 1/3. To reduce a rational number to lowest terms, we use the **fundamental property of rational numbers**.

If a/b is a rational number, and if $k \neq 0$, then

$$\frac{ak}{bk} = \frac{a}{b}.$$

Example 2 (a) $\frac{28}{35} = \frac{4(7)}{5(7)} = \frac{4}{5}.$

(b) $\frac{15}{45} = \frac{1(15)}{3(15)} = \frac{1}{3}.$

(c) $-\frac{9}{12} = \frac{-3(3)}{4(3)} = \frac{-3}{4}.$

Example 3 Find, and express in lowest terms, the product of 3/8 and 4/9.

$$\frac{3}{8} \cdot \frac{4}{9} = \frac{3 \cdot 4}{8 \cdot 9} = \frac{12}{72} = \frac{1 \cdot 12}{6 \cdot 12} = \frac{1}{6}.$$

Later we shall need to rewrite rational numbers as equivalent rational numbers with larger denominators. We can also use the fundamental property for this purpose. For example, to write 5/6 as a rational number with denominator 24, proceed as follows. Since the desired denominator 24 is $4 \cdot 6$, we multiply both the numerator and denominator of the fraction 5/6 by 4, using the fundamental property.

$$\frac{5}{6} = \frac{5(4)}{6(4)} = \frac{20}{24}.$$

Example 4 Write 5/8 as a rational number having denominator 72.

$$\frac{5}{8} = \frac{5(9)}{8(9)} = \frac{45}{72}.$$

Example 5 Which of the following two rational numbers is larger?

$$\frac{10}{11} \quad \text{or} \quad \frac{13}{14}.$$

We can compare two rational numbers if their denominators are alike by the size of their numerators. Therefore, we must first represent 10/11 and 13/14 as fractions with the same (common) denominator. Since 11 and 14 have no common factor, the least common denominator is $(11)(14) = 154$. Now we use the fundamental property to write each rational number as a number with denominator 154.

$$\frac{10}{11} = \frac{10(14)}{11(14)} = \frac{140}{154} \quad \text{and} \quad \frac{13}{14} = \frac{13(11)}{14(11)} = \frac{143}{154}.$$

By comparing numerators we see that 143/154 is larger than 140/154 so that 13/14 is larger than 10/11.

Example 6 Write the following rational numbers in order from smallest to largest: $\frac{2}{3}, \frac{3}{4}, \frac{5}{7}$.

As in Example 5, first write each of the three rational numbers as fractions with the same denominator.

$$\frac{2}{3} = \frac{2(28)}{3(28)} = \frac{56}{84} \qquad \frac{3}{4} = \frac{3(21)}{4(21)} = \frac{63}{84} \qquad \frac{5}{7} = \frac{5(12)}{7(12)} = \frac{60}{84}.$$

By comparing numerators, we see that the proper order is

$$\frac{2}{3} < \frac{5}{7} < \frac{3}{4}.$$

There are three signs associated with a rational number, as shown below by the parentheses.

$$(\ )\frac{(\ )a}{(\ )b}.$$

Either a plus or a minus sign can be placed in any of the three parentheses. It can be shown that *any two* of these signs can be changed without affecting the value of the rational number. For example, the following are all equal.

$$\frac{-4}{5}, \qquad \frac{4}{-5}, \qquad -\frac{4}{5}, \qquad -\frac{-4}{-5}.$$

The forms $4/-5$ and $-(-4/-5)$ are seldom used.

In Chapter 1 of this book, we defined an inverse for multiplication called the *reciprocal.* The reciprocal of the number a is the number that can be multiplied by a to give the product 1. From the definition of multiplication of rational numbers and the fundamental property, we see that

$$\frac{a}{b} \cdot \frac{b}{a} = 1 \qquad \left(\frac{a}{b} \neq 0\right).$$

Therefore, the reciprocal of a/b is b/a. (There is no reciprocal for zero.) Thus the reciprocal of $-9/10$ is $10/-9$, the reciprocal of 5 (or 5/1) is 1/5, and so on. In Chapter 1 we defined division in terms of multiplication, as follows.

If a and b are any two numbers, with $b \neq 0$, then

$$a \div b = \frac{a}{b} = a \cdot \frac{1}{b}.$$

(To divide a by b, multiply a by the reciprocal of b.) We can apply this same definition to rational numbers:

If a/b and c/d are any two rational numbers, with $c/d \neq 0$, then their quotient is

$$\frac{a}{b} \div \frac{c}{d} = \frac{\frac{a}{b}}{\frac{c}{d}} = \frac{a}{b} \cdot \frac{d}{c} = \frac{ad}{bc}.$$

Example 7 $\frac{3}{4} \div \frac{-5}{8} = \frac{3}{4} \cdot \frac{8}{-5} = \frac{24}{-20} = \frac{6}{-5} = \frac{-6}{5}.$

Example 8 $\frac{\frac{-2}{3}}{\frac{-6}{15}} = \frac{-2}{3} \cdot \frac{15}{-6} = \frac{-30}{-18} = \frac{5}{3}.$

EXERCISES

In Exercises 1–24, find the products or quotients. Reduce each answer to lowest terms.

1. $\frac{3}{4} \cdot \frac{-9}{5}$
2. $\frac{3}{8} \cdot \frac{2}{7}$
3. $\frac{1}{10} \cdot \frac{-6}{-3}$
4. $2 \cdot \frac{1}{3}$
5. $\frac{9}{-4} \cdot \frac{-8}{6}$
6. $\frac{3}{5} \cdot \frac{5}{3}$
7. $\frac{3}{8} \div \frac{5}{4}$
8. $\frac{9}{16} \div \frac{3}{8}$
9. $\frac{5}{12} \div \frac{15}{4}$
10. $\frac{15}{16} \div \frac{30}{8}$
11. $\frac{15}{32} \cdot \frac{8}{25}$
12. $\frac{24}{25} \cdot \frac{50}{3}$
13. $\frac{-2}{3} \cdot \frac{-5}{8}$
14. $-\frac{2}{4} \cdot \frac{3}{9}$
15. $\left(\frac{13}{2}\right)\left(\frac{2}{3}\right)$
16. $\left(\frac{-9}{4}\right)\left(\frac{8}{7}\right)$
17. $\left(\frac{-28}{3}\right)\left(\frac{6}{2}\right)$
18. $\left(\frac{-121}{9}\right)\left(\frac{18}{11}\right)$
19. $\frac{3}{8} \cdot \frac{5}{6} \div \frac{-8}{9} \cdot \frac{4}{3}$

(*Hint*: See the discussion on order of operations in Chapter 1.)

20. $\frac{2}{11} \div \frac{3}{22} \cdot \frac{5}{7} \div \frac{2}{3} \cdot \frac{9}{8}$

21. $\frac{2}{5} \div \frac{1}{5} \cdot \frac{3}{4} \div \frac{2}{7}$

22. $\frac{6}{7} \cdot \frac{2}{3} \div \frac{4}{9} \cdot \frac{3}{5}$

23. $\frac{5}{8} \cdot \frac{9}{5} \cdot \frac{2}{3} \div \frac{3}{4}$

24. $\frac{6}{4} \cdot \frac{4}{3} \div \frac{5}{8} \div \frac{1}{2} \cdot \frac{4}{5}$

Write the fractions in each of Exercises 25–32 in order from smallest to largest.

25. $\frac{12}{25}, \frac{13}{30}$

26. $\frac{23}{24}, \frac{14}{15}$

27. $\frac{4}{5}, \frac{6}{7}, \frac{7}{9}$

28. $\frac{5}{8}, \frac{6}{11}, \frac{3}{5}$

29. $\frac{3}{8}, \frac{1}{3}, \frac{2}{7}, \frac{3}{10}$

30. $\frac{9}{10} \cdot \frac{6}{7}, \frac{5}{6}, \frac{10}{11}$

31. $\frac{4}{7}, \frac{1}{2}, \frac{5}{12}, \frac{8}{15}$

32. $\frac{3}{4}, \frac{2}{3}, \frac{11}{15}, \frac{12}{17}$

For each of the following, first write an equation, and then solve it.

33. The product of a quantity n and 3/4 is 5/6. Find n.

34. If a certain number is multiplied by $-8/9$, the product is 2/3. Find the number.

35. If $-3/5$ is divided by a certain number, n, then 2/3 is the result. Find n.

36. The quotient of 5/8 and a number n is 3/8. Find n.

37. The quotient of 1/3 and a number n is 2. Find n.

38. The quotient of 3/4 and the product of a number n and 5/8 is 2/3. Find n.

5.2 ADDITION AND SUBTRACTION OF RATIONAL NUMBERS

To see how to add rational numbers, consider two rational numbers which have the same denominator, a/c and b/c. By the definition of division,

$$\frac{a}{c} = a \cdot \frac{1}{c} \quad \text{and} \quad \frac{b}{c} = b \cdot \frac{1}{c}.$$

Then we can write

$$\frac{a}{c} + \frac{b}{c} = a \cdot \frac{1}{c} + b \cdot \frac{1}{c}.$$

Use the distributive property on the right-hand side of the above equation.

$$\frac{a}{c} + \frac{b}{c} = (a + b)\frac{1}{c}.$$

By the definition of division, the right-hand side can be written as

$$\frac{a}{c} + \frac{b}{c} = \frac{a + b}{c}.$$

Example 1 Add 3/5 and $-8/5$.

Since 3/5 and $-8/5$ have the same denominators, the sum is

$$\frac{3}{5} + \frac{-8}{5} = \frac{3 + (-8)}{5} = \frac{-5}{5} = -1.$$

To add two rational numbers with different denominators, we must first find a common denominator. Then we use the fundamental property of rational numbers to convert each rational number to an equivalent fraction which has the common denominator for its denominator.

It is practical to find the *least common denominator*, the smallest whole number that both denominators divide into. We abbreviate the term "Least Common Denominator" to LCD. To find the LCD, factor each denominator into its prime factors. (The whole number a is a *prime number* if its only positive factors are the number a itself and the number 1. The number 1 is not a prime number.)

For example, to add $-5/6$ and 3/4 we first find the LCD. Factor each denominator into its prime factors.

$$6 = 2 \cdot 3 \qquad \text{and} \qquad 4 = 2 \cdot 2.$$

The LCD must be large enough so that both 6 and 4 will divide into it. Thus the LCD must have the factors $2 \cdot 3$ for the denominator 6, and the factors $2 \cdot 2$ for the denominator 4.

$$\text{LCD} = 2 \cdot 3 \cdot 2 = 12.$$

It is not necessary to have more than two factors of 2 in the LCD. Now use the fundamental property of rational numbers to write each number as a fraction with a denominator of 12.

$$\frac{-5}{6} + \frac{3}{4} = \frac{-5(2)}{6(2)} + \frac{3(3)}{4(3)}$$

$$= \frac{-10}{12} + \frac{9}{12}.$$

Finally, we apply the rule for adding two fractions with the same denominator.

$$\frac{-5}{6} + \frac{3}{4} = \frac{-1}{12}.$$

Example 2 Add 3/5 and 3/4.

To find the LCD, we find the prime factors of each denominator.

$$5 = 5 \qquad 4 = 2 \cdot 2.$$

Since the denominators have no common factors, the LCD is

$$(5)(2)(2) = 20.$$

Now convert each rational number to a fraction with denominator of 20, and add.

$$\begin{aligned}\frac{3}{5} + \frac{3}{4} &= \frac{3(4)}{5(4)} + \frac{3(5)}{4(5)}\\ &= \frac{12}{20} + \frac{15}{20}\\ &= \frac{12 + 15}{20}\\ &= \frac{27}{20}.\end{aligned}$$

To subtract two rational numbers, change the sign of the second number, and add, using the definition of subtraction given in Chapter 1. The difference of the rational numbers a/b and c/d is

$$\frac{a}{b} - \frac{c}{d} = \frac{a}{b} + \left(-\frac{c}{d}\right).$$

In the previous section of this chapter we noted that $-(c/d)$ could also be written as either $-c/d$ or $c/-d$. For the purpose of subtraction, the best form is $-c/d$.

$$\frac{a}{b} - \frac{c}{d} = \frac{a}{b} + \frac{-c}{d}.$$

Example 3

$$\begin{aligned}\frac{5}{6} - \frac{11}{5} &= \frac{5(5)}{6(5)} - \frac{11(6)}{5(6)}\\ &= \frac{25}{30} - \frac{66}{30}\\ &= \frac{25}{30} + \frac{-66}{30}\\ &= \frac{25 + (-66)}{30}\\ &= \frac{-41}{30}.\end{aligned}$$

A number such as $2\frac{1}{3}$, which represents the sum of an integer and a rational number, is called a **mixed number**. In algebra it is inconvenient to work with mixed numbers, so such numbers are usually rewritten as single fractions. For example, let us write $2\frac{1}{3}$ as a single fraction. First, we write

$$2\frac{1}{3} = 2 + \frac{1}{3}.$$

Then we express 2 as a fraction with a denominator of 3.

$$2 = 2\left(\frac{3}{3}\right) = \frac{2(3)}{3} = \frac{6}{3}.$$

Thus, we have

$$2\frac{1}{3} = \frac{6}{3} + \frac{1}{3} = \frac{7}{3}.$$

Example 4 Add $4\frac{3}{5}$ and 2/3 and express the answer in lowest terms.

First write $4\frac{3}{5}$ as a single fraction.

$$4\frac{3}{5} = 4 + \frac{3}{5} = 4\left(\frac{5}{5}\right) + \frac{3}{5} = \frac{4(5)}{5} + \frac{3}{5} = \frac{20}{5} + \frac{3}{5} = \frac{23}{5}.$$

Then add the two fractions.

$$4\frac{3}{5} + \frac{2}{3} = \frac{23}{5} + \frac{2}{3} = \frac{69}{15} + \frac{10}{15} = \frac{79}{15}.$$

The fraction 79/15 is in lowest terms.

Now that we have discussed the operations of multiplication, division, addition, and subtraction for rational numbers, we can work problems in which several of these operations appear.

Example 5 Find $\left(\frac{2}{3} + \frac{3}{4} \cdot \frac{1}{2}\right) \div \frac{1}{5}$, and reduce the answer to lowest terms.

We begin inside the parentheses by first finding the product

$$\frac{3}{4} \cdot \frac{1}{2} = \frac{3}{8}.$$

Then we add 2/3 to the resulting number.

$$\frac{2}{3} + \frac{3}{8} = \frac{2 \cdot 8}{3 \cdot 8} + \frac{3 \cdot 3}{8 \cdot 3} = \frac{16}{24} + \frac{9}{24} = \frac{25}{24}.$$

Finally, we divide by 1/5.

$$\frac{25}{24} \div \frac{1}{5} = \frac{25}{24} \cdot \frac{5}{1} = \frac{125}{24}.$$

Since 125 and 24 have no common factor, the answer is in lowest terms.

Example 6 Evaluate the following, and express in lowest terms.

$$\frac{2 - \frac{1}{4}}{\frac{1}{2} + \frac{3}{4}}.$$

First, simplify the numerator.

$$2 - \frac{1}{4} = \frac{2 \cdot 4}{1 \cdot 4} - \frac{1}{4} = \frac{8}{4} - \frac{1}{4} = \frac{7}{4}.$$

Next, simplify the denominator.

$$\frac{1}{2} + \frac{3}{4} = \frac{1 \cdot 2}{2 \cdot 2} + \frac{3}{4} = \frac{2}{4} + \frac{3}{4} = \frac{5}{4}.$$

Now perform the division.

$$\frac{7}{4} \div \frac{5}{4} = \frac{7}{4} \cdot \frac{4}{5} = \frac{7 \cdot 4}{4 \cdot 5} = \frac{7}{5}.$$

EXERCISES

In Exercises 1–22, find the sums or differences. Express each answer in lowest terms.

1. $\frac{1}{12} + \frac{3}{12}$

2. $\frac{2}{3} + \frac{2}{3}$

3. $\frac{1}{10} + \frac{6}{10}$

4. $\frac{3}{4} + \frac{8}{4}$

5. $\frac{4}{9} + \frac{2}{3}$

6. $\frac{3}{5} + \frac{2}{15}$

7. $\frac{8}{11} + \frac{3}{22}$

8. $\frac{9}{10} - \frac{3}{5}$

9. $\frac{2}{3} - \frac{4}{5}$

10. $\frac{8}{12} - \frac{5}{9}$

11. $\frac{5}{6} - \frac{3}{10}$

12. $\frac{11}{4} - \frac{11}{8}$

13. $3\frac{1}{4} + 6\frac{1}{8}$

14. $5\frac{2}{3} + \frac{3}{4}$

15. $4\frac{1}{2} + \frac{2}{3}$

16. $7\frac{1}{8} + 3\frac{1}{4}$

17. $6\frac{2}{3} - 5\frac{1}{4}$

18. $8\frac{2}{9} - 7\frac{4}{5}$

19. $\frac{2}{5} + \frac{1}{3} + \frac{9}{10}$

20. $\frac{3}{8} + \frac{5}{6} + \frac{2}{3}$

21. $\frac{5}{7} + \frac{2}{8} - \frac{1}{2}$

22. $\frac{2}{3} - \frac{1}{6} - \frac{1}{2}$

In Exercises 23–36, perform the indicated operations.

23. $\left(\frac{3}{4}\cdot\frac{2}{9}\right)+\frac{1}{3}$

24. $\left(\frac{1}{3}\div\frac{1}{2}\right)+\left(\frac{5}{6}-\frac{3}{4}\right)$

25. $\left(\frac{2}{5}-\frac{3}{8}\right)+\left(\frac{1}{5}\cdot\frac{3}{4}\right)$

26. $\left(\frac{6}{8}\div\frac{3}{16}\right)-\left(\frac{2}{3}\cdot\frac{3}{4}\right)$

27. $\left(\frac{5}{6}-\frac{3}{8}+\frac{2}{3}\cdot\frac{3}{5}\right)\div\frac{5}{2}$

28. $\frac{3}{4}\div\frac{2}{3}\cdot\frac{5}{8}+\frac{1}{5}\cdot\frac{2}{8}-\frac{5}{6}$

29. $\dfrac{1+\frac{3}{4}}{\frac{3}{2}+\frac{2}{3}}$

30. $\dfrac{5-\frac{11}{3}}{\frac{1}{4}+\frac{2}{3}}$

31. $\dfrac{\frac{1}{4}+\frac{2}{3}}{\frac{1}{4}-\frac{2}{3}}$

32. $\dfrac{3+\frac{2}{3}}{1-\frac{1}{3}}$

33. $\dfrac{\frac{5}{6}-\frac{2}{3}}{\frac{3}{4}+\frac{1}{2}}$

34. $\dfrac{\frac{1}{5}+\frac{3}{2}}{\frac{5}{2}-\frac{2}{5}}$

35. $1+\dfrac{1}{1+\frac{1}{2}}$

36. $2-\dfrac{1}{1-\frac{1}{2}}$

Write an equation for each of the following, and then solve it.

37. The sum of a certain quantity n and 3/8 is 5/6. Find n.

38. The difference of a number and 4 is 5/8. Find the number.

39. If a number is divided into the sum of the number and 3/4, the result is 5/3. Find the number.

40. When three times the difference of a number and 5/6 is divided by half the number, the result is 2. Find the number.

5.3 RATIONAL EXPRESSIONS

The quotient of two integers is called a *rational number*. Similarly, the quotient of two polynomials is called a *rational expression*.

If P and Q are polynomials, with $Q \neq 0$, then P/Q is a **rational expression**.

Examples of rational expressions include

$$\frac{-6x}{x^3+8}, \quad \frac{9x}{y+3}, \quad \frac{2m^3}{8}.$$

However, we do not consider $8x/0$ a rational expression since division by zero is not possible. For that reason, you must be careful when substituting a number for the variable in the denominator of a rational expression. For example, in the rational expression

$$\frac{8x^2}{x-3},$$

x can take on any numerical value except 3. When $x = 3$, the denominator of the rational expression becomes 0, so that the expression is meaningless. The set of all permissible numerical values for the variable in a rational expression is called the **domain of the rational expression**. The domain of

$$\frac{8x^2}{x-3}$$

is all real numbers except 3, as we saw above.

In this book, unless stated otherwise, the domain of a rational expression will be the set of all real numbers except those which make the denominator zero.

Example 1 The domain of the rational expression

$$\frac{p+5}{p+2}$$

includes all real numbers except -2, since $p + 2 = 0$ when $p = -2$.

Example 2 The domain of the rational expression

$$\frac{9m^2}{m^2-5m+6}$$

is all real numbers except 2 and 3. To see this, find all values of m for which $m^2 - 5m + 6$ is zero. That is, solve the equation $m^2 - 5m + 6 = 0$.

$$m^2 - 5m + 6 = 0$$
$$(m-2)(m-3) = 0$$
$$m - 2 = 0 \quad \text{or} \quad m - 3 = 0$$
$$m = 2 \qquad\qquad m = 3.$$

Example 3 Find the numerical value of the following rational expression **(a)** when $x = 1$; **(b)** when $x = 2$:

$$\frac{3x+5}{2x-4}.$$

(a) We can find the value of the rational expression for $x = 1$ by substitution.

$$\frac{3x+5}{2x-4} = \frac{3(1)+5}{2(1)-4} = \frac{3+5}{2-4} = \frac{8}{-2} = -4.$$

(b) If we substitute 2 for x in the rational expression, we see that the denominator, $2x - 4$, becomes 0. Since 2 is not in the domain, there is no value for the rational expression when $x = 2$.

Since a rational expression represents a number for each numerical value assigned to its variable, the properties of rational numbers also hold for rational expressions. The fundamental property of rational numbers has a parallel for rational expressions, the **fundamental property of rational expressions.**

If P/Q is a rational expression, and if K represents any rational expression, $K \neq 0$, then

$$\frac{PK}{QK} = \frac{P}{Q}.$$

The fundamental property can be used to reduce a rational expression to lowest terms, the form in which numerator and denominator have no common factor other than 1. For example,

$$\begin{aligned}\frac{6x^2+18x-24}{3x^2-3} &= \frac{6(x^2+3x-4)}{3(x^2-1)}\\ &= \frac{3 \cdot 2(x+4)(x-1)}{3(x+1)(x-1)}\\ &= \frac{2(x+4)}{x+1}.\end{aligned}$$

Note that the two rational expressions are equal only for values of x which are in both domains, that is, for $x \neq 1$ or $x \neq -1$.

Example 4 Reduce $\dfrac{x-y}{y-x}$ to lowest terms.

At first glance, there does not seem to be any way in which we can factor $x - y$ and $y - x$ to get a common factor. However, note that

$$y - x = -1(-y + x) = -1(x - y).$$

With these factors, the rational expression can be simplified.

$$\frac{x-y}{y-x} = \frac{1(x-y)}{-1(x-y)} = \frac{1}{-1} = -1.$$

Example 5 Reduce to lowest terms:

$$\frac{3p + 3q}{p^2 - q^2}.$$

The first step is to factor both numerator and denominator as in our work with rational numbers.

$$\frac{3p + 3q}{p^2 - q^2} = \frac{3(p + q)}{(p + q)(p - q)}.$$

Since there is a common factor, we can write the rational expression in lowest terms:

$$\frac{3(p + q)}{(p + q)(p - q)} = \frac{3}{p - q}.$$

The fundamental property can also be used, as it was with rational numbers, to convert a rational expression into an equivalent expression with a different denominator. Thus to convert

$$\frac{x(x - 3)}{(x + 2)(x - 1)}$$

into a rational expression having denominator $(x + 2)(x - 1)(x + 1)$ we multiply numerator and denominator by $x + 1$:

$$\frac{x(x - 3)}{(x + 2)(x - 1)} = \frac{x(x - 3)(x + 1)}{(x + 2)(x - 1)(x + 1)}.$$

Example 6 Reduce to lowest terms:

$$\frac{m^2 + 2m - 8}{2m^2 - m - 6}.$$

Always begin by factoring both numerator and denominator, if possible.

$$\frac{m^2 + 2m - 8}{2m^2 - m - 6} = \frac{(m + 4)(m - 2)}{(2m + 3)(m - 2)} = \frac{m + 4}{2m + 3}.$$

EXERCISES

In Exercises 1–6, find the numerical value of each rational expression when $x = -3$.

1. $\dfrac{x^3}{2x^2}$

2. $\dfrac{-5x + 1}{2x}$

3. $\dfrac{4x^2 - 2x}{3x}$

4. $\dfrac{(-8x)^2}{3x + 9}$

5. $\dfrac{2x + 5}{3 + x}$

6. $\dfrac{x^2 - 1}{x}$

Reduce each of the following to lowest terms. It may be necessary to factor first in some of the exercises.

7. $\frac{2y}{3y}$

8. $\frac{5m}{10}$

9. $\frac{12k^2}{6k}$

10. $\frac{9m^3}{3m}$

11. $\frac{-8y^6}{6y^3}$

12. $\frac{16x^4}{-8x^2}$

13. $\frac{12m^2p}{9mp^2}$

14. $\frac{6a^2b^3}{24a^3b^2}$

15. $\frac{6(y+2)}{-8(y+2)^2}$

16. $\frac{9(m+2)}{5(m+2)}$

17. $\frac{(x+1)(x-1)}{(x+1)^2}$

18. $\frac{3(t+5)}{(t+5)(t-1)}$

19. $\frac{m^2-n^2}{m+n}$

20. $\frac{a^2-b^2}{a-b}$

21. $\frac{5m^2-5m}{10m-10}$

22. $\frac{3y^2-3y}{2(y-1)}$

23. $\frac{16r^2-4s^2}{4r-2s}$

24. $\frac{11s^2-22s^3}{6-12s}$

25. $\frac{m^2-4m+4}{m^2+m-6}$

26. $\frac{a^2-a-6}{a^2+a-12}$

27. $\frac{x^2+3x-4}{x^2-1}$

28. $\frac{8m^2+6m-9}{16m^2-9}$

29. $\frac{6y^2+11y+4}{3y^2+7y+4}$

30. $\frac{z^2-5z+6}{2z^2-8}$

31. $\frac{-a+b}{b-a}$

32. $\frac{b-a}{a-b}$

33. $\frac{x^2-1}{1-x}$

34. $\frac{p^2-q^2}{q-p}$

35. $\frac{m^2-4m}{4m-m^2}$

36. $\frac{s^2-r^2}{r^2-s^2}$

5.4 MULTIPLICATION AND DIVISION OF RATIONAL EXPRESSIONS

To multiply two rational numbers, we multiply the numerators and multiply the denominators. Multiplication of rational expressions is very similar.

The product of the rational expressions P/Q and R/S is

$$\frac{P}{Q}\cdot\frac{R}{S}=\frac{PR}{QS}.$$

For example, let us use the definition to find the product of the rational expressions $6/x$ and $x^2/12$. Using the definition, we multiply the numerators and multiply the denominators.

$$\frac{6}{x}\cdot\frac{x^2}{12}=\frac{6\cdot x^2}{x\cdot 12}=\frac{6x^2}{12x}=\frac{x}{2}.$$

In the last step we used the fundamental property to reduce to lowest terms.

Example 1 Find the product: $\dfrac{x+y}{2x} \cdot \dfrac{x^2}{(x+y)^2}$.

By the definition of multiplication, we have

$$\frac{x+y}{2x} \cdot \frac{x^2}{(x+y)^2} = \frac{(x+y)x^2}{2x(x+y)^2}$$

$$= \frac{(x+y)x^2}{2x(x+y)(x+y)}$$

$$= \frac{x}{2(x+y)}.$$

In the last two steps we factored and used the fundamental property of rational expressions to reduce the answer to lowest terms.

Example 2 Find the product:

$$\frac{x^2+3x}{x^2-3x-4} \cdot \frac{x^2-5x+4}{x^2+2x-3}.$$

We use the definition of multiplication, but before multiplying the numerators and denominators, we factor wherever possible, then reduce by eliminating any pairs of common factors in the denominator and numerator as shown below. By doing this step first, the work is greatly simplified.

$$\frac{x^2+3x}{x^2-3x-4} \cdot \frac{x^2-5x+4}{x^2+2x-3} = \frac{x(x+3)}{(x-4)(x+1)} \cdot \frac{(x-4)(x-1)}{(x+3)(x-1)}$$

$$= \frac{x(x+3)(x-4)(x-1)}{(x-4)(x+1)(x+3)(x-1)}$$

$$= \frac{x}{x+1}.$$

To divide the rational number a/b by the non-zero rational number c/d, we multiply by the reciprocal of c/d, which is d/c. The reciprocal of a rational expression P/Q should be the quantity whose product with P/Q is 1. By the definition of multiplication, and the fundamental property, $P/Q \cdot Q/P = 1$, so that the reciprocal of P/Q is Q/P. We can now define division of rational expressions in a manner similar to division of rational numbers as follows:

If P/Q and R/S are any two rational expressions, with $R/S \neq 0$, then their quotient is

$$\frac{P}{Q} \div \frac{R}{S} = \frac{P}{Q} \cdot \frac{S}{R} = \frac{PS}{QR}.$$

We need the restriction $R/S \neq 0$ since we cannot divide by zero.

Example 3
$$\frac{y}{y+3} \div \frac{y}{y+5} = \frac{y}{y+3} \cdot \frac{y+5}{y}$$
$$= \frac{y(y+5)}{(y+3)y}$$
$$= \frac{y+5}{y+3}.$$

Example 4
$$\frac{(3m)^2}{(2n)^3} \div \frac{6m^3}{16n^2} = \frac{9m^2}{8n^3} \div \frac{6m^3}{16n^2}$$
$$= \frac{9m^2}{8n^3} \cdot \frac{16n^2}{6m^3}$$
$$= \frac{9 \cdot 16m^2n^2}{8 \cdot 6n^3m^3}$$
$$= \frac{3}{mn}.$$

Example 5 Divide $\dfrac{x^2-4}{(x+3)(x-2)}$ by $\dfrac{(x+2)(x+3)}{2x}$.

To find the quotient of these rational expressions, we use the definition of division, and then simplify the resulting answer.

$$\frac{x^2-4}{(x+3)(x-2)} \div \frac{(x+2)(x+3)}{2x} = \frac{(x+2)(x-2)}{(x+3)(x-2)} \cdot \frac{2x}{(x+2)(x+3)}$$

Recall that $x^2 - 4$ factors as $(x+2)(x-2)$.

$$= \frac{(x+2)(x-2)(2x)}{(x+3)(x+3)(x-2)(x+2)}$$
$$= \frac{2x}{(x+3)^2}.$$

Example 6 Divide $\dfrac{m^2-4}{m^2-1} \div \dfrac{2m^2+4m}{m-1}$.

$$\frac{m^2-4}{m^2-1} \div \frac{2m^2+4m}{m-1} = \frac{m^2-4}{m^2-1} \cdot \frac{m-1}{2m^2+4m}$$
$$= \frac{(m+2)(m-2)}{(m+1)(m-1)} \cdot \frac{m-1}{2m(m+2)}$$
$$= \frac{m-2}{2m(m+1)}.$$

EXERCISES

Find the following products or quotients. Write each answer in lowest terms.

1. $\frac{9m^2}{16} \cdot \frac{4}{3m}$

2. $\frac{21z^4}{8y} \cdot \frac{4y^3}{7z^5}$

3. $\frac{4p^2}{8p} \cdot \frac{3p^3}{16p^4}$

4. $\frac{6x^3}{9x} \cdot \frac{12x}{x^2}$

5. $\frac{8a^4}{12a^3} \cdot \frac{9a^5}{3a^2}$

6. $\frac{14p^5}{2p^2} \cdot \frac{8p^6}{28p^9}$

7. $\frac{3r^2}{9r^3} \div \frac{8r^4}{6r^5}$

8. $\frac{15m^{10}}{9m^5} \div \frac{6m^6}{10m^4}$

9. $\frac{3m^2}{(4m)^3} \div \frac{9m^3}{32m^4}$

10. $\frac{5x^3}{(4x)^2} \div \frac{15x^2}{8x^4}$

11. $\frac{a+b}{2} \cdot \frac{12}{(a+b)^2}$

12. $\frac{3(x-1)}{y} \cdot \frac{2y}{5(x-1)}$

13. $\frac{a-3}{16} \div \frac{a-3}{32}$

14. $\frac{9}{8-2y} \div \frac{3}{4-y}$

15. $\frac{2k+8}{6} \div \frac{3k+12}{2}$

16. $\frac{5m+25}{10} \cdot \frac{12}{6m+30}$

17. $\frac{9y-18}{6y+12} \cdot \frac{3y+6}{15y-30}$

18. $\frac{12p+24}{36p-36} \div \frac{6p+12}{8p-8}$

19. $\frac{3r+12}{8} \cdot \frac{16r}{9r+36}$

20. $\frac{2r+2p}{8z} \div \frac{r^2y+ryp}{72}$

21. $\frac{y^2-16}{y+3} \div \frac{y-4}{y^2-9}$

22. $\frac{9(y-4)^2}{8(z+3)^2} \cdot \frac{16(z+3)}{3(y-4)}$

23. $\frac{6(m+2)}{3(m-1)^2} \div \frac{(m+2)^2}{9(m-1)}$

24. $\frac{4y+12}{2y-10} \div \frac{y^2-9}{y^2-y-20}$

25. $\frac{2-y}{8} \cdot \frac{7}{y-2}$

(*Hint:* Recall Example 4, Section 5.3.)

26. $\frac{9-2z}{3} \cdot \frac{9}{2z-9}$

27. $\frac{8-r}{8+r} \div \frac{r-8}{r+8}$

28. $\frac{6r-18}{9r^2+6r-24} \cdot \frac{12r-16}{4r-12}$

29. $\frac{k^2-k-6}{k^2+k-12} \div \frac{k^2+2k-3}{k^2+3k-4}$

30. $\frac{m^2+3m+2}{m^2+5m+4} \cdot \frac{m^2+10m+24}{m^2+5m+6}$

31. $\frac{n^2-n-6}{n^2-2n-8} \cdot \frac{n^2+7n+12}{n^2-9}$

32. $\dfrac{6n^2 - 5n - 6}{6n^2 + 5n - 6} \cdot \dfrac{12n^2 - 17n + 6}{12n^2 - n - 6}$

33. $\dfrac{16 - r^2}{r^2 + 2r - 8} \div \dfrac{r^2 - 2r - 8}{4 - r^2}$

34. $\dfrac{y^2 + y - 2}{y^2 + 3y - 4} \div \dfrac{y + 2}{y + 3}$

35. $\dfrac{2m^2 - 5m - 12}{m^2 - 10m + 24} \div \dfrac{4m^2 - 9}{m^2 - 9m + 18}$

36. $\dfrac{9z^2 + 27zm}{9m^2 + 27zm} \div \dfrac{8zm + 24m^2}{16zm + 48z^2}$

37. $\dfrac{21p^2 - 20pq - q^2}{p^2 + pq - 2q^2} \div \dfrac{21p^2 + 22pq + q^2}{p^2 + pq - 2q^2}$

38. $\dfrac{2m^2 + 7m + 3}{m^2 - 9} \cdot \dfrac{m^2 - 3m}{2m^2 + 11m + 5}$

39. $\dfrac{(x + 1)^3(x + 4)}{x^2 + 5x + 4} \div \dfrac{x^2 + 2x + 1}{x^2 + 3x + 2}$

40. $\dfrac{m^2 - m - 6}{3m^2 + 10m + 8} \cdot \dfrac{6m^2 + 17m + 12}{4m^2 - 9}$

41. $\left(\dfrac{x^2 + 10x + 25}{x^2 + 10x} \cdot \dfrac{10x}{x^2 + 15x + 50}\right) \div \dfrac{x + 5}{x + 10}$

42. $\left(\dfrac{m^2 - 12m + 32}{8m} \cdot \dfrac{m^2 - 8m}{m^2 - 8m + 16}\right) \div \dfrac{m - 8}{m - 4}$

5.5 ADDITION AND SUBTRACTION OF RATIONAL EXPRESSIONS

As discussed in Section 5.2, to add the rational numbers 1/4 and 2/5, we proceed as follows:

1. Decide on the least common denominator, if needed. The LCD for these rational numbers is 20.
2. Rewrite each rational number, using this common denominator. Here

$$\frac{1}{4} = \frac{1 \cdot 5}{4 \cdot 5} = \frac{5}{20}$$

and

$$\frac{2}{5} = \frac{2 \cdot 4}{5 \cdot 4} = \frac{8}{20}.$$

3. Add the numerators to get the numerator of the sum. The LCD is used as the denominator of the sum.

$$\frac{1}{4}+\frac{2}{5}=\frac{5}{20}+\frac{8}{20}$$
$$=\frac{5+8}{20}$$
$$=\frac{13}{20}.$$

We can follow a similar procedure to find the sum of two rational expressions.

If P/Q and R/Q are rational expressions, then

$$\frac{P}{Q}+\frac{R}{Q}=\frac{P+R}{Q}.$$

Example 1 Add $\frac{2}{3y}+\frac{1}{4y}$.

The LCD is $3 \cdot 4 \cdot y = 12y$. (We need only one factor of y.) Thus, we must multiply in the first term by $12y/3y = 4$, and in the second term by $12y/4y = 3$.

$$\frac{2}{3y}+\frac{1}{4y}=\frac{2(4)}{3y(4)}+\frac{1(3)}{4y(3)}=\frac{8}{12y}+\frac{3}{12y}=\frac{11}{12y}.$$

Example 2 Add the rational expressions $\frac{x}{x^2-1}+\frac{x}{x+1}$.

To find the LCD, factor both denominators.

$$x^2-1=(x+1)(x-1); \qquad x+1 \text{ cannot be factored.}$$

The LCD is $(x+1)(x-1)$, so we have

$$\frac{x}{x^2-1}+\frac{x}{x+1}=\frac{x}{(x+1)(x-1)}+\frac{x(x-1)}{(x+1)(x-1)}$$
$$=\frac{x+x(x-1)}{(x+1)(x-1)}$$
$$=\frac{x+x^2-x}{(x+1)(x-1)}$$
$$=\frac{x^2}{(x+1)(x-1)}.$$

Example 3 Find the sum of the rational expressions

$$\frac{2x}{x^2 + 5x + 6} \quad \text{and} \quad \frac{x + 1}{x^2 + 2x - 3}.$$

To begin, we factor the denominators completely.

$$\frac{2x}{x^2 + 5x + 6} + \frac{x + 1}{x^2 + 2x - 3} = \frac{2x}{(x + 2)(x + 3)} + \frac{x + 1}{(x + 3)(x - 1)}.$$

The LCD is $(x + 2)(x + 3)(x - 1)$. By the fundamental property,

$$\frac{2x}{(x + 2)(x + 3)} + \frac{x + 1}{(x + 3)(x - 1)}$$

$$= \frac{2x(x - 1)}{(x + 2)(x + 3)(x - 1)} + \frac{(x + 1)(x + 2)}{(x + 2)(x + 3)(x - 1)}.$$

Since the two rational expressions above have the same denominator, we add their numerators, just as with rational numbers.

$$= \frac{2x(x - 1) + (x + 1)(x + 2)}{(x + 2)(x + 3)(x - 1)}$$

$$= \frac{2x^2 - 2x + x^2 + 3x + 2}{(x + 2)(x + 3)(x - 1)}$$

$$= \frac{3x^2 + x + 2}{(x + 2)(x + 3)(x - 1)}.$$

It is often convenient to leave the denominator in factored form in a problem of this type.

To subtract two rational numbers, we use the fact that

$$\frac{a}{b} - \frac{c}{d} = \frac{a}{b} + \left(-\frac{c}{d}\right).$$

Again, a parallel procedure is used for rational expressions.

If P/Q and R/S are rational expressions, then their difference is

$$\frac{P}{Q} - \frac{R}{S} = \frac{P}{Q} + \left(-\frac{R}{S}\right).$$

Example 4 Find $\dfrac{12}{x^2} - \dfrac{-8}{x^2}$.

$$\frac{12}{x^2} - \frac{-8}{x^2} = \frac{12}{x^2} + \left(-\frac{-8}{x^2}\right) = \frac{12}{x^2} + \frac{8}{x^2} = \frac{20}{x^2}.$$

Example 5 Find $\dfrac{6x}{(x-1)^2} - \dfrac{2}{x^2-1}$.

$$\frac{6x}{(x-1)^2} - \frac{2}{x^2-1} = \frac{6x}{(x-1)(x-1)} + \frac{-2}{(x-1)(x+1)}$$

We changed subtraction to addition and changed the numerator of the second fraction from 2 to -2. We also factored the two denominators, so that we can now identify a common denominator as $(x-1)(x-1)(x+1)$. We must use the factor $x-1$ twice, since it appears twice in the first denominator.

$$= \frac{6x(x+1)}{(x-1)(x-1)(x+1)} + \frac{-2(x-1)}{(x-1)(x-1)(x+1)}$$

$$= \frac{6x(x+1) - 2(x-1)}{(x-1)(x-1)(x+1)}$$

$$= \frac{6x^2 + 6x - 2x + 2}{(x-1)(x-1)(x+1)}$$

$$= \frac{6x^2 + 4x + 2}{(x-1)(x-1)(x+1)}.$$

EXERCISES

Find the sums or differences.

1. $\dfrac{2}{p} + \dfrac{5}{p}$

2. $\dfrac{3}{r} + \dfrac{6}{r}$

3. $\dfrac{9}{k} - \dfrac{12}{k}$

4. $\dfrac{15}{z} - \dfrac{25}{z}$

5. $\dfrac{y}{y+1} + \dfrac{1}{y+1}$

6. $\dfrac{3m}{m-4} + \dfrac{-12}{m-4}$

7. $\dfrac{m^2}{m-n} - \dfrac{n^2}{m-n}$

8. $\dfrac{a+b}{2} - \dfrac{a-b}{2}$

9. $\dfrac{m^2}{m+6} + \dfrac{6m}{m+6}$

10. $\dfrac{y^2}{y-1} + \dfrac{-y}{y-1}$

11. $\dfrac{2}{2r+2} + \dfrac{2r}{2r+2}$

12. $\dfrac{a^2}{a+b} + \dfrac{ab}{a+b}$

13. $\dfrac{3}{m} + \dfrac{1}{2}$

14. $\dfrac{6}{p} - \dfrac{2}{3}$

15. $\dfrac{9}{m} + \dfrac{3}{2}$

16. $\dfrac{9}{10} + \dfrac{r}{2}$

17. $\frac{3}{5} - \frac{1}{y}$

18. $\frac{9y}{7} - \frac{2y}{8}$

19. $\frac{5m}{6} - \left(\frac{2m}{3} - \frac{m}{6}\right)$

20. $\frac{4 + 2k}{5} + \frac{2 + k}{10}$

21. $\frac{5 - 4r}{8} - \frac{2 - 3r}{6}$

22. $\left(\frac{3}{x} + \frac{4}{2x}\right) - \frac{5}{4x}$

23. $\frac{6}{y^2} - \frac{2}{y}$

24. $\frac{3}{p} + \frac{5}{p^2}$

25. $\frac{9}{2p} + \frac{4}{p^2}$

26. $\frac{15}{4k^2} - \frac{3}{k}$

27. $\frac{3m + n}{3} + \frac{m + 2n}{6}$

28. $\frac{5r + s}{3} - \frac{2r - s}{9}$

29. $\frac{-1}{x^2} + \frac{-3}{xy}$

30. $\frac{9}{p^2} + \frac{p}{x}$

31. $\frac{m + 2}{m} + \frac{m}{m + 2}$

32. $\frac{2x - 5}{x - 2} + \frac{x}{2x - 4}$

33. $\frac{8}{x - 2} - \frac{4}{x + 2}$

34. $\frac{6m}{m - n} + \frac{2n}{m + n}$

35. $\frac{2x}{x + y} - \frac{3x}{2x + 2y}$

36. $\frac{1}{a + b} - \frac{a}{a^2 - b^2}$

37. $\frac{1}{m^2 - 1} - \frac{1}{m^2 + 3m + 2}$

38. $\frac{2}{4y^2 - 16} + \frac{3}{4 + 2y}$

39. $\frac{1}{m^2 - 9} + \frac{1}{m + 3}$

40. $\frac{2}{y^2 - 4} - \frac{3}{2y + 4}$

41. $\frac{4}{2 - m} + \frac{7}{m - 2}$

42. $\frac{9}{8 - y} + \frac{6}{y - 8}$

43. $\frac{-1}{-3 + y} - \frac{2}{y - 3}$

44. $\frac{-8}{11 + p} - \frac{6}{p + 11}$

45. $\frac{5m}{m + 2n} - \frac{3m}{-m - 2n}$

46. $\frac{6k}{2k + 3m} - \frac{4k}{-2k - 3m}$

47. $\frac{x + 3y}{x^2 + 2xy + y^2} + \frac{x - y}{x^2 + 4xy + 3y^2}$

48. $\left(\frac{m + 1}{m^2 - 1} + \frac{m - 1}{m^2 + 2m + 1}\right) - \frac{3}{m - 1}$

49. $\left(\dfrac{a-b}{a^2+2ab+b^2}-\dfrac{a}{a^2-b^2}\right)-\dfrac{1}{a-b}$

50. $\left(\dfrac{m-n}{m^2+4mn+4n^2}+\dfrac{n}{m^2-4n^2}\right)-\dfrac{2}{m-2n}$

5.6 EQUATIONS INVOLVING RATIONAL EXPRESSIONS

To solve equations with fractions we can simplify our work by using the multiplication property of equality. We want to replace an equation that includes fractions by another equation which does not include fractions. To do this, we choose as a multiplier the LCD of all denominators in the fractions of the equation. For example, let us solve

$$\frac{x}{3}+\frac{x}{4}=10+x.$$

The LCD of the two fractions is 12. Therefore, we begin by multiplying both sides of the equation by 12.

$$12\left(\frac{x}{3}+\frac{x}{4}\right)=12(10+x)$$

$$12\left(\frac{x}{3}\right)+12\left(\frac{x}{4}\right)=12(10)+12x$$

$$4x+3x=120+12x$$

$$7x=120+12x$$

$$-5x=120$$

$$x=-24.$$

Check the solution $x=-24$ by substituting in each side of the original equation separately.

$$\frac{x}{3}+\frac{x}{4}=\frac{-24}{3}+\frac{-24}{4}=-8+(-6)=-14$$

and $\qquad 10+x=10+(-24)=-14.$

Therefore, -24 is the solution.

When solving equations which have a variable in the denominator, we must remember that the number 0 cannot be used as a denominator. Therefore, the solution cannot be a number which is excluded from the domain of any rational expression in the equation. The following example illustrates this.

Example 1 Solve $\dfrac{x}{x-2} = \dfrac{2}{x-2} + 2.$

The common denominator is $x - 2$. Multiply both sides of the equation by $x - 2$.

$$(x-2)\left(\frac{x}{x-2}\right) = (x-2)\left(\frac{2}{x-2}\right) + (x-2)(2)$$

$$x = 2 + 2x - 4$$

$$x = -2 + 2x$$

$$0 = -2 + x$$

$$2 = x.$$

The solution to the final equation in the sequence of equations is 2. However, we cannot have a solution of 2 in this equation because this value makes both denominators zero, and the equation is meaningless. Therefore, this equation has no solution.

When we first mentioned the multiplication property of equality, we said that the equation $A = B$ has the same solution as the equation $AC = BC$, only if $C \neq 0$. Here we multiplied our equation by $x - 2$, which has the value 0 when $x = 2$. Thus the multiplication property has not failed. We just neglected to note the restriction.

Example 2 Solve $\dfrac{2}{x^2 - x} = \dfrac{1}{x^2 - 1}.$

To solve the equation, we begin by finding a common denominator. Since $x^2 - x$ can be factored as $x(x - 1)$, while $x^2 - 1$ can be factored as $(x + 1)(x - 1)$, the least common denominator of the two rational expressions is $x(x + 1)(x - 1)$. We multiply both sides of the equation by $x(x + 1)(x - 1)$.

$$x(x+1)(x-1)\frac{2}{x(x-1)} = x(x+1)(x-1)\frac{1}{(x+1)(x-1)}$$

$$2(x+1) = x$$

$$2x + 2 = x$$

$$x + 2 = 0$$

$$x = -2.$$

To be sure that $x = -2$ is a solution, check the proposed solution in the original equation by substitution.

$$\frac{2}{x^2 - x} = \frac{2}{(-2)^2 - (-2)} = \frac{2}{4+2} = \frac{2}{6} = \frac{1}{3}$$

$$\frac{1}{x^2 - 1} = \frac{1}{(-2)^2 - 1} = \frac{1}{4-1} = \frac{1}{3}.$$

Since -2 is in the domain and satisfies the equation, the solution is -2.

Example 3 Solve $\dfrac{1}{x-1} + \dfrac{1}{2} = \dfrac{2}{x^2-1}$.

The least common denominator is $2(x+1)(x-1)$. We multiply both sides of the equation by this common denominator.

$$2(x+1)(x-1)\left(\frac{1}{x-1} + \frac{1}{2}\right) = 2(x+1)(x-1)\frac{2}{(x+1)(x-1)}$$

$$2(x+1)(x-1)\frac{1}{x-1} + 2(x+1)(x-1)\frac{1}{2} = 2(x+1)(x-1)\frac{2}{(x+1)(x-1)}$$

$$2(x+1) + (x+1)(x-1) = 4$$

$$2x + 2 + x^2 - 1 = 4$$

$$x^2 + 2x + 1 = 4$$

$$x^2 + 2x - 3 = 0.$$

Upon factoring, we have

$$(x+3)(x-1) = 0.$$

Therefore, it seems $x = -3$ or $x = 1$. But 1 is not in the domain of the original equation. (Why?) Therefore, the only solution is -3. This can be verified by checking both proposed solutions in the original equation.

Example 4 Solve the equation

$$\frac{2m}{m^2-4} + \frac{1}{m-2} = \frac{2}{m+2}.$$

Since $m^2 - 4 = (m+2)(m-2)$, use $(m+2)(m-2)$ as common denominator:

$$(m+2)(m-2)\left(\frac{2m}{m^2-4} + \frac{1}{m-2}\right) = (m+2)(m-2)\frac{2}{m+2}$$

$$(m+2)(m-2)\frac{2m}{m^2-4} + (m+2)(m-2)\frac{1}{m-2} = (m+2)(m-2)\frac{2}{m+2}$$

$$2m + m + 2 = 2(m-2)$$

$$3m + 2 = 2m - 4$$

$$m = -6.$$

Check that -6 is in the domain of the equation so that the solution is -6.

Example 5 Solve the equation

$$4 + \frac{x+3}{x-3} - \frac{4x^2}{x^2-9} = \frac{x-3}{x+3}.$$

The least common denominator is $(x+3)(x-3)$. After multiplying both sides of the equation by the common denominator, we have

$$\begin{aligned} 4(x+3)(x-3) + (x+3)(x+3) - 4x^2 &= (x-3)(x-3) \\ 4x^2 - 36 + x^2 + 6x + 9 - 4x^2 &= x^2 - 6x + 9 \\ -27 + 6x &= -6x + 9 \\ 12x &= 36 \\ x &= 3. \end{aligned}$$

There is no solution. The only proposed solution, 3, is not in the domain.

EXERCISES

Solve each equation.

1. $\frac{1}{4} = \frac{x}{2}$
2. $\frac{2}{m} = \frac{5}{12}$
3. $\frac{9}{k} = \frac{3}{4}$
4. $\frac{p}{15} = \frac{4}{15}$
5. $\frac{3}{4} - m = 2m$
6. $3r - \frac{1}{2} = \frac{11}{2}$
7. $\frac{6}{x} - \frac{4}{x} = 5$
8. $\frac{3}{x} + \frac{2}{x} = 5$
9. $\frac{x}{2} - \frac{x}{4} = 6$
10. $\frac{4}{y} + \frac{2}{3} = 1$
11. $\frac{9}{m} = 5 - \frac{1}{m}$
12. $\frac{3x}{5} + 2 = \frac{1}{4}$
13. $\frac{2t}{7} - 5 = t$
14. $\frac{1}{2} + \frac{2}{m} = 1$
15. $\frac{x+1}{2} = \frac{x+2}{3}$
16. $\frac{t-4}{3} = t + 2$
17. $\frac{3m}{2} + m = 5$
18. $\frac{9}{x+1} = 3$
19. $\frac{9}{x-2} = 3$
20. $\frac{2y-1}{y} + 2 = \frac{1}{2}$
21. $\frac{2k+3}{k} = \frac{3}{2}$
22. $\frac{a}{2} - \frac{17+a}{5} = 2a$
23. $\frac{5-y}{y} + \frac{3}{4} = \frac{7}{y}$
24. $\frac{x}{x-4} = \frac{2}{x-4} + 5$
25. $\frac{a-4}{4} = \frac{a+8}{16}$
26. $\frac{m-2}{5} = \frac{m+8}{10}$
27. $\frac{2p+8}{9} = \frac{10p+4}{27}$
28. $\frac{5r-3}{7} = \frac{15r-2}{28}$
29. $\frac{8x-1}{6x+8} = \frac{3}{4}$
30. $\frac{6m+9}{5m+10} = \frac{3}{5}$
31. $\frac{2}{y} = \frac{y}{5y-12}$
32. $\frac{8x+3}{x} = 3x$

33. $\dfrac{m}{2m+2} = \dfrac{-2m}{4m+4} + \dfrac{m}{m+1}$

34. $\dfrac{5p+1}{3p+3} = \dfrac{5p-5}{5p+5} + \dfrac{3p-1}{p+1}$

35. $\dfrac{1}{x^2+5x+6} + \dfrac{1}{x^2-2x-8} = \dfrac{-1}{12(x+2)}$

36. $\dfrac{x+4}{x^2-3x+2} - \dfrac{5}{x^2-4x+3} = \dfrac{x-4}{x^2-5x+6}$

37. $\dfrac{3y}{y^2+5y+6} = \dfrac{5y}{y^2+2y-3} - \dfrac{2}{y^2+y-2}$

38. $\dfrac{3}{r^2+r-2} - \dfrac{1}{r^2-1} = \dfrac{7}{2(r^2+3r+2)}$

5.7 APPLICATIONS OF RATIONAL EXPRESSIONS

Now we can discuss some applications of our work with rational expressions.

Example 1 If the same number is added to both the numerator and denominator of the rational number 3/4, the result is 1/12 larger than 3/4. Find the number.

If x represents the number that is added to numerator and denominator, we can write

$$\frac{3+x}{4+x}$$

to represent the result of adding the same number to both numerator and denominator. Since this result is 1/12 larger than 3/4, we can write

$$\frac{3+x}{4+x} = \frac{3}{4} + \frac{1}{12}.$$

Adding 1/12 and 3/4 gives

$$\frac{3+x}{4+x} = \frac{9}{12} + \frac{1}{12}$$

$$\frac{3+x}{4+x} = \frac{10}{12}$$

$$\frac{3+x}{4+x} = \frac{5}{6}.$$

If we multiply both sides of the equation by the common denominator $6(4 + x)$, we have

$$6(4+x)\frac{3+x}{4+x} = 6(4+x)\frac{5}{6}$$

$$6(3+x) = 5(4+x)$$

$$18 + 6x = 20 + 5x$$

$$x = 2.$$

Checking the solution in the words of the original problem, we add 2 to the numerator and denominator of 3/4.

$$\frac{3+2}{4+2} = \frac{5}{6}.$$

Since $5/6 = 1/12 + 3/4$, 2 is the number we want.

Problems involving *percent* occur frequently in business and in everyday matters. Percent means "per hundred," so that 60% means 60/100 or .60. Algebra is useful in solving percent problems.

Example 2 What percent of 80 is 20?

The question can be restated as "20 is what percent of 80?" or "Find a number x such that 20 is x percent of 80." Since percent means "per hundred," write x percent as $x/100$. Then the statement can be translated as

$$20 = \frac{x}{100}(80)$$

$$2000 = 80x$$

$$x = 25.$$

We see that 20 is 25 percent of 80.

Example 3 If 34% of a certain number is 38.08, find the number.

In other words, "Find a number x such that 34% of x is 38.08." First change 34% to .34 (since 38.08 is a decimal). Then translate the words into symbols.

$$.34x = 38.08.$$

Multiply both sides of the equation by 1/.34.

$$x = \frac{38.08}{.34} = 112.$$

Check that 34% of 112 is 38.08.

Example 4 Working alone, John can cut his lawn in 8 hours. If John's pet sheep is released to eat the grass, the lawn can be cut in 14 hours. If both John and the sheep work on the lawn, how long will it take to cut it?

Let x be the number of hours that it takes John and the sheep to cut the lawn, working together. Certainly x will be less than 8, since John alone can cut the lawn in 8 hours. In one hour, John can do 1/8 of the lawn, and in one hour the sheep can do 1/14 of the lawn. Since it takes them x hours to cut the lawn when working together, in one hour together they can do $1/x$ of the lawn. The amount of the lawn cut by John in one hour plus the amount cut by the sheep in one hour must equal the amount they can do together in one hour. In symbols, we have

$$\frac{1}{8} + \frac{1}{14} = \frac{1}{x}.$$

The quantity $56x$ is the least common denominator for 8, 14, and x, and so we multiply both sides of the equation by $56x$.

$$56x\left(\frac{1}{8} + \frac{1}{14}\right) = 56x \cdot \frac{1}{x}$$

$$56x \cdot \frac{1}{8} + 56x \cdot \frac{1}{14} = 56x \cdot \frac{1}{x}$$

$$7x + 4x = 56$$

$$11x = 56$$

$$x = \frac{56}{11}.$$

Working together, John and his sheep cut the lawn in 56/11 hours, or $5\frac{1}{11}$ hours, about 5 hours and 5 minutes.

Example 5 The Big Muddy has a current of 3 miles per hour. A motorboat takes as long to go 12 miles downstream as to go 8 miles upstream. What is the speed of the boat in still water?

This requires the distance formula, $d = rt$ (distance = rate · time). For our problem, let's use x to represent the speed of the boat in still water. Since the current pushes the boat when the boat is going downstream, the speed of the boat downstream will be the sum of the speed of the boat and the speed of the current, or $x + 3$ miles per hour. Similarly, the speed of the boat upstream is given by $x - 3$ miles per hour.

We can summarize the information in the problem in a chart.

	d	r	t
downstream	12	$x + 3$	
upstream	8	$x - 3$	

	d	r	t
downstream	12	$x + 3$	
upstream	8	$x - 3$	

To fill in the last column, representing time, we solve the formula $d = rt$ for t.

$$d = rt, \qquad \frac{d}{r} = t.$$

Then the time upstream is

$$\frac{d}{r} = \frac{8}{x - 3},$$

while the time downstream is

$$\frac{d}{r} = \frac{12}{x + 3}.$$

Now we can complete the chart.

	d	r	t
downstream	12	$x + 3$	$\frac{12}{x + 3}$
upstream	8	$x - 3$	$\frac{8}{x - 3}$

The problem states that the time upstream equals the time downstream. Thus the two times from the chart are equal.

$$\frac{12}{x + 3} = \frac{8}{x - 3}.$$

To solve this equation, multiply both sides by $(x + 3)(x - 3)$.

$$(x + 3)(x - 3)\frac{12}{x + 3} = (x + 3)(x - 3)\frac{8}{x - 3}$$

$$12(x - 3) = 8(x + 3)$$

$$12x - 36 = 8x + 24$$

$$4x = 60$$

$$x = 15.$$

The speed of the boat in still water is 15 miles per hour.

To check, note that the speed of the boat downstream would be $15 + 3 = 18$ miles per hour, and to go 12 miles would take

$$12 = 18t,$$

$$t = \frac{2}{3} \text{ hour.}$$

On the other hand, the speed of the boat upstream is $15 - 3 = 12$ miles per hour, and to go 8 miles would take

$$8 = 12t,$$

$$t = \frac{2}{3} \text{ hour.}$$

The time upstream equals the time downstream, as required.

EXERCISES

1. One-half of a number is three more than one-sixth of the same number. What is the number?
2. The numerator of the fraction 4/7 is increased by an amount so that the value of the resulting fraction is 27/21. By what amount was the numerator increased?
3. In a certain fraction, the denominator is 5 larger than the numerator. If 3 is added to both the numerator and the denominator, the result is 3/4. Find the original fraction.
4. The denominator of a certain fraction is three times the numerator. If 1 is added to the numerator and subtracted from the denominator, the result equals 1/2. Find the original rational number.
5. One number is three more than another. If the smaller is added to two-thirds the larger, the result is half the sum of the original numbers. Find the numbers.
6. The sum of a number and its reciprocal is 5/2. Find the number.
7. If twice the reciprocal of a number is subtracted from the number, the result is $-7/3$. Find the number.
8. The sum of the reciprocals of two consecutive integers is 5/6. Find the integers.
9. If three times a number is added to twice its reciprocal, the answer is 5. Find the number.
10. A man and his son worked four days at a job. The son's daily wage was 2/5 that of the father. If together they earned $168, what were their daily wages?
11. The profits from a student show are to be given to two scholarships in the ratio 2/3. If the fund receiving the larger amount was given $390, how much was given to the other fund?
12. If 19% of a certain number is 38, find the number.
13. Eight is what percent of 12?

14. Six is what percent of four?

15. When 28% of a certain number is added to 25, the result is 81. Find the number.

16. The distance from Sacramento to Yosemite is about 180 miles. About 120 miles of the distance is freeway. What percent of the distance is freeway?

17. It is estimated that a thermostat setting 6° higher on an air conditioner saves about 24% in the amount of electricity consumed. Assume that a thermostat setting is changed from 72° to 78°. Find the decrease in the bill if the cost of using the machine was $48 last summer.

18. Ms. Lopez earned $9000 last year. She received an 8% increase in salary this year. Inflation this year was 7%. Find the net change in her income, taking her raise and inflation both into account.

19. The owner of a plant boutique desires to sell a certain plant at $10.00. The owner must have a 40% "markup" on the wholesale purchase price to cover overhead and profit. (That is, the owner must take 40% of the purchase price—the markup—and add this markup to the purchase price. The result is the retail selling price, which is $10.00 here.) How much can the owner pay for the plant wholesale?

Selling price (retail)	
Purchase price (wholesale)	Markup

20. If a 25% markup is applied to an item, and it sells for $15.00, how much did it cost the merchant?

21. George signs a note with the bank, agreeing to pay them $2000 at the end of two years. The bank then gives him $1800. What percent simple annual interest is he paying? (See inside the back cover for the formula for simple interest.)

22. In a certain state, the state sales tax is 5%. At the end of business one day, John discovered that he had taken in $1200 during the day. (This amount includes his sales and the sales tax he collected.) He multiplied this figure by 5%: (.05)(1200) = $60, and then sent $60 to the state as sales tax he had collected. An algebra student could have told him that he had sent in too much. Tell why.

23. On Monday John is told he will receive a 10% raise in salary. On Tuesday, it is discovered that John was entitled to no raise after all, and so his salary is then cut 10%. John ends up with less money than he started with. Describe how this happened.

24. Sam can row four miles per hour in still water. It takes as long rowing 8 miles upstream as 24 miles downstream. How fast is the current?

25. Mary flew from here to there at 180 miles per hour, and from there to here at 150 miles per hour. The trip from there to here took one hour longer than the other trip. Find the distance from here to there. (Assume there was no wind in either direction.)

26. On a business trip, Joann traveled to her destination at an average speed of 60 m.p.h. Coming home, her average speed was 50 m.p.h. and the trip took 1/2 hour longer. How far did she travel?

27. Jenny flew her airplane 500 miles against the wind in the same time it took her to fly it 600 miles with the wind. If the speed of the wind was 10 m.p.h., what was the average speed of her plane?

28. George can paint a room, working alone, in 8 hours. Hortense can paint the same room, working alone, in 6 hours. How long will it take them if they work together?

29. Machine A can do a certain job in 7 hours, while machine B takes 12 hours. How long will it take the two machines working together?

30. One pipe can fill a swimming pool in 6 hours, while another pipe can do it in 9 hours. How long will it take the two pipes working together to fill the pool 3/4 full?

31. Dennis can do a job in 4 days. When Dennis and Sue work together, the job takes $3\frac{1}{3}$ days. How long would the job take Sue if she worked alone?

32. An inlet pipe can fill a swimming pool in nine hours, while an outlet pipe can empty the pool in 12 hours. Through an error, both pipes are left open. How long will it take to fill the pool?

33. A cold water faucet can fill a sink in 12 minutes, and a hot water faucet in 15. The drain can empty the sink in 25 minutes. If both faucets and the drain are open, how long will it take to fill the sink?

34. In Exercise 32, assume the error was discovered after both pipes had been running for 3 hours, and the outlet pipe was then closed. How much more time would then be required to fill the pool? (*Hint*: How much of the job had been done when the error was discovered?)

5.8 COMPLEX FRACTIONS

A fraction containing rational numbers or rational expressions in the numerator, denominator, or both is called a **complex fraction**. Examples of complex fractions include

$$\frac{3 + \frac{4}{x}}{5}, \qquad \frac{\frac{3x^2 - 5x}{6x^2}}{2x - \frac{1}{x}}, \qquad \frac{3 + x}{5 - \frac{2}{x}}$$

In Section 5.2 we simplified the complex fraction

$$\frac{\frac{1}{5} - \frac{1}{4}}{\frac{2}{3} + \frac{3}{4}}$$

by first rewriting the numerator and then the denominator as single fractions. Then we performed the division.

Numerator: $\frac{1}{5} - \frac{1}{4} = \frac{4}{20} - \frac{5}{20} = \frac{-1}{20}.$

Denominator: $\frac{2}{3} + \frac{3}{4} = \frac{8}{12} + \frac{9}{12} = \frac{17}{12}.$

Divide numerator by denominator:

$$-\frac{1}{20} \div \frac{17}{12} = -\frac{1}{20} \cdot \frac{12}{17} = \frac{-1 \cdot 4 \cdot 3}{4 \cdot 5 \cdot 17} = \frac{-3}{85}.$$

The same procedure can be used to simplify the complex fraction

$$\frac{6 + \frac{3}{x}}{\frac{2x + 1}{8}}.$$

First write the numerator as a rational expression.

$$\begin{aligned} 6 + \frac{3}{x} &= \frac{6}{1} + \frac{3}{x} \\ &= \frac{6x}{x} + \frac{3}{x} \\ &= \frac{6x + 3}{x}. \end{aligned}$$

Using this result, the complex fraction can be written as

$$\frac{\dfrac{6x+3}{x}}{\dfrac{2x+1}{8}}.$$

Next, use the rule for division and the fundamental property.

$$\begin{aligned}\frac{6x+3}{x} \div \frac{2x+1}{8} &= \frac{6x+3}{x} \cdot \frac{8}{2x+1}\\ &= \frac{3(2x+1)}{x} \cdot \frac{8}{2x+1}\\ &= \frac{24}{x}.\end{aligned}$$

Example 1 Simplify the complex fraction

$$\frac{\dfrac{x^2-4}{3x+2}}{\dfrac{x+2}{6x+4}}.$$

$$\begin{aligned}\frac{x^2-4}{3x+2} \div \frac{x+2}{6x+4} &= \frac{x^2-4}{3x+2} \cdot \frac{6x+4}{x+2}\\ &= \frac{(x+2)(x-2)}{3x+2} \cdot \frac{2(3x+2)}{x+2}\\ &= 2(x-2).\end{aligned}$$

The result may be written as $2x - 4$.

Example 2 Simplify $2 + \dfrac{2}{2 + \dfrac{2}{2+2}}$.

We begin by simplifying the fraction in the denominator.

$$2 + \frac{2}{2+\dfrac{2}{2+2}} = 2 + \frac{2}{2+\dfrac{2}{4}} = 2 + \frac{2}{2+\dfrac{1}{2}} = 2 + \frac{2}{\dfrac{4}{2}+\dfrac{1}{2}} = 2 + \frac{2}{\dfrac{5}{2}}$$

$$= 2 + \frac{2}{1}\cdot\frac{2}{5} = 2 + \frac{4}{5} = \frac{10}{5} + \frac{4}{5} = \frac{14}{5}.$$

EXERCISES

Simplify each complex fraction.

1. $\dfrac{\frac{9}{10}}{\frac{3}{5}}$

2. $\dfrac{\frac{5}{8}}{\frac{25}{16}}$

3. $\dfrac{1+\frac{3}{4}}{\frac{5}{4}}$

4. $\dfrac{2-\frac{3}{8}}{\frac{3}{4}}$

5. $\dfrac{-6+\frac{5}{6}}{\frac{31}{5}}$

6. $\dfrac{3+\frac{1}{3}}{2-\frac{1}{3}}$

7. $\dfrac{\frac{x}{y}}{\frac{x^2}{y}}$

8. $\dfrac{\frac{a}{b+1}}{\frac{a^2}{b}}$

9. $\dfrac{\frac{pq}{r}}{\frac{p^2q}{r^2}}$

10. $\dfrac{\frac{x+1}{y}}{\frac{y+1}{x}}$

11. $\dfrac{m+\frac{1}{m}}{\frac{3}{m}-m}$

12. $\dfrac{y+\frac{2}{y}}{\frac{y^2+2}{3}}$

13. $\dfrac{\frac{3}{y}+1}{\frac{3+y}{2}}$

14. $\dfrac{\frac{1}{x}}{\frac{1+x}{1-x}}$

15. $\dfrac{x+\frac{1}{y}}{\frac{1}{x}+y}$

16. $\dfrac{y-\frac{1}{y}}{y+\frac{1}{y}}$

17. $\dfrac{\frac{1}{x}+\frac{1}{y}}{\frac{1}{x+y}}$

18. $\dfrac{\frac{a}{a-b}}{\frac{1}{a}-\frac{1}{b}}$

19. $\dfrac{\frac{p+q}{p}}{\frac{1}{p}+\frac{1}{q}}$

20. $\dfrac{r+\frac{1}{r}}{\frac{1}{r}-r}$

21. $\dfrac{\frac{1}{m+n}-\frac{1}{m-n}}{\frac{4}{m^2-n^2}}$

22. $\dfrac{\frac{a+1}{a-1}-\frac{a-1}{a+1}}{\frac{a-1}{a+1}-\frac{a+1}{a-1}}$

23. $1-\dfrac{1}{1+\dfrac{1}{1-\dfrac{1}{1+1}}}$

24. $3-\dfrac{2}{\dfrac{1}{1+\dfrac{1}{2+\frac{1}{2}}}}$

CHAPTER 5 SUMMARY

Key Words

Rational number
Lowest terms
Mixed number
Complex fraction
Least common denominator
Rational expression
Domain of a rational expression
Rational equation

Fundamental Property of Rational Numbers

If a/b is a rational number, and if $k \neq 0$, then

$$\frac{ak}{bk} = \frac{a}{b}.$$

Addition of Rational Numbers

If a/c and b/c are rational numbers, then their sum is

$$\frac{a}{c} + \frac{b}{c} = \frac{a+b}{c}.$$

Subtraction of Rational Numbers

If a/b and c/d are rational numbers, then their difference is

$$\frac{a}{b} - \frac{c}{d} = \frac{a}{b} + \frac{-c}{d}.$$

Multiplication of Rational Numbers

If a/b and c/d are rational numbers, then their product is

$$\frac{a}{b} \cdot \frac{c}{d} = \frac{ac}{bd}.$$

Division of Rational Numbers

If a/b and c/d are rational numbers, with $c/d \neq 0$, then

$$\frac{a}{b} \div \frac{c}{d} = \frac{a}{b} \cdot \frac{d}{c} = \frac{ad}{bc}.$$

Fundamental Property of Rational Expressions

If P/Q is a rational expression, and if K is a rational expression, $K \neq 0$, then

$$\frac{PK}{QK} = \frac{P}{Q}.$$

Addition of Rational Expressions

If P/Q and R/Q are rational expressions, then their sum is

$$\frac{P}{Q} + \frac{R}{Q} = \frac{P+R}{Q}.$$

Subtraction of Rational Expressions

If P/Q and R/S are rational expressions, then their difference is

$$\frac{P}{Q} - \frac{R}{S} = \frac{P}{Q} + \left(-\frac{R}{S}\right).$$

Multiplication of Rational Expressions

If P/Q and R/S are rational expressions, then their product is

$$\frac{P}{Q} \cdot \frac{R}{S} = \frac{PR}{QS}.$$

Division of Rational Expressions

If P/Q and R/S are rational expressions, with $R/S \neq 0$, then

$$\frac{P}{Q} \div \frac{R}{S} = \frac{P}{Q} \cdot \frac{S}{R} = \frac{PS}{QR}.$$

CHAPTER 5 TEST

1. Write the three numbers 5/8, 9/16, and 7/12 in order, from smallest to largest.

Perform the following operations. Write all answers in lowest terms.

2. $\frac{7}{9} + \frac{1}{3} - \frac{5}{12}$

3. $\frac{3}{4} \cdot \frac{8}{5} \div \frac{5}{6}$

4. $\frac{x^6y}{x^3} \cdot \frac{y^2}{x^2y^3}$

5. $\frac{m^3}{n} \div \frac{m^2}{n^3}$

6. $\frac{5}{x} - \frac{6}{x}$

7. $\frac{1}{a+1} + \frac{5}{6a+6}$

8. $1 - \frac{t-1}{t+1}$

9. $\frac{1}{a^2 + a - 6} + \frac{1}{a^2 + 2a - 8}$

10. $\frac{6m^2 - m - 2}{8m^2 + 10m + 3} \cdot \frac{4m^2 + 7m + 3}{3m^2 + 5m + 2}$

11. $\frac{5a^2 + 7a - 6}{2a^2 + 3a - 2} \div \frac{5a^2 + 17a - 12}{2a^2 + 5a - 3}$

12. $\dfrac{1 + \frac{1}{3}}{1 - \frac{1}{3}}$

13. $\dfrac{\frac{3}{x}}{\frac{1}{1+x}}$

Solve the following equations

14. $\frac{1}{8} = \frac{x}{12}$

15. $\frac{3}{t-1} + \frac{1}{t+1} = \frac{6}{5}$

For each problem, write an equation, and then solve it.

16. Eight is what percent of five?

17. If four times a number is added to the reciprocal of twice the number, the result is 3. Find the number.

18. If the numerator of $3/x$ is decreased by x, the result is 2/3. Find x.

19. Harry can paint his house, working alone, in five hours. His wife Gertie can do it in four hours. How long will it take them working together?

20. When Jane walks to school, a distance of three miles, it takes her three-fifths of an hour. If she can travel three times as fast on her bike, how long does it take her to get to school when she rides?

6 LINEAR EQUATIONS

6.1 LINEAR EQUATIONS IN TWO VARIABLES

In a faraway city a taxi driver, who is a big fan of mathematics, charges riders by the rule

$$y = 25x + 50$$

where y is the price in cents of a taxi ride of x miles. For example, to find the price of a ride of 5 miles, the driver substitutes 5 for x in the equation $y = 25x + 50$.

$$\begin{aligned} y &= 25x + 50 & & \\ y &= 25(5) + 50 & &\text{Let } x = 5 \\ y &= 125 + 50 & & \\ y &= 175. & &\text{If } x = 5\text{, then } y = 175 \end{aligned}$$

A taxi ride of 5 miles costs 175¢, which is \$1.75.

A ride of $x = 8$ miles costs

$$\begin{aligned} y &= 25(8) + 50 & &\text{Let } x = 8 \\ y &= 250. & &\text{If } x = 8\text{, then } y = 250 \end{aligned}$$

A taxi ride of 8 miles costs 250¢ (\$2.50). Check that a ride of 11 miles costs \$3.25.

It is not convenient to write a solution in the form

$$\text{if } x = 11\text{, then } y = 325$$

Instead, we write

$$(11, 325).$$

The abbreviation gives the x-value and the y-value as a pair of numbers, enclosed by parentheses. The x-value is written first. A pair of numbers written in this order is called an **ordered pair**. (If other letters are used, the values are usually written inside the parentheses in alphabetical order.)

Example 1 Write solutions to the taxi price equation as ordered pairs.

	Solution	*Ordered Pair*
(a)	If $x = 5$, then $y = 175$.	$(5, 175)$
(b)	If $x = 8$, then $y = 250$.	$(8, 250)$
(c)	If $x = 11$, then $y = 325$.	$(11, 325)$

Other ordered pairs can be found that represent solutions of $y = 25x + 50$. Let $x = 0$, $x = 1$, $x = 2$, and so on. The corresponding ordered pairs are $(0, 50)$, $(1, 75)$, $(2, 100)$, and so on. In fact, there is an infinite number of ordered pairs for this equation. Each ordered pair whose x and y values make a given equation true is a **solution** of the equation. Such an ordered pair is said to **satisfy** the equation.

Example 2 Decide in each case whether the given equation is satisfied by the given ordered pair.

(a) $2x + 3y = 12$; $(3, 2)$.

To see whether or not the ordered pair $(3, 2)$ satisfies the equation $2x + 3y = 12$, substitute 3 for x and 2 for y in the given equation.

$$2x + 3y = 12$$
$$2(3) + 3(2) = 12 \qquad \text{Let } x = 3; \text{ let } y = 2$$
$$6 + 6 = 12$$
$$12 = 12. \qquad \text{True}$$

This result is true, so that $(3, 2)$ satisfies the equation $2x + 3y = 12$.

(b) $2x + 3y = 12$; $(-6, 8)$.

Substitute -6 for x and 8 for y.

$$2x + 3y = 12$$
$$2(-6) + 3(8) = 12 \qquad \text{Let } x = -6; \text{ let } y = 8$$
$$-12 + 24 = 12$$
$$12 = 12. \qquad \text{True}$$

Since this result is true, $(-6, 8)$ satisfies $2x + 3y = 12$.

(c) $2x + 3y = 12$; $(-2, -7)$.

$$2x + 3y = 12$$
$$2(-2) + 3(-7) = 12 \qquad \text{Let } x = -2; \text{ let } y = -7$$
$$-4 - 21 = 12$$
$$-25 = 12. \qquad \text{False}$$

This result is false, and $(-2, -7)$ does *not* satisfy the equation $2x + 3y = 12$.

Most equations with two variables are satisfied by an infinite number of ordered pairs. There is no completely satisfactory way to write the set of all solutions of such equations. However, we can draw a graph of the solutions, as shown in the next section.

Example 3 Given the equation $5x - y = 24$, complete the ordered pairs.

Equation	*Ordered pairs*
$5x - y = 24$	$(5, \quad)$
	$(-3, \quad)$
	$(0, \quad)$.

To find the y-value for the first ordered pair, replace x with 5 in the equation, and solve the resulting equation for y.

$$\begin{aligned} 5x - y &= 24 \\ 5(5) - y &= 24 && \text{Let } x = 5 \\ 25 - y &= 24 \\ -y &= -1 \\ y &= 1. && \text{Value of } y \text{ when } x = 5. \end{aligned}$$

The correct ordered pair is $(5, 1)$.

We can complete the ordered pair $(-3, \quad)$ by letting $x = -3$ in the equation. To complete $(0, \quad)$, let $x = 0$.

$$\begin{aligned} &\text{If} & x &= -3, \\ &\text{then} & 5x - y &= 24 \\ &\text{becomes} & 5(-3) - y &= 24 \\ && -15 - y &= 24 \\ && -y &= 39 \\ && y &= -39. \end{aligned} \qquad \begin{aligned} &\text{If} & x &= 0, \\ &\text{then} & 5x - y &= 24 \\ &\text{becomes} & 5(0) - y &= 24 \\ && 0 - y &= 24 \\ && -y &= 24 \\ && y &= -24. \end{aligned}$$

Thus, the completed ordered pairs are

Equation	*Ordered pairs*
$5x - y = 24$	$(5, 1)$
	$(-3, -39)$
	$(0, -24)$.

Example 4 Complete the following ordered pairs.

Equation	*Ordered pairs*	
$x - 2y = 8$	$(2, \quad)$	$(\quad , 0)$
	$(10, \quad)$	$(\quad , -2)$.

Complete the two ordered pairs on the left by letting $x = 2$ and $x = 10$, respectively.

$$\begin{aligned} &\text{If} & x &= 2, \\ &\text{then} & x - 2y &= 8 \\ &\text{becomes} & 2 - 2y &= 8 \\ && -2y &= 6 \\ && y &= -3. \end{aligned} \qquad \begin{aligned} &\text{If} & x &= 10, \\ &\text{then} & x - 2y &= 8 \\ &\text{becomes} & 10 - 2y &= 8 \\ && -2y &= -2 \\ && y &= 1. \end{aligned}$$

To finish the problem, let $y = 0$ and $y = -2$.

If $y = 0$,	If $y = -2$,
then $x - 2y = 8$	then $x - 2y = 8$
becomes $x - 2(0) = 8$	becomes $x - 2(-2) = 8$
$x - 0 = 8$	$x + 4 = 8$
$x = 8$.	$x = 4$.

The completed ordered pairs are as follows.

Equation	*Ordered pairs*	
$x - 2y = 8$	(2, −3)	(8, 0)
	(10, 1)	(4, −2).

Example 5 Complete the following ordered pairs.

Equation	*Ordered pairs*		
$x = 5$	(, −2)	(, 6)	(, 3).

The equation we are given here is $x = 5$. Therefore, no matter which value of y we might choose, we always have the same value of x, namely, 5. Therefore, each ordered pair can be completed by placing 5 in the first position.

Equation	*Ordered pairs*		
$x = 5$	(5, −2)	(5, 6)	(5, 3).

When an equation such as $x = 5$ is discussed along with equations of two variables, it is customary to think of $x = 5$ as an equation in two variables by rewriting $x = 5$ as $x + 0 = 5$, or $x + 0 \cdot y = 5$. This last form shows that for any value of y, we have $x = 5$.

The equations we worked with in this section all fit the pattern

$$ax + by = c,$$

where a, b, and c are real numbers. Such an equation is called a **linear equation in two variables.** Examples of linear equations in two variables include

$$2x + 3y = 12 \qquad y = 3x - 5 \qquad x = 5 \;\; (x + 0y = 5).$$

EXERCISES

In Exercises 1–12, decide whether or not the given ordered pair satisfies the given equation.

1. $x + y = 9$; (2, 7)

2. $3x + y = 8$; (0, 8)

3. $2x - y = 6$; (2, −2)

4. $2x + y = 5$; (2, 1)

5. $4x - 3y = 6$; $(1, 2)$

6. $5x - 3y = 1$; $(0, 1)$

7. $y = 3x$; $(1, 3)$

8. $x = -4y$; $(8, -2)$

9. $x = -6$; $(-6, 8)$

10. $y = 2$; $(9, 2)$

11. $x + 4 = 0$; $(-5, 1)$

12. $x - 6 = 0$; $(5, -1)$

In Exercises 13–32, complete the given ordered pairs.

Equation	*Ordered pairs*		
13. $y = 2x + 1$	$(3, \quad)$	$(0, \quad)$	$(-1, \quad)$
14. $y = 3x - 5$	$(2, \quad)$	$(0, \quad)$	$(-3, \quad)$
15. $y = 8 - 3x$	$(2, \quad)$	$(0, \quad)$	$(-3, \quad)$
16. $y = -2 - 5x$	$(4, \quad)$	$(0, \quad)$	$(-4, \quad)$
17. $2x + y = 9$	$(0, \quad)$	$(3, \quad)$	$(12, \quad)$
18. $-3x + y = 4$	$(1, \quad)$	$(0, \quad)$	$(-2, \quad)$
19. $2x + 3y = 6$	$(0, \quad)$	$(\quad, 0)$	$(\quad, 4)$
20. $4x + 3y = 12$	$(0, \quad)$	$(\quad, 0)$	$(\quad, 8)$
21. $3x - 5y = 15$	$(0, \quad)$	$(\quad, 0)$	$(\quad, -6)$
22. $4x - 9y = 36$	$(\quad, 0)$	$(0, \quad)$	$(\quad, 4)$
23. $x = -4$	$(\quad, 6)$	$(\quad, 2)$	$(\quad, -3)$
24. $x = 8$	$(\quad, 3)$	$(\quad, 8)$	$(\quad, 0)$
25. $y = 3$	$(8, \quad)$	$(4, \quad)$	$(-2, \quad)$
26. $y = -8$	$(4, \quad)$	$(0, \quad)$	$(-4, \quad)$
27. $x + 9 = 0$	$(\quad, 8)$	$(\quad, 3)$	$(\quad, 0)$
28. $y + 4 = 0$	$(9, \quad)$	$(2, \quad)$	$(0, \quad)$
29. $4x + 5y = 10$	$(0, \quad)$	$(\quad, 0)$	$(\quad, 3)$
30. $2x - 3y = 4$	$(\quad, 0)$	$(0, \quad)$	$(\quad, 3)$
31. $6x - 4y = 5$	$(0, \quad)$	$(\quad, 0)$	$(2, \quad)$
32. $4x - 3y = 7$	$(\quad, 0)$	$(2, \quad)$	$(\quad, -1)$

6.2 GRAPHING ORDERED PAIRS

When we graph the solution of a linear equation in *one* variable, we locate a point on a number line. The coordinate of the point is the value of x that satisfies the equation.

Now we want to graph the solutions of a linear equation in *two* variables. The solutions are ordered pairs in the form (x, y). We have to take values of both x and y into account. We need two number lines, one for each variable. The two number lines are placed at right angles, crossing at the 0 point on each line. (See Figure 6.1.)

The horizontal number line is called the x-**axis.** The vertical line is called the y-**axis.** Together, the x-axis and y-axis are called the **coordinate axes.** They form a **coordinate system** in the plane. (A plane is a flat surface—like this page.) A coordinate system is like the map of a town showing the grid of streets. The coordinate axes divide the plane into four **quadrants.** These are numbered in Figure 6.1, counterclockwise. The axes are not part of any quadrant. The coordinate axes cross at the zero point of the two number lines. This point is called the **origin.** It represents the ordered pair (0, 0), where the x and y values are both zero.

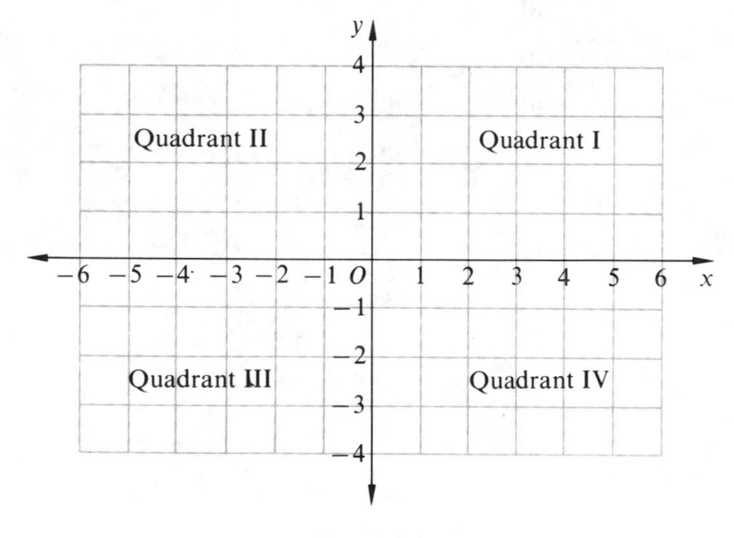

Figure 6-1

We can graph any ordered pair (x, y) on the coordinate system by locating the x and y coordinates of a point. This is called **plotting** the point. To plot the point representing the ordered pair (2, 3), start at the origin. Go 2 units (x-value) to the right, along the x-axis. Then, turn and go up 3 units (y-value), on a line parallel to the y-axis. This locates the point representing the ordered pair (2, 3). The point is shown in Figure 6.2. From now on, the statement "The point representing the ordered pair (2, 3)" will be abbreviated as simply "The point (2, 3)."

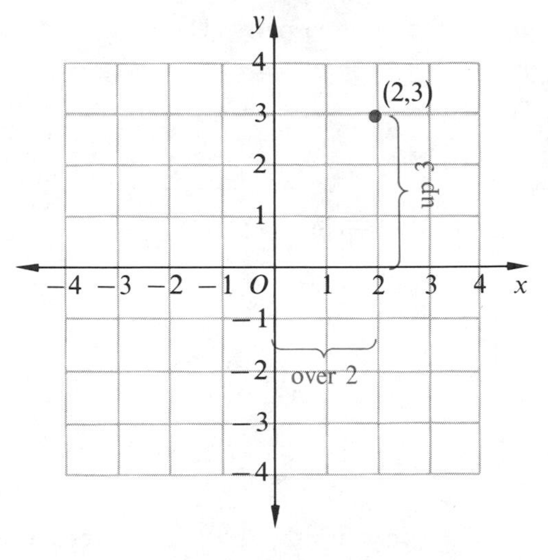

Figure 6-2

Example 1 Plot the following points on a coordinate system.

(a) $(1, 5)$ **(d)** $(7, -2)$

(b) $(-2, 3)$ **(e)** $(3/2, 2)$.

(c) $(-1, -4)$

To locate the point $(-1, -4)$, for example, go 1 unit to the *left* along the x-axis. Then turn and go 4 units *down*, parallel to the y-axis. To plot the point $(3/2, 2)$, go $3/2$ (or $1\frac{1}{2}$) units to the right along the x-axis. Then turn and go 2 units up. Figure 6.3 shows the graphs of the points in this example.

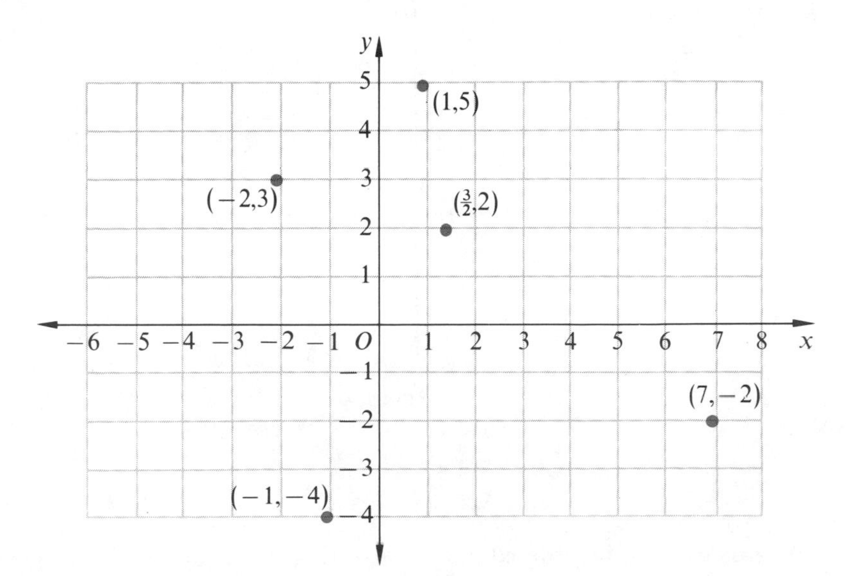

Figure 6-3

Example 2 Complete the ordered pairs. Then graph the ordered pairs on a coordinate system.

Equation	*Ordered pairs*	
$x + 2y = 7$	(1,)	(3,)
	$(-3,$)	(7,).

To complete the ordered pairs, substitute the given x-values into the equation $x + 2y = 7$.

If	$x = 1,$	If	$x = -3,$
then	$x + 2y = 7$	then	$x + 2y = 7$
becomes	$1 + 2y = 7$	becomes	$-3 + 2y = 7$
	$2y = 6$		$2y = 10$
	$y = 3.$		$y = 5.$

In the same way, if $x = 3$ then $y = 2$; and if $x = 7$ then $y = 0$.

Equation	*Ordered pairs*	
$x + 2y = 7$	(1, 3)	(3, 2)
	$(-3, 5)$	(7, 0).

The graph of these ordered pairs is in Figure 6.4.

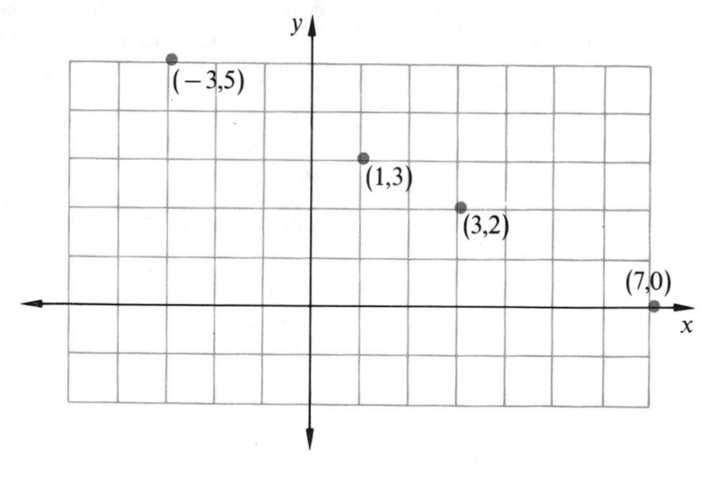

Figure 6-4

Example 3 Some ordered pairs that satisfy the equation $3x - 2y = 6$ are given below. Graph the ordered pairs on a coordinate system.

Equation	*Ordered pairs*	
$3x - 2y = 6$	$(0, -3)$	$(4, 3)$
	$(2, 0)$	$(6, 6)$
	$(-2, -6)$	$(-3, -15/2)$
	$(-4, -9)$	$(1, -3/2)$.

The graph of these ordered pairs is in Figure 6.5.

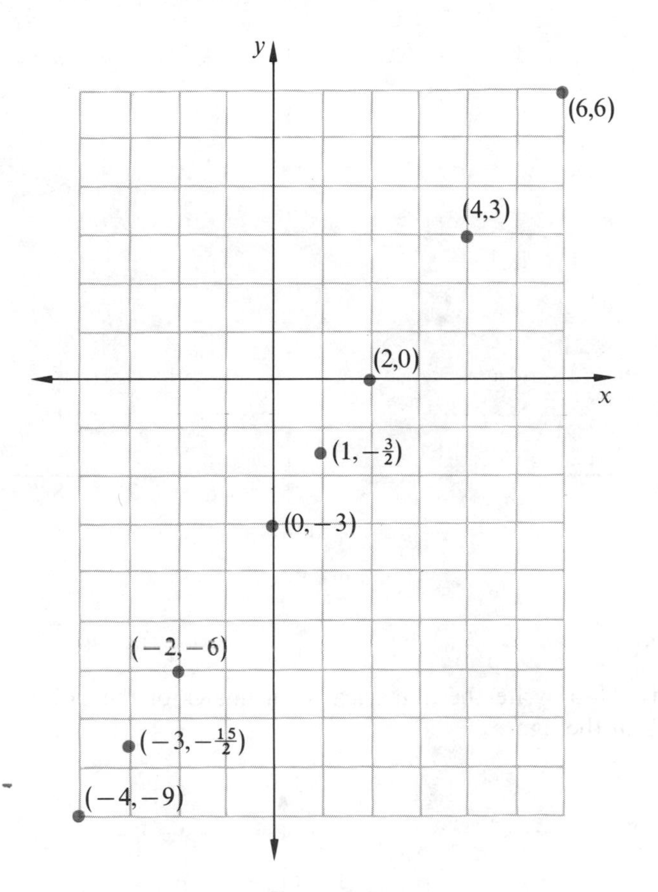

Figure 6-5

Example 4 At the beginning of this chapter we mentioned the taxi driver who charges riders by the rule

$$y = 25x + 50,$$

where y represents the cost in cents of a ride of x miles. Some ordered pairs satisfying this equation are listed.

(1, 75)	(4, 150)	(7, 225)
(2, 100)	(5, 175)	(8, 250)
(3, 125)	(6, 200)	(9, 275).

The above ordered pairs are graphed in Figure 6.6. The y-axis has a shorter unit measure than the x-axis in order to fit the figure better on the page.

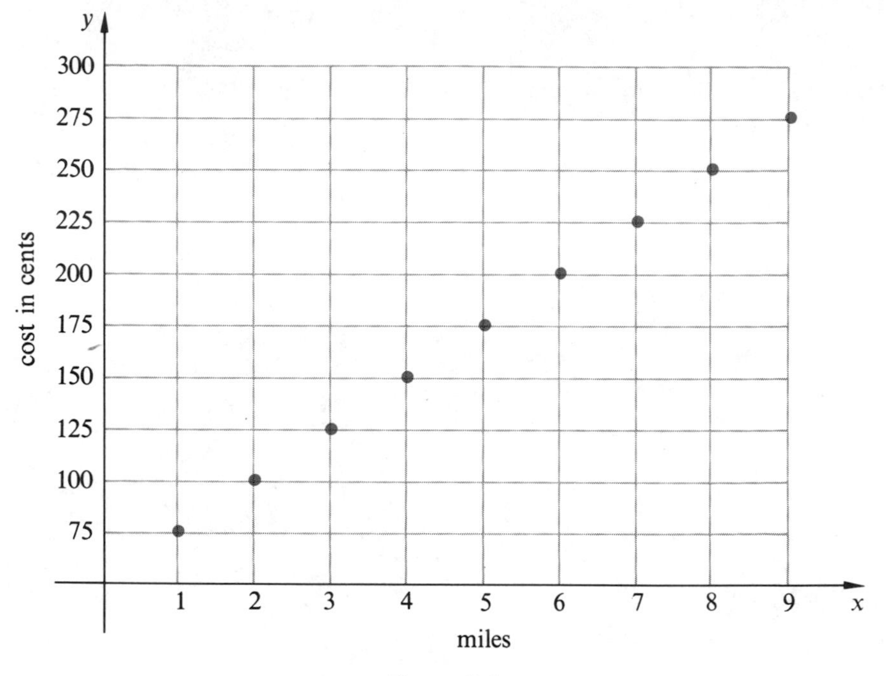

Figure 6-6

EXERCISES

In Exercises 1–6, write the numerical coordinates of the points labeled A through F in the figure.

1. A

2. B

3. C

4. D

5. E

6. F

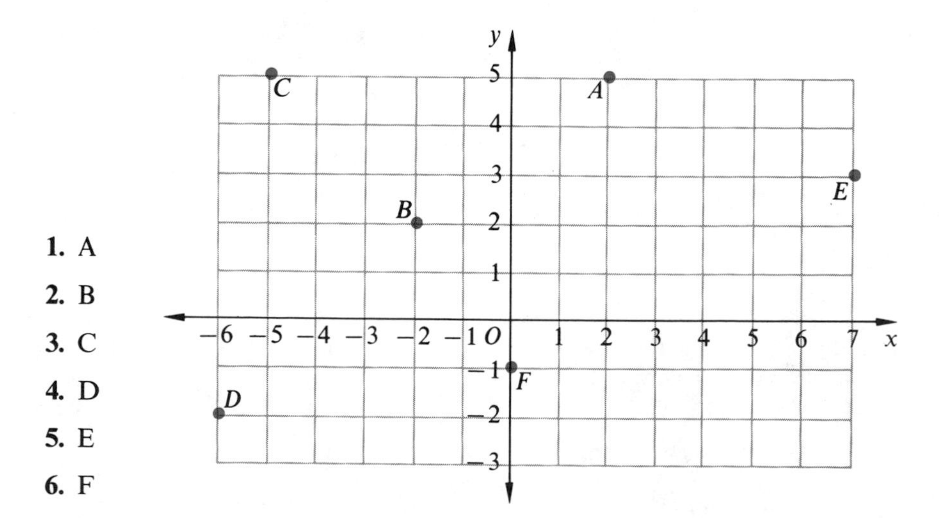

For Exercises 7–22, plot the given points on the same coordinate system.

7. $(6, 1)$

8. $(4, -2)$

9. $(3, 5)$

10. $(-4, -5)$

11. $(-2, 4)$

12. $(-5, -1)$

13. $(-3, 5)$

14. $(3, -5)$

15. $(4, 0)$

16. $(1, 0)$

17. $(-2, 0)$

18. $(-5, 0)$

19. $(0, 3)$

20. $(0, 6)$

21. $(0, -5)$

22. $(0, 0)$

Without plotting, state the quadrant in which each point of Exercises 23–34 will be found.

23. $(2, 3)$

24. $(2, -3)$

25. $(-2, 3)$

26. $(-2, -3)$

27. $(-1, -1)$

28. $(4, 7)$

29. $(-3, 6)$

30. $(1, 5)$

31. $(5, -4)$

32. $(9, -1)$

33. $(0, 0)$

34. $(-2, 0)$

In Exercises 35–44, complete the given ordered pairs. Then graph the ordered pairs.

Equation	*Ordered pairs*	
35. $y = 2x + 6$	$(0, \quad)$	$(2, \quad)$
	$(\quad, 0)$	$(\quad, 2)$
36. $y = 8 - 4x$	$(0, \quad)$	$(3, \quad)$
	$(\quad, 0)$	$(\quad, 16)$
37. $3x + 5y = 15$	$(0, \quad)$	$(10, \quad)$
	$(\quad, 0)$	$(\quad, 6)$
38. $2x - 5y = 10$	$(0, \quad)$	$(10, \quad)$
	$(\quad, 0)$	$(\quad, -6)$
39. $y = 3x$	$(0, \quad)$	$(-2, \quad)$
	$(4, \quad)$	$(\quad, -3)$
40. $x + 2y = 0$	$(0, \quad)$	$(\quad, 3)$
	$(4, \quad)$	$(\quad, -1)$
41. $x = 3$	$(\quad, 2)$	$(\quad, 5)$
	$(\quad, 0)$	$(\quad, -3)$
42. $y = -5$	$(2, \quad)$	$(-3, \quad)$
	$(0, \quad)$	$(-1, \quad)$
43. $y + 2 = 0$	$(5, \quad)$	$(0, \quad)$
	$(-3, \quad)$	$(-2, \quad)$
44. $x - 4 = 0$	$(\quad, 7)$	$(\quad, 0)$
	$(\quad, -4)$	$(\quad, 4)$

6.3 GRAPHING LINEAR EQUATIONS

As we have seen, there are an infinite number of ordered pairs which are solutions of a linear equation such as $x + 2y = 7$. For example, if $x = 1$, we can find a value for y by substituting 1 for x in the equation.

$$\begin{aligned} x + 2y &= 7 \\ 1 + 2y &= 7 \qquad \text{Let } x = 1 \\ 2y &= 6 \\ y &= 3. \qquad \text{Value of } y \text{ when } x = 1 \end{aligned}$$

Thus the ordered pair (1, 3) is one solution to the given equation. This pair and some others that satisfy the equation $x + 2y = 7$ are listed below. These pairs are plotted in Figure 6.7.

Equation	*Ordered pairs*			
$x + 2y = 7$	$(-3, 5)$	$(1, 3)$	$(6, 1/2)$	$(7, 0)$
	$(-1, 4)$	$(3, 2)$	$(0, 7/2)$	$(-2, 9/2)$.

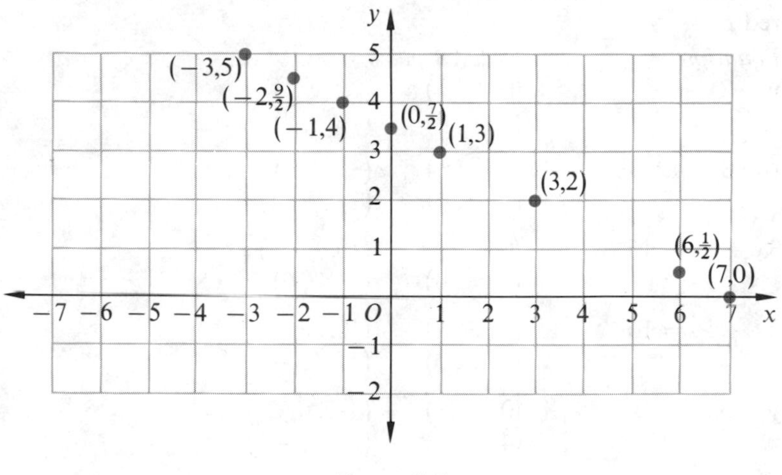

Figure 6-7

The points plotted in Figure 6.7 seem to lie on a straight line. In fact, all the points whose coordinates satisfy $x + 2y = 7$ do lie on a straight line. Thus the graph of all solutions of $x + 2y = 7$ is a line of infinite length. A portion of this line is shown in Figure 6.8. (A portion is all we can show, but remember that the line goes on forever in both directions.)

The graph of any linear equation in two variables is a straight line. A straight line is completely determined if we know two different points that lie on the line. Knowing these facts, we need to plot only a few points representing ordered pairs that satisfy a given linear equation. We graph the equation by drawing a line through these points. It is a good idea to find at least

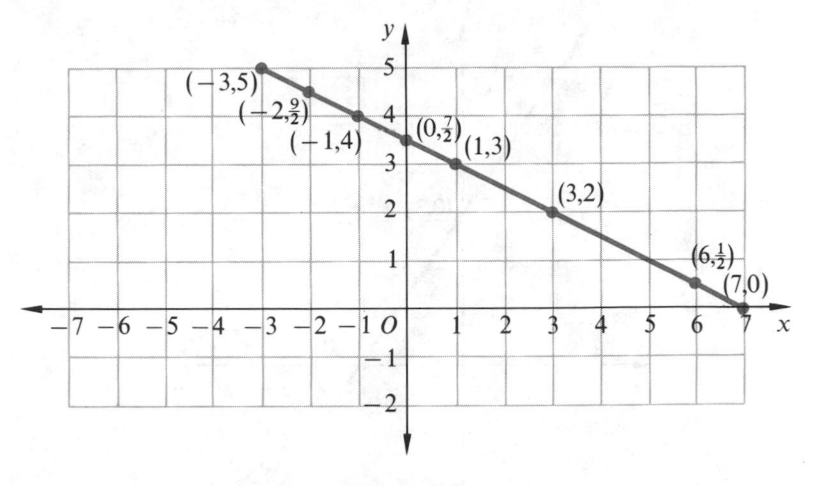

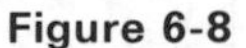
Figure 6-8

three points on the line. The third point serves as a check. Two of the three points can usually be found by letting $x = 0$ and by letting $y = 0$. To get the third point, let x or y equal some other number.

Example 1 Graph $3x + 2y = 6$.

We need to find three points representing ordered pairs that satisfy the equation $3x + 2y = 6$. For two of the points, let $x = 0$ and then let $y = 0$.

If	$x = 0,$	If	$y = 0,$
then	$3x + 2y = 6$	then	$3x + 2y = 6$
becomes	$3(0) + 2y = 6$	becomes	$3x + 2(0) = 6$
	$0 + 2y = 6$		$3x + 0 = 6$
	$2y = 6$		$3x = 6$
	$y = 3.$		$x = 2.$
Ordered pair:	$(0, 3)$.	Ordered pair:	$(2, 0)$.

To get the third point, let x or y equal some other number. For example, let $x = -2$. Check that y is then equal to 6. This gives the ordered pair $(-2, 6)$. Plot the three ordered pairs $(0, 3)$, $(2, 0)$, and $(-2, 6)$, as in Figure 6.9 (over). Draw a straight line through the three points. This straight line is the graph of $3x + 2y = 6$.

Example 2 Graph $4x - 5y = 20$.

Let $x = 0$. Then $y = -4$, giving the ordered pair $(0, -4)$. Also, if $y = 0$, then $x = 5$, which leads to $(5, 0)$. Let $y = 2$, say, to get the third point. To find x, substitute 2 for y in the equation $4x - 5y = 20$. Check that $x = 15/2$. The ordered pairs $(15/2, 2)$, $(0, -4)$, and $(5, 0)$ are plotted in Figure 6.10. The line through these three points is the graph of $4x - 5y = 20$.

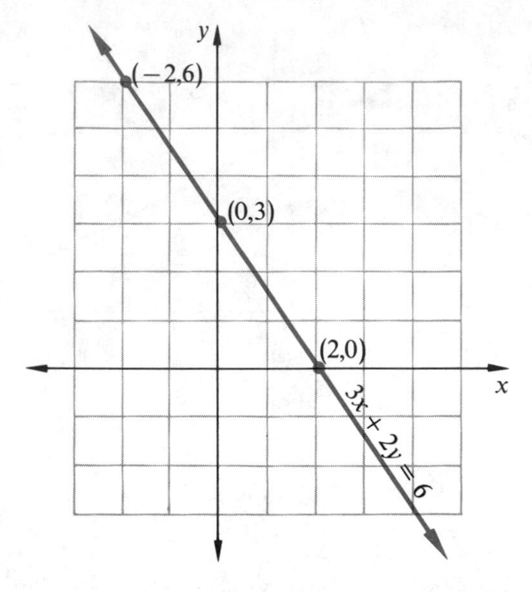

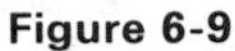
Figure 6-9

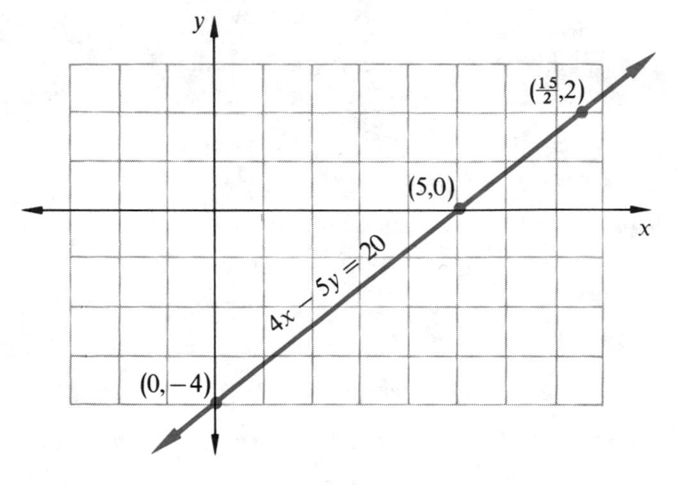

Figure 6-10

Example 3 Graph $x = 3y$.

Here, if we let $x = 0$, we get $y = 0$. Also, if $y = 0$ then $x = 0$. Both procedures lead to the same ordered pair, (0, 0). We need to find two more ordered pairs for the graph. We can do this by selecting other values for x or y. For example, if we let $y = 2$, then $x = 6$, which gives the ordered pair (6, 2). Another ordered pair is $(-6, -2)$. These ordered pairs were used to get the graph shown in Figure 6.11.

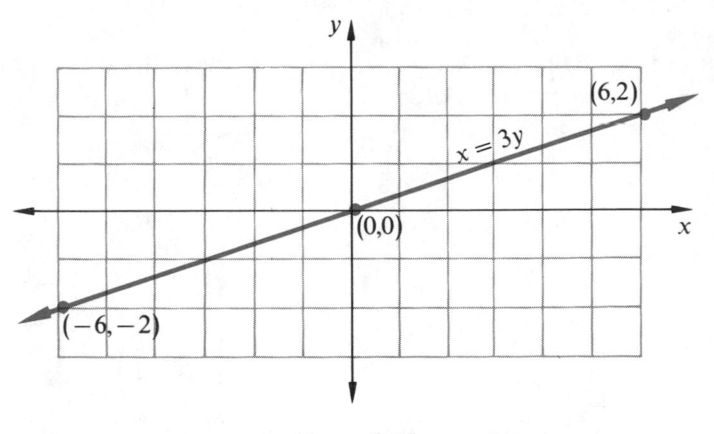

Figure 6-11

Example 4 Graph $y = -4$.

Think of $y = -4$ as $0 + y = -4$ or $0 \cdot x + y = -4$. For any value of x that we might choose, we always have $y = -4$. Some ordered pairs that satisfy $y = -4$ are $(-2, -4)$, $(0, -4)$, and $(3, -4)$. Figure 6.12 shows the graph of these ordered pairs, with a line through them. This line is parallel to the x-axis, and is horizontal.

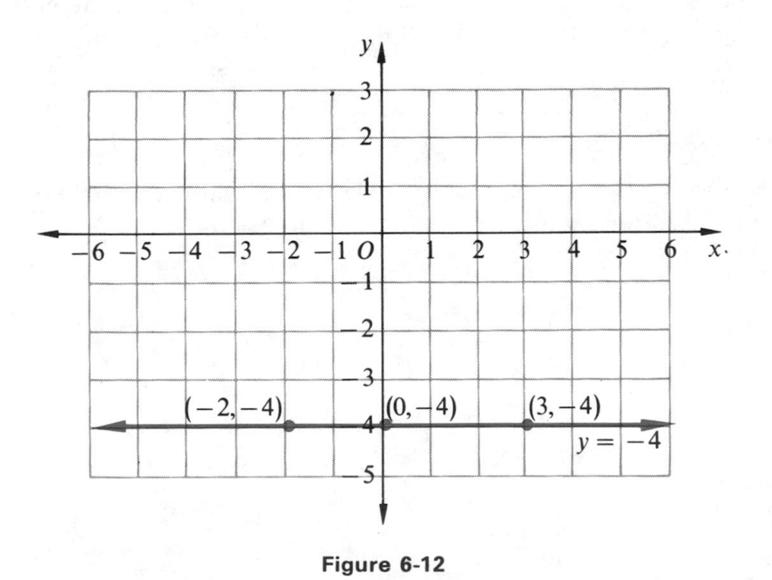

Figure 6-12

In general, the graph of a linear equation $y = k$, where k is a real number, is a horizontal line going through the point $(0, k)$.

Example 5 Graph $x = 3$.

All ordered pairs that satisfy this equation have an x-value of 3. To graph $x = 3$, we can use the ordered pairs (3, 3), (3, 0), and (3, −2). (See Figure 6.13.)

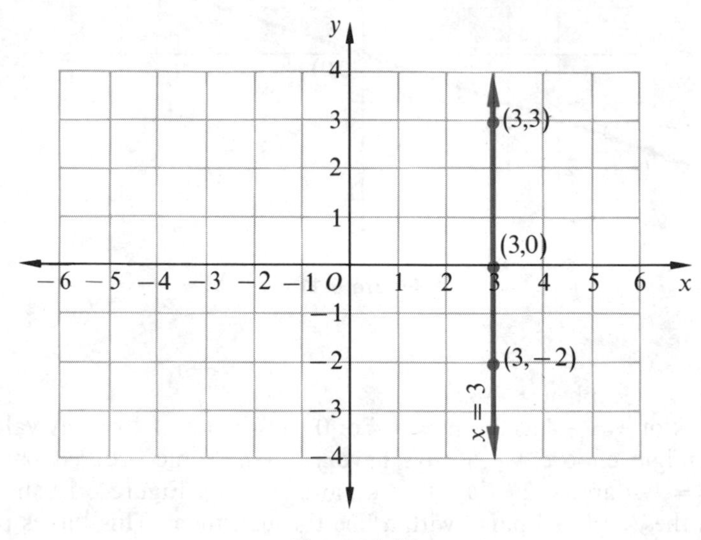

Figure 6-13

In general, every linear equation $x = k$, where k is a real number, has a vertical line for its graph. This vertical line goes through the point $(k, 0)$.

EXERCISES

In Exercises 1–10, first complete the ordered pairs for each equation. Then graph the equation, using the ordered pairs to plot points.

1. $x + y = 5$; (0,) (, 0) (2,)

2. $y = x - 3$; (0,) (, 0) (5,)

3. $y = x + 4$; (0,) (, 0) (−2,)

4. $y + 5 = x$; (0,) (, 0) (6,)

5. $y = 3x - 6$; (0,) (, 0) (3,)

6. $y = 5x - 10$; (0,) (, 0) (3,)

7. $2x + 5y = 20$; (0,) (, 0) (5,)

8. $3x - 4y = 12$; (0,) (, 0) (8,)

9. $x + 5 = 0$; (, 2) (, 0) (, −3)

10. $y - 4 = 0$; (3,) (0,) (−2,)

In Exercises 11–36, graph each of the linear equations.

11. $x - y = 2$

12. $x + y = 6$

13. $y = x + 2$

14. $y = x - 1$

15. $y = 2x - 4$

16. $y = 3x + 9$

17. $x = 3y - 12$

18. $x = 2y - 10$

19. $3x - 2y = 6$

20. $2x + 3y = 12$

21. $2x - 7y = 14$

22. $3x + 5y = 15$

23. $3x + 7y = 21$

24. $6x - 5y = 30$

25. $y = 2x$

26. $y = -3x$

27. $y + 6x = 0$

28. $y - 4x = 0$

29. $x + 2 = 0$

30. $y - 3 = 0$

31. $y = 6$

32. $x = 2$

33. $y = -1$

34. $x = 4$

35. $x = 0$

36. $y = 0.$

Translate each of the statements of Exercises 37–42 into an equation. Then graph the equation.

37. The x-value is 2 more than the y-value.

38. The y-value is 3 less than the x-value.

39. The y-value is 3 less than twice the x-value.

40. The x-value is 4 more than three times the y-value.

41. If 3 is added to the y-value, the result is 4 less than twice the x-value.

42. If 6 is subtracted from 4 times the y-value, the result is three times the x-value.

43. As a rough estimate, the weight of a man taller than about 60 inches is approximated by $y = 5.5x - 220$, where x is the height of the person in inches, and y is the weight in pounds. Estimate the weights of men whose heights are
(a) 62 inches
(b) 64 inches
(c) 68 inches
(d) 70 inches
(e) 72 inches
(f) Graph $y = 5.5x - 220$. Use only the numbers 62 through 76 on the x-axis.

6.4 GRAPHING LINEAR INEQUALITIES

In the preceding section, we discussed methods of graphing linear equations, such as $2x + 3y = 6$. In this section we shall extend this discussion to **linear inequalities**, such as $2x + 3y \le 6$. The symbol $\le$ is read "is less than or equal to," so that the points of the line $2x + 3y = 6$ satisfy the inequality $2x + 3y \le 6$. Thus the graph of $2x + 3y \le 6$ will include the graph of $2x + 3y = 6$.

The first step in graphing $2x + 3y \le 6$ is to graph the line $2x + 3y = 6$, as shown in Figure 6.14. This line divides the plane into three regions: one region below the line; one region above the line; and the third region the points of the line itself.

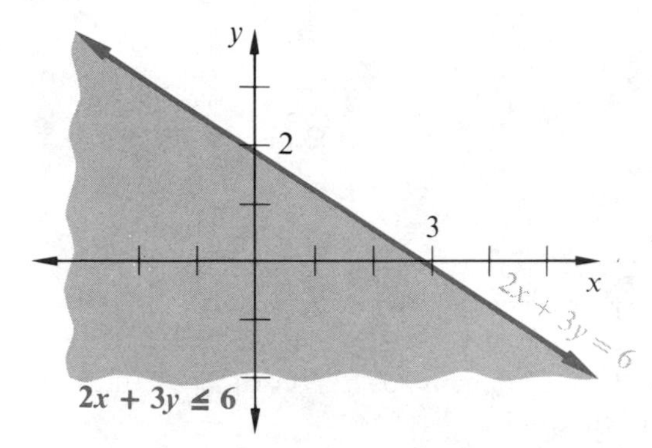

Figure 6-14

The graph of the linear inequality $2x + 3y \le 6$ includes either all the points below the line, or all the points above the line. To decide which, choose as a test point any point not on the line $2x + 3y = 6$. When it is not on the line, the origin, (0, 0), is a good choice. Substitute 0 for x and 0 for y in the *original inequality*, to see if the resulting statement is true or false. Here we have

$$2x + 3y \le 6$$

$$2(0) + 3(0) \le 6 \qquad \text{Let } x = 0, \text{ and } y = 0$$

$$0 + 0 \le 6$$

$$0 \le 6. \qquad \text{True}$$

The statement is true. Therefore, the graph of the inequality includes the region containing (0, 0). This region is shaded in Figure 6.14.

Example 1 Graph $x - y > 5$.

This time the inequality is restricted to "greater than." Thus the points of the line $x - y = 5$ do *not* satisfy the inequality. But we still need a boundary line to establish the two regions. One of the regions is the graph of $x - y > 5$.

Begin by graphing the line $x - y = 5$. Instead of a solid line, draw a dashed line. The dashed line in Figure 6.15 shows that the boundary line is *not* part of the graph of the inequality $x - y > 5$.

To see which side of the line to shade, choose the point (0, 0) as a test point. Substitute 0 for x and 0 for y in the inequality $x - y > 5$.

$$x - y > 5$$
$$0 - 0 > 5 \qquad \text{Let } x = 0 \text{ and } y = 0$$
$$0 > 5. \qquad \text{False}$$

The statement is false this time. Therefore, the graph of the inequality does *not* include the region containing (0, 0). The graph must include the other region. Shade the region below the line $x - y = 5$, as in Figure 6.15. The graph of the inequality $x - y > 5$ contains only the shaded region, and excludes the dashed line, which serves only as a boundary.

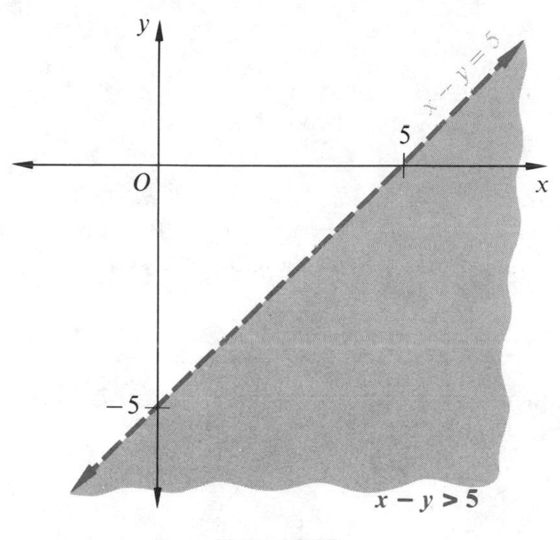

Figure 6-15

Example 2 Graph $2x - 5y \geq 10$.

First graph the straight line $2x - 5y = 10$ by finding three ordered pairs that satisfy the equation. This is the solid line in Figure 6.16. Choose any point not on the line $2x - 5y = 10$, say (0, 0), and substitute into the original inequality.

$$2x - 5y \geq 10$$
$$2(0) - 5(0) \geq 10 \qquad \text{Let } x = 0 \text{ and } y = 0$$
$$0 - 0 \geq 10$$
$$0 \geq 10. \qquad \text{False}$$

Thus (0, 0) is *not* part of the graph of $2x - 5y \geq 10$. Because of this, shade the region on the other side of the line, as in Figure 6.16. The line $2x - 5y = 10$ is part of the graph, along with the shaded region.

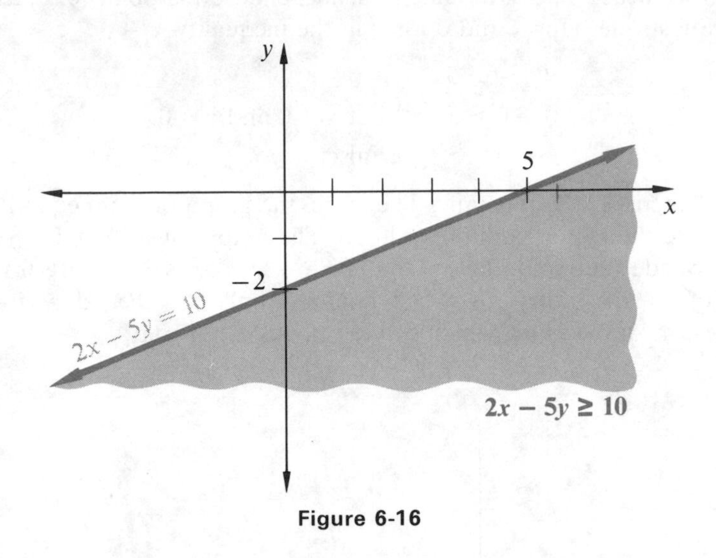

Figure 6-16

Example 3 Graph $x \leq 3$.

The line $x = 3$ is vertical, and goes through the point (3, 0). If we choose (0, 0) to substitute into the given inequality, we get

$$x \leq 3$$
$$0 \leq 3. \qquad \text{True}$$

Since the result is true, shade the region of the plane to the left of the line $x = 3$, as in Figure 6.17.

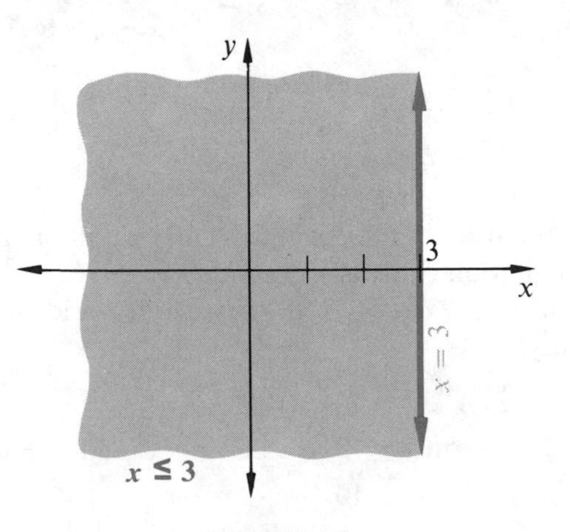

Figure 6-17

Example 4 Graph $x \leq 2y$.

First graph the line $x = 2y$, as in Figure 6.18. To test for the region, choose any point not on the line $x = 2y$, say (0, 3). We cannot choose (0, 0) in this case, since it is on the line $x = 2y$.

$$x \leq 2y$$

$$0 \leq 2(3) \qquad \text{Let } x = 0 \text{ and let } y = 3$$

$$0 \leq 6. \qquad \text{True}$$

The region including the point (0, 3) is shaded in Figure 6.18. The graph of $x \leq 2y$ includes this region as well as the boundary line $x = 2y$.

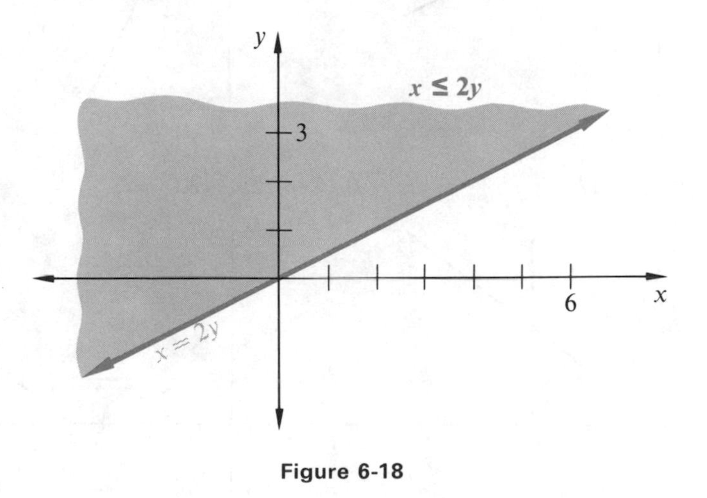

Figure 6-18

EXERCISES

In Exercises 1–10, the figure for each inequality shows the first step in graphing the inequality. The required straight line has been graphed. Complete each graph by shading the correct region.

1. $x + y \leq 4$

2. $x + y \geq 2$

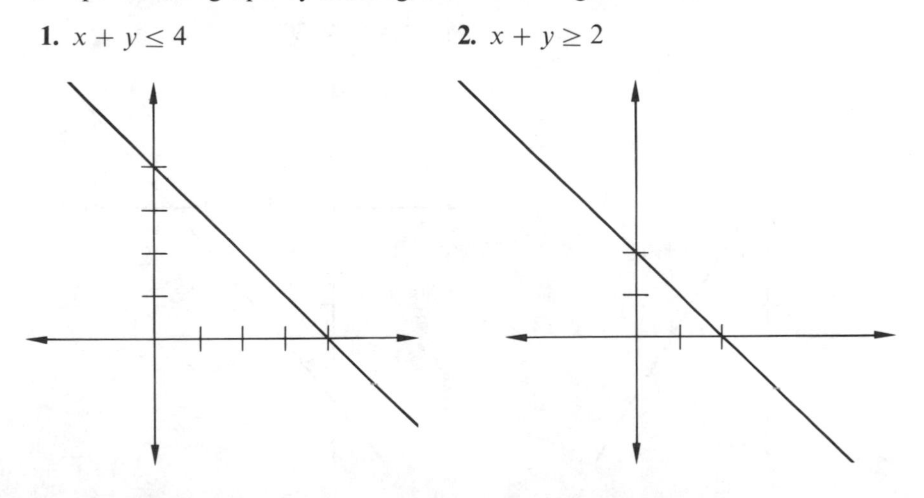

3. $x + 2y \leq 7$

4. $2x + y \leq 5$

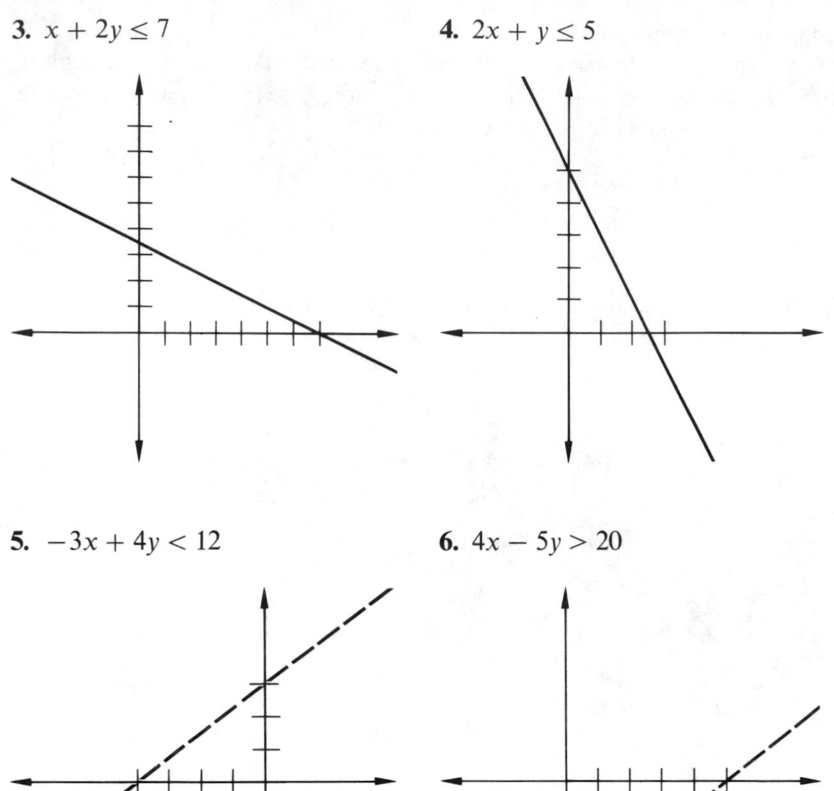

5. $-3x + 4y < 12$

6. $4x - 5y > 20$

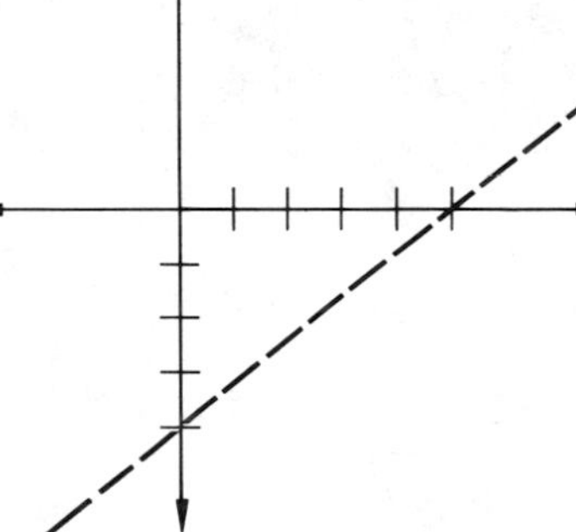

7. $5x + 3y > 15$

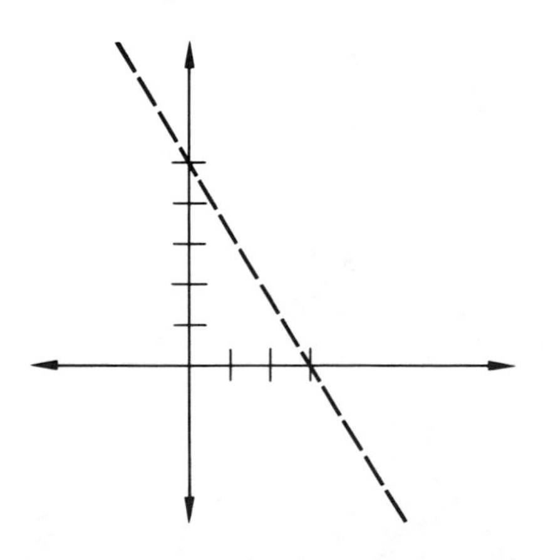

8. $6x - 5y < 30$

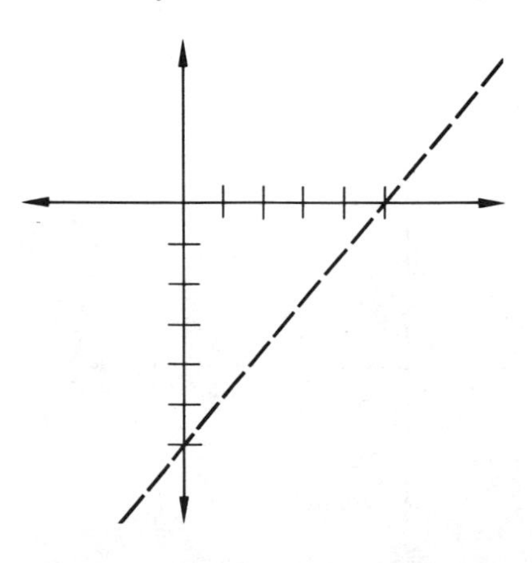

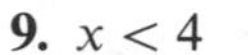

9. $x < 4$

10. $y > -1$

In Exercises 11–40, graph the inequalities.

11. $x + y \leq 8$

12. $x + y \geq 2$

13. $x - y \leq -2$

14. $x - y \leq 3$

15. $x + 2y \geq 4$

16. $x + 3y \leq 6$

17. $2x + 3y > 6$

18. $3x + 4y > 12$

19. $3x - 4y < 12$

20. $2x - 3y < -6$

21. $3x + 7y \geq 21$

22. $2x + 5y \geq 10$

23. $4x - 5y \geq 20$

24. $3x + 5y \leq 15$

25. $4x + 7y \leq 14$

26. $3x - 5y \leq 10$

27. $x \leq 3y$

28. $x \leq 5y$

29. $x \geq -2y$

30. $x > -4y$

31. $2x + 3y \geq 0$

32. $3x + 4y \geq 0$

33. $x < 4$

34. $x < -2$

35. $y \leq 2$

36. $y \leq -3$

37. $x \geq -2$

38. $x \geq 5$

39. $x \geq 0$

40. $y \geq 0$

6.5 RELATIONS AND FUNCTIONS (OPTIONAL)

An equation or inequality that describes a relationship between two variables is called a **relation.** The taxi driver at the beginning of this chapter described the relationship between the cost of a ride and the miles driven by a linear equation, $y = 25x + 50$.

The equation $y = 3x - 1$ is also a relation between two variables, x and y. The two variables are related such that y is 1 less than 3 times x.

There are an infinite number of ordered pairs that satisfy $y = 3x - 1$. Some of them are $(0, -1)$, $(2, 5)$, $(3, 8)$, $(-1, -4)$, and $(-2, -7)$. The relation $y = 3x - 1$ can be written as a set, $\{(x, y) | y = 3x - 1\}$, which reads "the set of all ordered pairs (x, y) such that $y = 3x - 1$."

In the relation $y = 3x - 1$, we could choose any value we like as a replacement for x. The set of all possible x-values is called the **domain** of the relation. The domain of $y = 3x - 1$ is the set of all real numbers.

The set of all possible y-values is called the **range** of the relation. Since y can also take on any value, the range of $y = 3x - 1$ is also the set of all real numbers.

In the relation $y = x^2$ the equation relating x and y is $y = x^2$. We can choose any value we want as a replacement for x. Thus the domain of $y = x^2$ is the set of all real numbers. However, the value of y (or of x^2) will never be negative. Therefore, the set of all possible y-values, the range of $y = x^2$, is restricted to the set of all non-negative real numbers.

Example 1 Find the domain and range of the relation $x = |y|$.

The symbol $|y|$ denotes the absolute value of y. For example, $|2| = 2$, $|-4| = 4$, and $|0| = 0$. Since the absolute value of a number is never negative, the domain is the set of all non-negative numbers. However, we can find the absolute value of any number. Thus the range is the set of all real numbers.

A **function** is a relation where each value of x yields exactly one value of y.

For example, the relation $y = 3x - 1$ is a function. If we choose a value for x, we will find exactly one value for y. On the other hand, the relation $x = |y|$ is not a function. If we choose $x = 5$, then $x = |y|$ becomes $5 = |y|$, which leads to two y-values, $y = 5$ and $y = -5$. Since one x-value results in two y-values, $x = |y|$ is not a function.

Example 2 Identify any of the following relations which are also functions.

(a) $y = x^2$.

If we choose a value of x, we will find exactly one value of y (by squaring our value of x). Thus, this relation is a function.

(b) $x = y^2$.

If we choose the x-value $x = 36$, we get *two* y-values, $y = 6$ and $y = -6$. For this reason, this relation is not a function.

It is common to use the letters f, g, and h to name functions. For example, suppose we use set symbols to write the function f:

$$f = \{(x, y) | y = 2x + 5\}.$$

For this function f, if $x = 5$, then $y = 2 \cdot 5 + 5 = 15$. This result is often written

$$f(5) = 15,$$

read "f of 5 equals 15." Also for the function f, if $x = -3$, then $y = -1$. Thus,

$$f(-3) = -1.$$

In the same way, $f(0)=5$, $f(2)=9$, $f(-5)=-5$, $f(8)=21$, and so on.

The function notation $f = \{(x, y) | y = 2x + 5\}$ is often abbreviated as

$$f(x) = 2x + 5$$

or as

$$y = 2x + 5.$$

Example 3 For the function $f(x) = x^2 - 3$, find **(a)** $f(2)$ **(b)** $f(0)$ **(c)** $f(-3)$.

Using the definition of $f(x)$, we can find the necessary values as follows.

(a) To find $f(2)$, substitute 2 for x. We have

$$f(x) = x^2 - 3$$
$$f(2) = 2^2 - 3 \qquad \text{Let } x = 2$$
$$f(2) = 4 - 3$$
$$f(2) = 1.$$

(b) In the same way, $f(0) = 0^2 - 3 = 0 - 3 = -3$.

(c) $f(-3) = (-3)^2 - 3 = 9 - 3 = 6$.

Example 4 Let $P(x) = 5x^2 - 4x + 3$. Find **(a)** $P(0)$ **(b)** $P(-2)$ **(c)** $P(3)$.

The letter P is used to show that this function is a polynomial.

(a) To find $P(0)$, substitute 0 for x.

$$P(x) = 5x^2 - 4x + 3$$
$$P(0) = 5 \cdot 0^2 - 4 \cdot 0 + 3 \qquad \text{Let } x = 0$$
$$P(0) = 3.$$

(b) Let $x = -2$.

$$P(-2) = 5 \cdot (-2)^2 - 4(-2) + 3$$
$$P(-2) = 20 + 8 + 3$$
$$P(-2) = 31.$$

(c) $P(3) = 5 \cdot 3^2 - 4 \cdot 3 + 3 = 36$.

Example 5 Define a function f as follows: Let $f(x)$ be the total charge for the rental of a chain saw, where this charge is \$2 per hour, or any fraction of an hour. For example, the charge for a four-hour rental is \$8, and the charge for a rental of 4.1 hours is \$10. Graph this function.

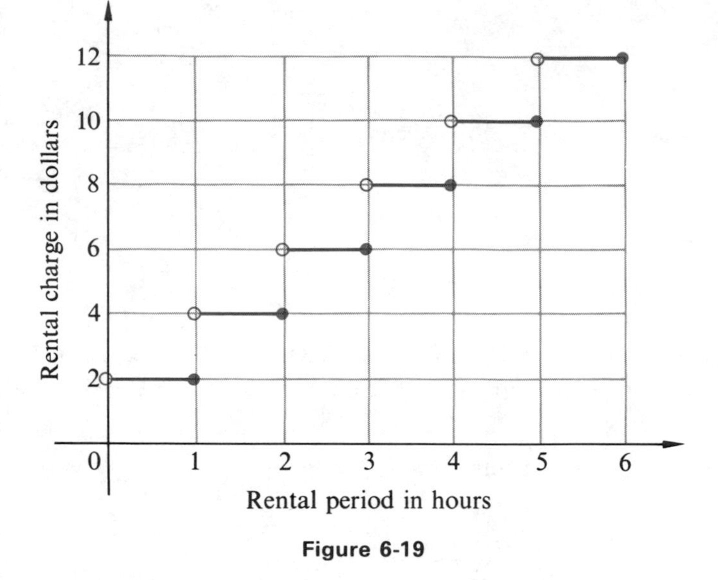

Figure 6-19

The charge for a rental of 0 to 1 hour is \$2, while the charge is \$4 if the rental is for more than 1 hour, but not more than 2 hours, and so on. By finding the rental charges for various numbers of hours, we get the graph in Figure 6.19. The solid dots are used to show that the right-hand endpoints of the line segments are included as part of the graph. The graph shows that the range of the function is the set $\{2, 4, 6, 8, 10, \ldots\}$. This is an example of a function (or relation) that does not have an equation which defines the relationship between the variables.

If we are given the graph of a relation, the definition of a function can be used to determine whether or not the given relation is a function.

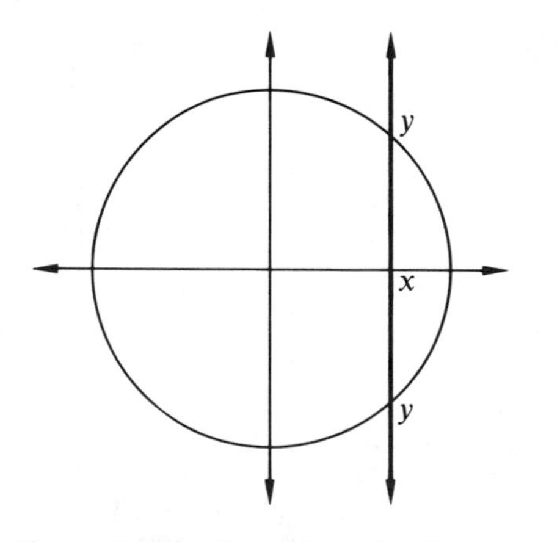

Figure 6-20A One value of x leads to two values of y

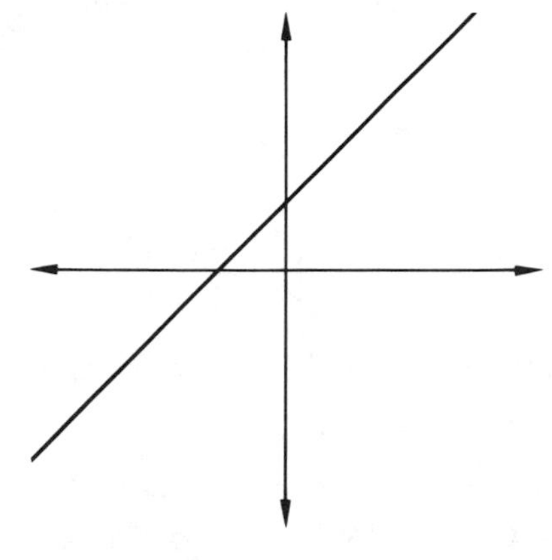

Figure 6-20B

By the definition of function, for each value of x there must be exactly one value of y. As shown in Figure 6.20A, there are two values of y for the indicated value of x, so that this graph is not the graph of a function. We can draw a vertical line that cuts the graph in more than one point.

On the other hand, in Figure 6.20B any vertical line will cut the graph in no more than one point. Thus, the graph in Figure 6.20B is the graph of a function.

We call this procedure the **vertical line test for a function.**

If any vertical line cuts the graph of a relation in more than one point, then the graph is not the graph of a function.

EXERCISES

In Exercises 1–12, state whether or not each relation is a function.

1. $y = 5x - 1$
2. $y = 4x + 5$
3. $2x + 3y = 6$
4. $4x - 3y = 12$
5. $y = x^2 + 3$
6. $y = 5 - x^2$
7. $x = y^2 - 4$
8. $x = y^2 + 6$
9. $x = |y + 2|$
10. $x = |y - 5|$
11. $2x + y < 6$
12. $3x - 4y > 2$

In Exercises 13–26, find the domain and range for each function.

13. $y = 2x + 5$
14. $y = 5x - 6$
15. $2x - y = 6$
16. $3x + 2y = 12$
17. $y = x^2 + 4$
18. $y = x^2 - 3$
19. $y = (x + 4)^2$
20. $y = (x - 2)^2$
21. $y = -(x - 3)^2$
22. $y = -(x + 5)^2$
23. $y = |x| + 2$
24. $y = |x| - 3$
25. $y = |x - 4|$
26. $y = |x + 3|$

In Exercises 27–40, find (a) $f(2)$ (b) $f(0)$ (c) $f(-3)$.

27. $f(x) = 3x + 2$
28. $f(x) = 4x - 1$
29. $f(x) = 4 - x$
30. $f(x) = 2 - 3x$
31. $f(x) = -4 - 4x$
32. $f(x) = -5 - 6x$
33. $f(x) = x^2 + 2$
34. $f(x) = x^2 - 5$
35. $f(x) = (x - 3)^2$
36. $f(x) = (x + 5)^2$
37. $f(x) = -(x + 2)^2$
38. $f(x) = -(x - 4)^2$
39. $f(x) = -|x + 2|$
40. $f(x) = -|x - 3|$

For each function in Exercises 41–46, find (a) $P(0)$ (b) $P(-3)$ (c) $P(2)$.

41. $P(x) = x^2 + 2x$

42. $P(x) = 2x^2 + 3x - 6$

43. $P(x) = -x^2 - 8x + 9$

44. $P(x) = -3x^2 + 4x - 2$

45. $P(x) = x^3 - 4x^2 + 1$

46. $P(x) = x^3 + 5x^2 - 11x + 2$

Write *function* or *not a function* for each graph.

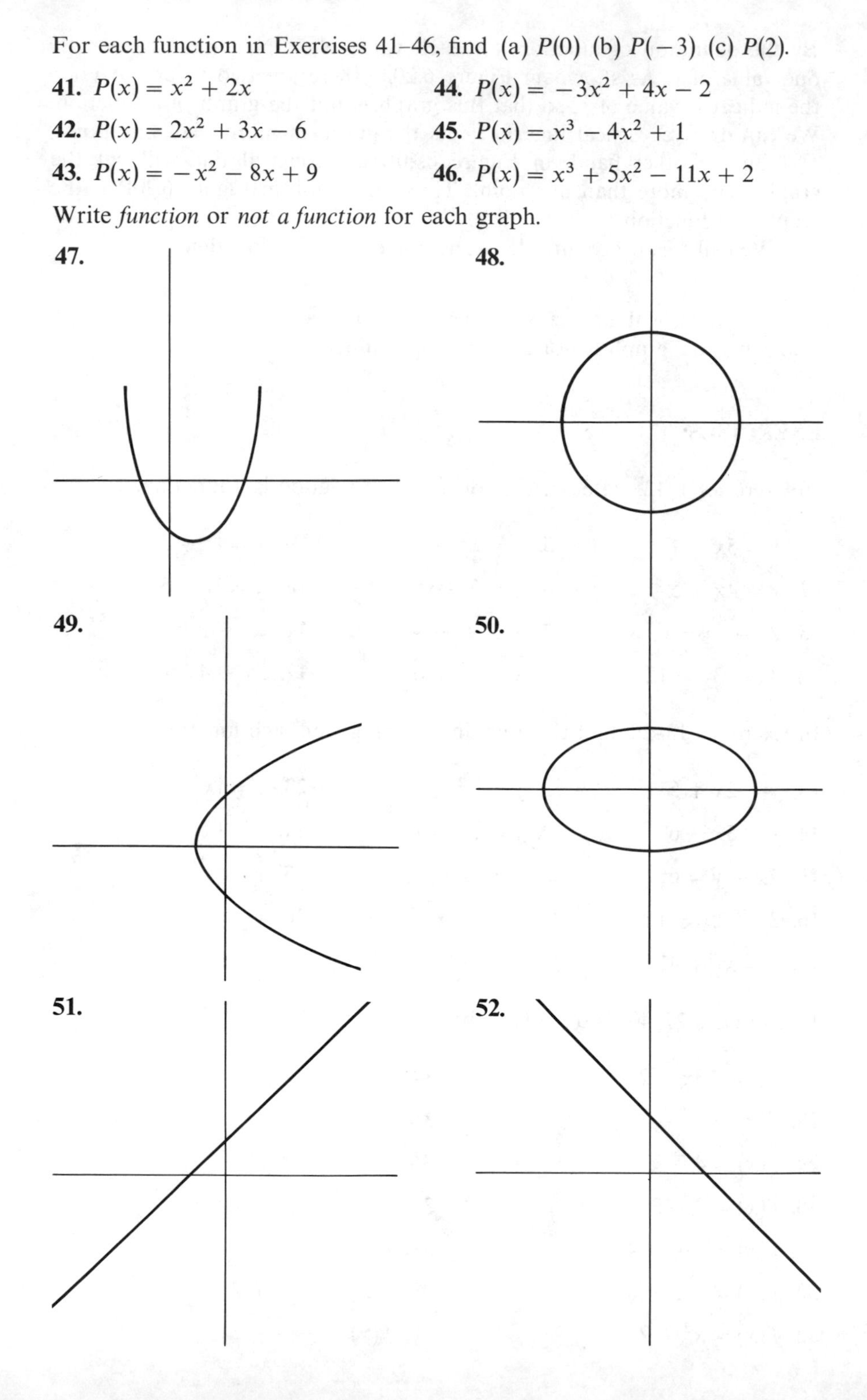

53.

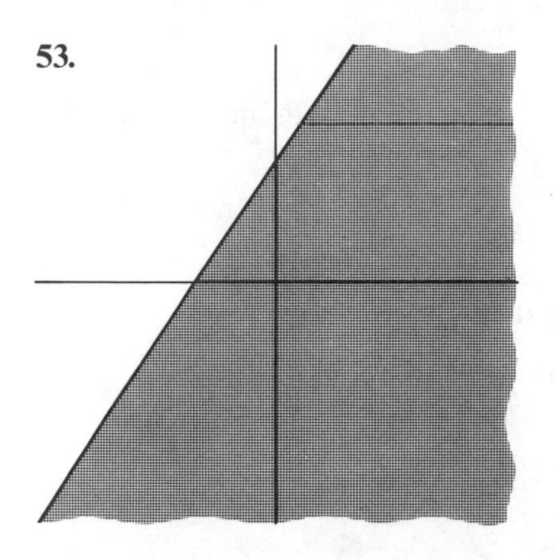

54.

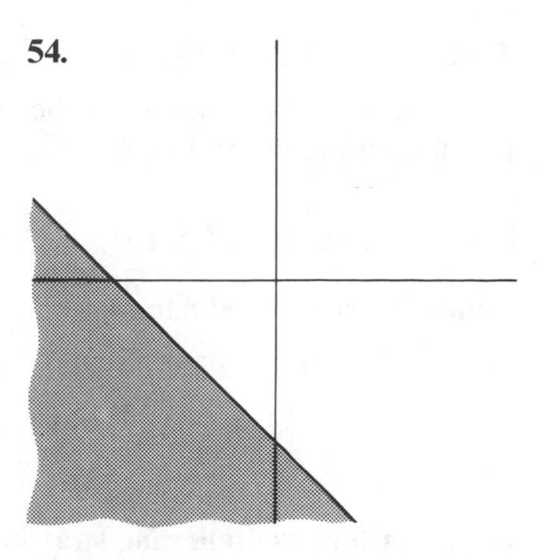

CHAPTER 6 SUMMARY

Key Words

Coordinate system	Relation
Coordinate axes	Domain
x-axis	Range
y-axis	Linear equation in two variables
Origin	Linear inequality
Quadrant	Function
Ordered pair	

Graphing a Straight Line

1. The graph of an equation in the form $ax + by = c$, where $a \neq 0$ and $b \neq 0$, is a straight line. To graph the equation, find three ordered pairs that satisfy the equation $ax + by = c$. Plot these three points. Then draw a straight line through them.

2. The graph of $x = k$, where k is a real number, is a vertical line through the point $(k, 0)$.

3. The graph of $y = k$, where k is a real number, is a horizontal line through the point $(0, k)$.

Graphing a Linear Inequality

1. To graph a linear inequality of the form $ax + by \leq c$ or $ax + by \geq c$, first graph the straight line $ax + by = c$. Then choose any point not on the line. Use the point as a test in the original inequality to see which region to shade.

2. To graph a linear inequality of the form $ax + by < c$ or $ax + by > c$, proceed as in (1), except that the graph of the straight line $ax + by = c$ should be dashed.

Vertical Line Test for a Function

If any vertical line cuts the graph of a relation in more than one point, then the graph is not the graph of a function.

CHAPTER 6 TEST

Complete the ordered pairs for each equation.

1. $y = 5x - 6$	$(0, \quad)$	$(-2, \quad)$	$(\quad, 14)$
2. $3x - 5y = 30$	$(0, \quad)$	$(\quad, 0)$	$(5, \quad)$
3. $x + 4 = 0$	$(\quad, 2)$	$(\quad, 0)$	$(\quad, -3)$

Graph each of the following straight lines.

4. $x + y = 9$

5. $2x + y = 6$

6. $3x - 2y = 18$

7. $4x + 5y = 10$

8. $x = 4y$

9. $x + 5 = 0$

10. $y = 2$

Graph each of the following linear inequalities.

11. $x + y \leq 6$

12. $3x - 4y > 12$

13. $5x + 6y < 30$

14. $y + 1 \geq 0$

Write *function* **or** *not a function* **for each of the following.**

15. $2x = y - 6$

16. $x^2 = y - 2$

17. $y = -|x + 6|$

18. $3x - 2y < 4$

19.

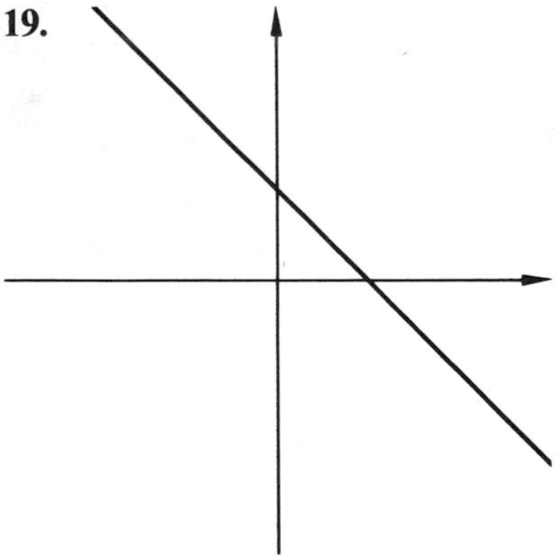

20.

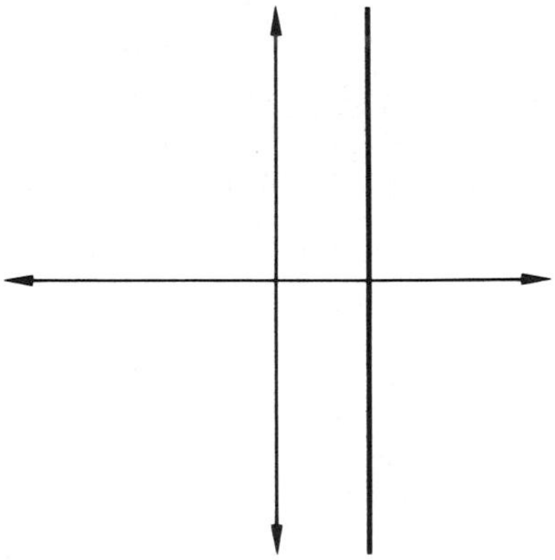

Let $f(x) = 6x - 2$. Find each of the following.

21. $f(0)$

22. $f(4)$

23. $f(-3)$

Let $g(x) = 5x^2 + |x|$. Find each of the following.

24. $g(-2)$

25. $g(-1)$

7 LINEAR SYSTEMS OF EQUATIONS

7.1 SOLVING SYSTEMS OF LINEAR EQUATIONS BY GRAPHING

In this book, a **system** of linear equations is two linear equations, each of which contains the same variables. Examples of systems of two linear equations include

System A	*System B*	*System C*
$2x + 3y = 4$	$x + 3y = 1$	$x - y = 1$
$3x - y = -5$	$-y = 4 - 2x$	$y = 3.$

In system C, you can think of $y = 3$ as an equation in two variables by writing $0x + y = 3$.

To solve a system of two linear equations, you must find all ordered pairs of numbers that make both equations true *at the same time.* Such a solution is called a **simultaneous solution** of the system.

Suppose someone says that the ordered pair $(-1, 2)$ is a simultaneous solution of the system

$$\begin{aligned} 2x + 3y &= 4 \\ 3x - y &= -5. \end{aligned}$$

How would you check this? Check by making sure that $(-1, 2)$ is a solution of both of the equations of the system. Substitute -1 for x and 2 for y in both equations of the system.

$$\begin{aligned} 2x + 3y &= 4 \\ 2(-1) + 3(2) &= 4 \\ -2 + 6 &= 4 \\ 4 &= 4, \quad \text{True} \end{aligned} \qquad \begin{aligned} 3x - y &= -5 \\ 3(-1) - (2) &= -5 \\ -3 - 2 &= -5 \\ -5 &= -5. \quad \text{True} \end{aligned}$$

Both equations are satisfied by $(-1, 2)$, so that $(-1, 2)$ is a simultaneous solution of the system.

Example 1 Is $(4, -3)$ a simultaneous solution for the systems in (a) and (b)?

(a) $x + 4y = -8$
$3x + 2y = 6.$

(b) $2x + 5y = -7$
$3x + 4y = 2.$

To find out whether or not $(4, -3)$ is a simultaneous solution for the given systems, substitute 4 for x and -3 for y in both equations of the two given systems.

(a)

$$\begin{aligned} x + 4y &= -8 \\ 4 + 4(-3) &= -8 \\ 4 + (-12) &= -8 \\ -8 &= -8, \quad \text{True} \end{aligned} \qquad \begin{aligned} 3x + 2y &= 6 \\ 3(4) + 2(-3) &= 6 \\ 12 + (-6) &= 6 \\ 6 &= 6. \quad \text{True} \end{aligned}$$

Since both equations are satisfied by the ordered pair $(4, -3)$, this ordered pair is a simultaneous solution for system (a).

(b)

$$\begin{aligned} 2x + 5y &= -7 \\ 2(4) + 5(-3) &= -7 \\ 8 + (-15) &= -7 \\ -7 &= -7, \quad \text{True} \end{aligned} \qquad \begin{aligned} 3x + 4y &= 2 \\ 3(4) + 4(-3) &= 2 \\ 12 + (-12) &= 2 \\ 0 &= 2. \quad \text{False} \end{aligned}$$

Since $(4, -3)$ does not satisfy both equations of system (b), this ordered pair is not a simultaneous solution for that system.

Let us now look for a method to use in finding the solution for a system of equations. We know that the graph of a linear equation is a straight line. Therefore, one way to solve a system of two equations would be to graph both equations together on the same axes. Then the coordinates of any common point of intersection on the graphs would give the simultaneous solution of the system. Since two different straight lines can cross in one point only, *there will never be more than one solution for such a system.*

Example 2 Solve the system of equations

$$\begin{aligned} 2x + 3y &= 4 \\ 3x - y &= -5 \end{aligned}$$

by graphing both equations on the same axes.

Figure 7.1 shows the graphs of both $2x + 3y = 4$ and $3x - y = -5$ on the same axes. The simultaneous solution can be read from the graph. The graphs are straight lines crossing in only one point, with coordinates $(-1, 2)$. This point gives the only solution for the system. The simultaneous solution is $(-1, 2)$. Check this by replacing x with -1 and y with 2 in both equations of the original system.

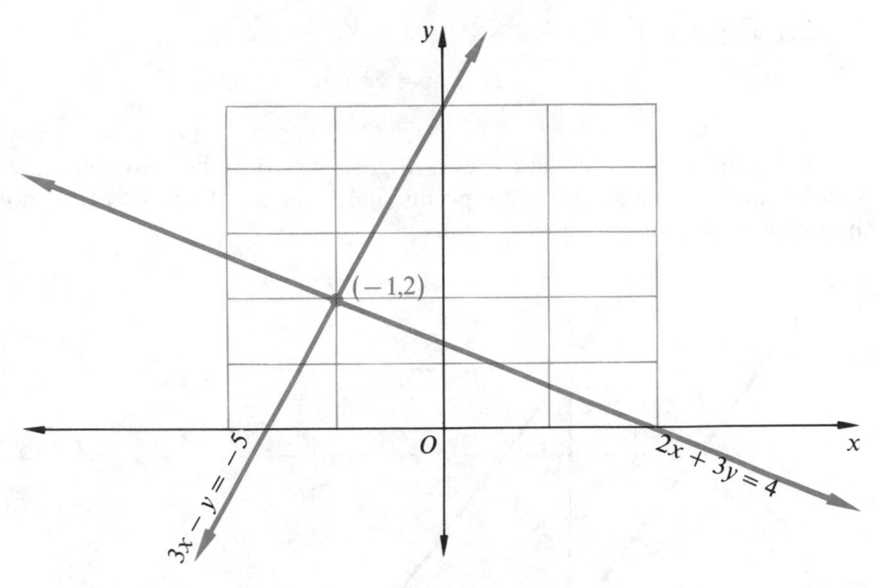

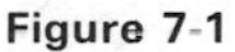
Figure 7-1

Example 3 Solve the system

$$3x + y = 5$$
$$2x - y = 10.$$

To find the simultaneous solution, graph both lines on the same axes, and find the point of intersection. As shown in Figure 7.2, the coordinates of the point of intersection are $(3, -4)$, which is the simultaneous solution.

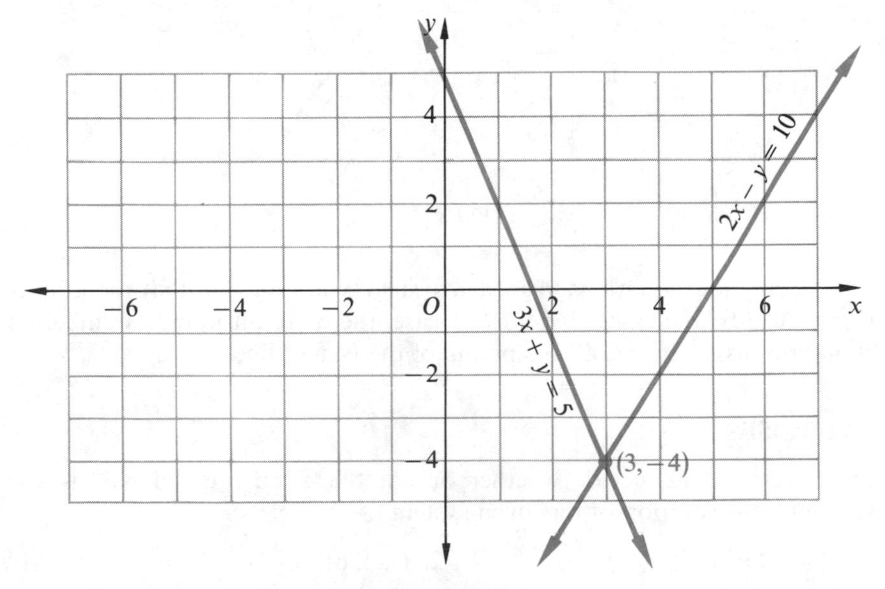

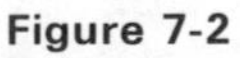
Figure 7-2

Example 4 Solve the system

$$2x + y = 2$$
$$2x + y = 8.$$

The graphs of these lines are in Figure 7.3. But the two lines are parallel, and therefore have no points in common. Thus there is no simultaneous solution for this system.

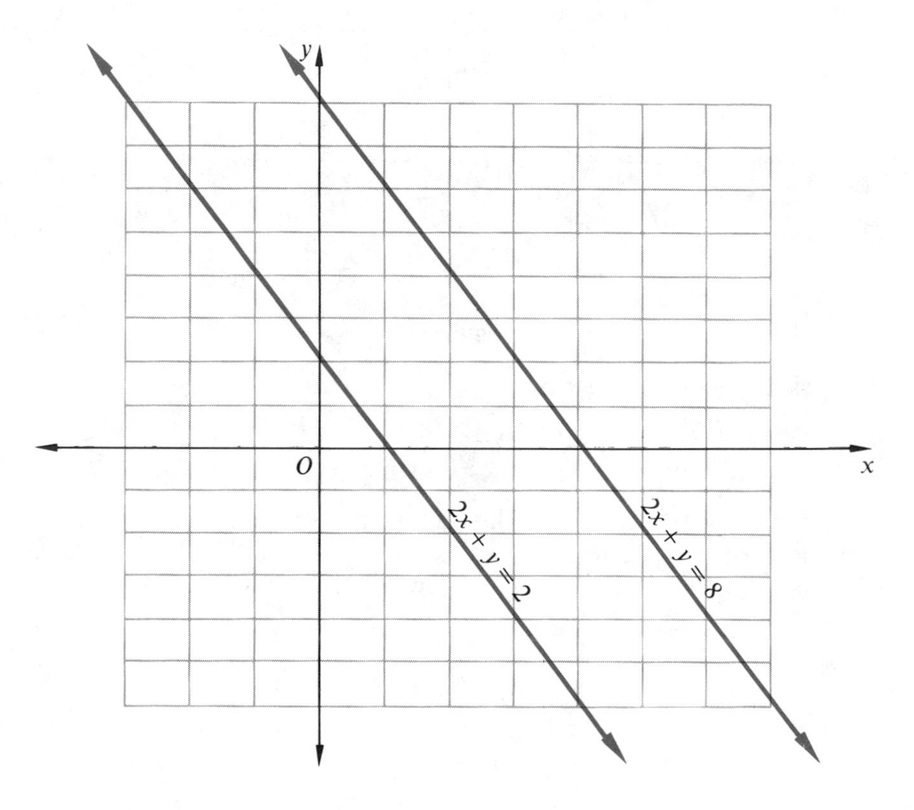

Figure 7-3

In graphing equations to solve a system, you may find that the graphs turn out to be the same line. In this case, there are an infinite number of simultaneous solutions, all the points of the (same) line.

EXERCISES

In Exercises 1–12, decide whether or not the given ordered pair is the simultaneous solution of the given system.

1. $(2, -5)$ $\quad 3x + y = 1$, $\quad 2x + 3y = -11$

2. $(-1, 6)$ $\quad 2x + y = 4$, $\quad 3x + 2y = 9$

3. $(4, -2)$ $\quad x + y = 2$
$2x + 5y = 2$

4. $(-6, 3)$ $\quad x + 2y = 0$
$3x + 5y = 3$

5. $(2, 0)$ $\quad 3x + 5y = 6$
$4x + 2y = 5$

6. $(0, -4)$ $\quad 2x - 5y = 20$
$3x + 6y = -20$

7. $(5, 2)$ $\quad 4x + 3y = 26$
$3x + 7y = 29$

8. $(9, 1)$ $\quad 2x + 5y = 23$
$3x + 2y = 29$

9. $(6, -8)$ $\quad x + 2y + 10 = 0$
$2x - 3y + 30 = 0$

10. $(-5, 2)$ $\quad 3x - 5y + 20 = 0$
$2x + 3y + 4 = 0$

11. $(5, -2)$ $\quad x - 5 = 0$
$y + 2 = 0$

12. $(-8, 3)$ $\quad x = 8$
$y = 3$

In Exercises 13–42, graph both equations on the same coordinate axis. Estimate the simultaneous solution of the given system, if there is one.

13. $x + y = 8$
$x - y = 2$

14. $x + y = -1$
$x - y = 3$

15. $x + y = 12$
$y - x = 4$

16. $y - x = -5$
$x + y = 1$

17. $2x + y = 6$
$3x + 2y = 8$

18. $4x + y = 10$
$2x - y = 8$

19. $2x - 5y = 17$
$3x + y = 0$

20. $2x + 3y = 11$
$3x - y = 11$

21. $5x + 4y = 7$
$2x - 3y = 12$

22. $2x + 5y = 17$
$3x - 4y = -9$

23. $4x + 5y = 3$
$2x - 5y = 9$

24. $2x + y = 1$
$3x - 4y = 29$

25. $3x + 2y = -12$
$x - 2y = -20$

26. $4x + y = -14$
$3x - 2y = -5$

27. $3x - 4y = -8$
$5x + 2y = -22$

28. $x - 2y = -3$
$4x + 3y = -34$

29. $3x - 2y = 15$
$4x + 3y = 20$

30. $-4x + 3y = 16$
$2x - 3y = -8$

31. $3x - 4y = 8$
$4x + 5y = -10$

32. $3x + 2y = 10$
$4x - 3y = -15$

33. $2x + 5y = 20$
$x - 2y = 1$

34. $2x + 3y = 3$
$x + y = 3$

35. $x + 2y - 5 = 0$
$2x + y - 1 = 0$

36. $2x - y - 2 = 0$
$3x + y - 3 = 0$

37. $5x + y - 7 = 0$
$2x + 2y + 2 = 0$

38. $2x + 5y - 20 = 0$
$x + y - 10 = 0$

39. $x = 6$
$y = -2$

40. $x = -1$
$y = 3$

41. $x + 2 = 4$
$y - 1 = 6$

42. $x - 3 = 2$
$y - 5 = 1$

In Exercises 43–50, graph each system. If the two equations produce parallel lines, write "no solution." If the two equations produce the same line, write "same line."

43. $2x + 3y = 5$
$4x + 6y = 9$

44. $5x - 4y = 5$
$10x - 8y = 23$

45. $3x = y + 5$
$6x - 2y = 5$

46. $4y + 1 = x$
$2x - 3 = 8y$

47. $2x - y = 4$
$4x = 2y + 8$

48. $3x = 5 - y$
$6x + 2y = 10$

49. $x - 4y + 6 = 0$
$8y - 2x = 12$

50. $3x - 4y = 8$
$8y - 6x + 16 = 0$

7.2 SOLVING SYSTEMS OF LINEAR EQUATIONS BY ADDITION

The graphical method for finding the simultaneous solution of a system of linear equations has one serious drawback. It can be very difficult to estimate accurately a solution involving fractions, such as $(1/3, \ -5/6)$, from a graph.

A way around this difficulty is an algebraic method of solving systems of equations which depends on the addition property of equality. First stated in Section 2.2, the addition property says that if two quantities are equal, then addition of the same quantity to each results in equal sums.

$$\text{If } A = B, \text{ then } A + C = B + C.$$

We can take this a step further—addition of *equal* quantities, rather than the *same* quantity, also results in equal sums.

$$\text{If } A = B \text{ and } C = D, \text{ then } A + C = B + D.$$

Example 1 Use the addition property of equality to find the simultaneous solution of the system

$$x + y = 5$$
$$x - y = 3.$$

Each equation of this system is a statement of equality, so, as discussed above, the sum of the right-hand sides equals the sum of the left-hand sides. Adding in this way, we have

$$(x + y) + (x - y) = 5 + 3.$$

Combining terms and simplifying gives

$$2x = 8$$
$$x = 4.$$

Thus, $x = 4$ gives the x-value of the simultaneous solution of the given system. To find the y-value of the solution, substitute 4 for x in either of the two equations of the system. Choose the first equation, $x + y = 5$.

$$x + y = 5$$
$$4 + y = 5 \qquad \text{Let } x = 4$$
$$y = 1.$$

Then $y = 1$ gives the y-value of the simultaneous solution. The solution is given by the ordered pair (4, 1).

This can be checked by substitution into the other equation of the given system, $x - y = 3$.

$$x - y = 3$$
$$4 - 1 = 3 \qquad \text{Let } x = 4\text{; let } y = 1$$
$$3 = 3. \qquad \text{True}$$

This result is true, so that the simultaneous solution of the given system is (4, 1).

Example 2 Solve the system

$$x - 2y = -11$$
$$-x + 5y = 26.$$

Add left-hand sides and right-hand sides of this system.

$$(x - 2y) + (-x + 5y) = -11 + 26$$
$$3y = 15$$
$$y = 5.$$

Substitute 5 for y in the first equation above.

$$x - 2y = -11$$
$$x - 2(5) = -11 \qquad \text{Let } y = 5$$
$$x - 10 = -11$$
$$x = -1.$$

The simultaneous solution for this system is thus $(-1, 5)$, as can be checked by substitution into both the original equations.

In solving the system

$$\begin{aligned} x - 2y &= -11 \\ -x + 5y &= 26, \end{aligned}$$

the work would be easier if we drew a line under the second equation and added vertically. To do this, the like terms must be lined up in columns.

$$\begin{array}{r} x - 2y = -11 \\ -x + 5y = 26 \\ \hline 3y = 15. \end{array}$$

Continue as before to obtain the solution.

Example 3 Solve the system

$$x + 3y = 7 \qquad (1)$$
$$2x + 5y = 12. \qquad (2)$$

If we add the two given equations, we get $3x + 8y = 19$, which does not help us find the solution. However, we can use the multiplication property of equality to rewrite either, or both, of the original equations so that we can add the equations and eliminate one of the variables.

If we multiply both sides of equation (1) by -2, the terms containing the variable x would drop out when we add. So multiply both sides of equation (1) by -2.

$$-2(x + 3y) = -2(7)$$
$$-2x - 6y = -14. \qquad (3)$$

Now add equations (3) and (2).

$$\begin{array}{rr} -2x - 6y = -14 & (3) \\ 2x + 5y = 12 & (2) \\ \hline -y = -2. & \end{array}$$

From this result, we get $y = 2$. Substituting back into equation (1) gives

$$\begin{aligned} x + 3y &= 7 \\ x + 3(2) &= 7 \qquad \text{Let } y = 2 \\ x + 6 &= 7 \\ x &= 1. \end{aligned}$$

The solution of this system is (1, 2).

Example 4 Solve the system

$$2x + 3y = -15 \qquad (4)$$
$$5x + 2y = 1. \qquad (5)$$

Here we use the multiplication property of equality with both equations. Multiply by numbers that will cause the coefficients of x (or of y) in the two equations to be negatives of each other. For example, multiply both sides of equation (4) by 5, and both sides of equation (5) by -2.

$$\begin{array}{r} 10x + 15y = -75 \\ -10x - 4y = -2 \\ \hline 11y = -77. \end{array}$$

This gives $y = -7$. By substituting -7 for y in equation (4), or in equation (5), we get $x = 3$. The solution of the system is thus $(3, -7)$.

We would have obtained the same result by multiplying equation (4) by 2 and equation (5) by -3. (Check this.)

Example 5 Solve the system

$$6x - 6y = 7 \qquad (6)$$

$$9x + 4y = 4. \qquad (7)$$

Multiply equation (6) by 9 and equation (7) by -6.

$$\begin{array}{r} 54x - 54y = 63 \\ -54x - 24y = -24 \\ \hline -78y = 39. \end{array}$$

From the equation $-78y = 39$, we get $y = -39/78 = -1/2$. Substitute $y = -1/2$ into equation (7) to find x.

$$9x + 4y = 4$$

$$9x + 4\left(-\frac{1}{2}\right) = 4 \qquad \text{Let } y = -\frac{1}{2}$$

$$9x - 2 = 4$$

$$9x = 6$$

$$x = \frac{2}{3}.$$

The solution of the system is $(2/3, -1/2)$.

The method of solution in this section is called the **addition method,** or **elimination method.** For most systems, this method is more efficient than graphing. The solution of a linear system of equations can be found by the addition method through the following steps.

1. Write both equations of the system in the form $ax + by = c$.
2. Multiply one or both equations by appropriate numbers so that the coefficients of x (or y) are additive inverses of each other.

3. Add the two equations to get an equation with only one variable.
4. Solve the equation from Step 3.
5. Substitute the solution from Step 4 into either of the original equations.
6. Solve the resulting equation from Step 5 for the remaining variable.
7. Check the answer.

EXERCISES

In Exercises 1–42, solve each system of equations. Use the addition method.

1. $x - y = 3$
$x + y = -1$

2. $x + y = 7$
$x - y = -3$

3. $x + y = 2$
$2x - y = 4$

4. $3x - y = 8$
$x + y = 4$

5. $2x + y = 14$
$x - y = 4$

6. $2x + y = 2$
$-x - y = 1$

7. $3x + 2y = 6$
$-3x - y = 0$

8. $5x - y = 9$
$-5x + 2y = -8$

9. $6x - y = 1$
$-6x + 5y = 7$

10. $6x + y = -2$
$-6x + 3y = -14$

11. $2x - y = 5$
$4x + y = 4$

12. $x - 4y = 13$
$-x + 6y = -18$

13. $5x - y = 15$
$7x + y = 21$

14. $x - 4y = 12$
$-x + 6y = -18$

15. $2x - y = 7$
$3x + 2y = 0$

16. $x + y = 7$
$-3x + 3y = -9$

17. $x + 3y = 16$
$2x - y = 4$

18. $4x - 3y = 8$
$2x + y = 14$

19. $x + 4y = -18$
$3x + 5y = -19$

20. $2x + y = 3$
$5x - 2y = -15$

21. $3x - 2y = -6$
$-5x + 4y = 16$

22. $-4x + 3y = 0$
$5x - 6y = 9$

23. $2x - y = -8$
$5x + 2y = -20$

24. $5x + 3y = -9$
$7x + y = -3$

25. $2x + y = 5$
$5x + 3y = 11$

26. $2x + 7y = -53$
$4x + 3y = -7$

27. $5x - 4y = -1$
$-7x + 5y = 8$

28. $6x + 2y = 0$
$-5x + 3y = 56$

29. $3x + 5y = 33$
$4x - 3y = 15$

30. $2x + 8y = 30$
$5x - 3y = 29$

31. $3x + 5y = -7$
$5x + 4y = 10$

32. $2x + 3y = -11$
$5x + 2y = 22$

33. $2x + 3y = -12$
$5x - 7y = -30$

34. $2x + 9y = 16$
$5x - 6y = 40$

35. $4x - 3y = 0$
$6x + 6y = 7$

36. $8x + 3y = 9$
$12x + 6y = 13$

37. $8x + 12y = 13$
$16x - 18y = -9$

38. $9x + 6y = -9$
$6x + 8y = -16$

39. $3x - 2y = 3$
$3x + 3y = 78$

40. $3x - 2y = 27$
$2x - 7y = -50$

41. $5x - 7y = 6$
$3x - 6y = 2$

42. $3x + 7y = -12$
$-4x + 3y = 16$

7.3 TWO SPECIAL CASES

When you graph each equation of a linear system to solve it, you usually obtain two lines that cross in exactly one point. The coordinates of the point give you the simultaneous solution of the system. Two other possibilities exist where the graphs do not intersect, and we consider these special cases in this section.

If the graphs are parallel lines, there is no simultaneous solution for the system of equations.

If both equations have the same line for a graph, then there are an infinite number of ordered pairs in the solution set.

Let us try the addition method on systems leading to parallel lines or to the same line.

Example 1 Solve the following system by the addition method.

$$2x + 4y = 5$$
$$4x + 8y = -9.$$

If we multiply both sides of $2x + 4y = 5$ by -2, and then add $4x + 8y = -9$, we get

$$\begin{array}{r} -4x - 8y = -10 \\ 4x + 8y = -9 \\ \hline 0 = -19. \end{array} \qquad \text{False}$$

The false statement, $0 = -19$, shows that the given system is self-contradictory. It has no simultaneous solution. This means that the graphs of the equations of this system are parallel lines, as shown in Figure 7.4.

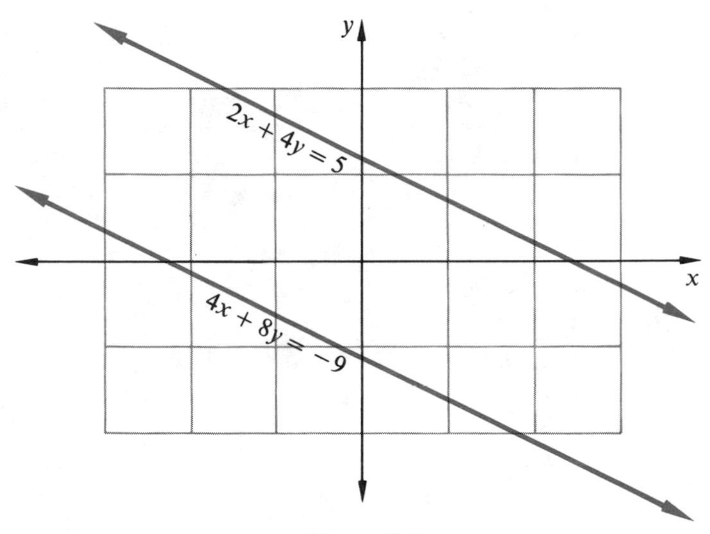

Figure 7-4

Example 2 Solve the system

$$\begin{aligned} 3x - y &= 4 \\ -9x + 3y &= -12. \end{aligned}$$

If we multiply both sides of the top equation by 3, and add, we get

$$\begin{array}{r} 9x - 3y = 12 \\ -9x + 3y = -12 \\ \hline 0 = 0. \end{array} \qquad \text{True}$$

This result, $0 = 0$, is true for any ordered pair (x, y) which satisfies either equation. Thus, any ordered pair which satisfies either equation is a simultaneous solution, and there are an infinite number of solutions. The solutions can be written as the solution set $\{(x, y) | 3x - y = 4\}$, or as the solution set $\{(x, y) | -9x + 3y = -12\}$. However, in the answers at the back of this book, such a solution is indicated by the words "same line."

There are three possibilities when using the addition method to solve a linear system of equations:

1. The result of the addition step is a statement such as $x = 2$ or $y = -3$. The solution will be exactly one ordered pair. The graphs of the equations of the system will cross in exactly one point.
2. The result of the addition step is a false statement, such as $0 = 4$. In this case, the graphs are parallel lines, and there is no simultaneous solution for the system.
3. The result of the addition step is a true statement, such as $0 = 0$. The graphs of both equations of the system are the same line, and there are an infinite number of ordered pairs which are solutions.

EXERCISES

Solve each system in Exercises 1–22.

1. $x + y = 4$
$x + y = -2$

2. $2x - y = 1$
$2x - y = 4$

3. $5x - 2y = 6$
$10x - 4y = 10$

4. $3x - 5y = 2$
$6x - 10y = 8$

5. $x + 3y = 5$
$2x + 6y = 10$

6. $6x - 2y = 12$
$-3x + y = -6$

7. $2x + 3y = 8$
$4x + 6y = 12$

8. $4x + y = 6$
$-8x - 2y = 21$

9. $5x = y + 4$
$5x = y - 4$

10. $4y = 3x - 2$
$4y = 3x + 5$

11. $6x + 3y = 0$
$-12x - 6y = 0$

12. $3x - 5y = 0$
$6x - 10y = 0$

13. $2x - 3y = 0$
$4x + 5y = 0$

14. $3x - 5y = 0$
$6x + 10y = 0$

15. $3x + 5y = 19$
$4x - 3y = 6$

16. $2x + 5y = 17$
$4x + 3y = -1$

17. $4x - 2y = 1$
$8x - 4y = 1$

18. $-2x + 3y = 5$
$4x - 6y = 5$

19. $3x - 2y = 8$
$-3x + 2y = -8$

20. $4x + y = 4$
$-8x - 2y = -8$

21. $$\begin{aligned} 4x - y &= 3 \\ -2x + \frac{1}{2}y &= -\frac{3}{2} \end{aligned}$$

22. $$\begin{aligned} 5x - 2y &= 8 \\ -\frac{5}{2}x + y &= -4 \end{aligned}$$

7.4 SOLVING SYSTEMS OF LINEAR EQUATIONS BY SUBSTITUTION

We have now looked at the graphical method and the elimination method for solving systems of linear equations. A third method is the **substitution method,** useful for systems where at least one equation gives one variable in terms of the other. However, the substitution method can be used to solve any system of linear equations.

Example 1 Solve the system

$$\begin{aligned} 3x + 5y &= 26 \\ y &= 2x. \end{aligned}$$

From the second of these two equations, we observe that $y = 2x$. Using this fact, we can substitute $2x$ for y in the first equation.

$$\begin{aligned} 3x + 5y &= 26 \\ 3x + 5(2x) &= 26 \qquad \text{Let } y = 2x \\ 3x + 10x &= 26 \\ 13x &= 26 \\ x &= 2. \end{aligned}$$

Since $y = 2x$, we have $y = 2(2) = 4$. The solution of the given system is thus $(2, 4)$.

Example 2 Use substitution to solve the system

$$\begin{aligned} 2x + 5y &= 7 \\ x &= -1 - y. \end{aligned}$$

The second equation gives x in terms of y. Substitute $-1 - y$ for x in the first equation.

$$\begin{aligned} 2x + 5y &= 7 \\ 2(-1 - y) + 5y &= 7 \qquad \text{Let } x = -1 - y \\ -2 - 2y + 5y &= 7 \\ -2 + 3y &= 7 \\ 3y &= 9 \\ y &= 3. \end{aligned}$$

Since $x = -1 - y$, we have $x = -1 - 3$, or $x = -4$. The solution of the given system is $(-4, 3)$.

Example 3 Use substitution to solve the system

$$x = 5 - 2y$$
$$2x + 4y = 6.$$

Substitute $5 - 2y$ for x in the second equation.

$$\begin{aligned} 2x + 4y &= 6 \\ 2(5 - 2y) + 4y &= 6 \qquad \text{Let } x = 5 - 2y \\ 10 - 4y + 4y &= 6 \\ 10 &= 6. \qquad \text{False} \end{aligned}$$

The false result means that the equations of the system have graphs that are parallel lines. Thus there is no simultaneous solution for this system.

Example 4 Solve the following system by substitution.

$$2x + 3y = 8$$
$$-4x - 2y = 0.$$

To use the substitution method, we need an equation giving x in terms of y (or y in terms of x). We can choose the first equation of the system, $2x + 3y = 8$, and solve the equation for x. This means that we need to have x alone on one side of the equation. To get this, we need to first add $-3y$ to both sides.

$$\begin{aligned} 2x + 3y &= 8 \\ 2x &= 8 - 3y. \end{aligned}$$

Now divide both sides of this equation by 2.

$$x = \frac{8 - 3y}{2}.$$

Finally, substitute this result for x in the second equation of the system.

$$\begin{aligned} -4x - 2y &= 0 \\ -4\left(\frac{8 - 3y}{2}\right) - 2y &= 0 \qquad \text{Let } x = \frac{8 - 3y}{2} \\ -2(8 - 3y) - 2y &= 0 \\ -16 + 6y - 2y &= 0 \\ -16 + 4y &= 0 \\ 4y &= 16 \\ y &= 4. \end{aligned}$$

Let $y = 4$ in $x = (8 - 3y)/2$.

$$x = \frac{8 - 3 \cdot 4}{2}$$

$$x = \frac{8 - 12}{2}$$

$$x = \frac{-4}{2}$$

$$x = -2.$$

The solution of the given system is $(-2, 4)$.

Example 5 Use substitution to solve the system

$$2x = 3 - 2y \qquad (1)$$
$$6 + 5y = 10 - 2x + 2y. \qquad (2)$$

To begin, simplify the second equation by adding -6 and $-2y$ to both sides. This gives the simplified system

$$2x = 3 - 2y \qquad (1)$$
$$3y = 4 - 2x. \qquad (3)$$

To use the substitution method, we must solve one of these equations for either x or y, and then substitute the result into the other equation. Let us solve equation (1) for x, as follows

$$2x = 3 - 2y$$

$$\frac{1}{2}(2x) = \frac{1}{2}(3 - 2y)$$

$$x = \frac{3}{2} - y.$$

Now substitute $3/2 - y$ for x in equation (3).

$$3y = 4 - 2x \qquad (3)$$

$$3y = 4 - 2\left(\frac{3}{2} - y\right) \qquad \text{Let } x = \frac{3}{2} - y$$

$$3y = 4 - 3 + 2y$$

$$y = 1.$$

Since $x = \frac{3}{2} - y$, and $y = 1$, we have $x = \frac{3}{2} - 1$, or $x = \frac{1}{2}$. The solution of the system is $(\frac{1}{2}, 1)$.

EXERCISES

In Exercises 1–22, solve each system by the substitution method.

1. $x + y = 6$
$y = 2x$

2. $x + 3y = -11$
$y = -4x$

3. $3x + 2y = 26$
$x = y + 2$

4. $4x + 3y = -14$
$x = y - 7$

5. $x + 5y = 3$
$x = 2y + 10$

6. $5x + 2y = 14$
$y = 2x - 11$

7. $5x + 7y = 40$
$x = 2y - 9$

8. $4x + 9y = -7$
$y = 2x - 13$

9. $3x - 2y = 14$
$2x + y = 0$

10. $2x - 5 = -y$
$x + 3y = 0$

11. $x + y = 6$
$x - y = 4$

12. $3x - 2y = 13$
$x + y = 6$

13. $3x - y = 6$
$y = 3x - 5$

14. $4x - y = 4$
$y = 4x + 3$

15. $6x - 8y = 4$
$3x = 4y + 2$

16. $12x + 18y = 12$
$2x = 2 - 3y$

17. $4x + 5y = 5$
$2x + 3y = 1$

18. $3x + 4y = 10$
$4x + 5y = 14$

19. $2x + 3y = 11$
$y = 1$

20. $3x + 4y = -10$
$x = -6$

21. $4x + y = 5$
$x - 2 = 0$

22. $5x + 2y = -19$
$y - 3 = 0$

In Exercises 23–36, solve by either the addition method or the substitution method. First simplify any equations where necessary.

23. $x + 4y = 34$
$y = 4x$

24. $3x - y = -14$
$x = -2y$

25. $4 + 4x - 3y = 34 + x$
$4x = -y - 2 + 3x$

26. $5x - 4y = 42 - 8y - 2$
$2x + y = x + 1$

27. $4x - 2y + 8 = 3x + 4y - 1$
$3x + y = x + 8$

28. $5x - 4y - 8x - 2 = 6x + 3y - 3$
$4x - y = -2y - 8$

29. $2x - 8y + 3y + 2 = 5y + 16$
$8x - 2y = 4x + 28$

30. $7x - 9 + 2y - 8 = -3y + 4x + 13$
$4y - 8x = -8 + 9x + 32$

31. $2x + 3y = 10$
$4x + 5y = 10 - y$

32. $10x + 21y = 90$
$5x + 11y = 10 - 5x - 10y$

33. $-2x + 3y = 12 + 2y$
$2x - 5y + 4 = -8 - 4y$

34. $2x + 5y = 7 + 4y - x$
$5x + 3y + 8 = 22 - x + y$

35. $y + 9 = 3x - 2y + 6$
$5 - 3x + 24 = -2x + 4y + 3$

36. $5x - 2y = 16 + 4x - 10$
$4x + 3y = 60 + 2x + y$

In Exercises 37–40, first clear fractions. Then solve the system.

Example. To clear $3x + \frac{1}{4}y = 2$ of fractions, multiply both sides of the equation by 4:

$$4\left(3x + \frac{1}{4}y\right) = 4 \cdot 2$$

$$4(3x) + 4 \cdot \frac{1}{4}y = 4 \cdot 2$$

$$12x + y = 8.$$

37. $x + \frac{1}{3}y = y - 2$

$\frac{1}{4}x - y = x - y$

38. $\frac{5}{3}x + 2y = \frac{1}{3} + y$

$2x - 3 + \frac{y}{3} = -2 + x$

39. $\frac{x}{6} + \frac{y}{6} = 1$

$-\frac{1}{2}x - \frac{1}{3}y = -5$

40. $\frac{x}{2} - \frac{y}{3} = \frac{5}{6}$

$\frac{x}{5} - \frac{y}{4} = \frac{1}{10}$

7.5 APPLICATIONS

Many practical problems are more easily translated into mathematical equations if *two* variables are used. However, we must have a system of two equations before we can get the desired solution.

The examples in this section illustrate this method of using two equations with two variables to solve practical problems. In each example, two variables are used. Two equations are written from the information in the problem. The two equations form a system of equations which is solved to find the solution for the problem.

Example 1 The sum of two numbers is 63. Their difference is 19. Find the two numbers.

Let x represent one number and y the other. Then, from the information of the problem we set up a system of equations.

$$x + y = 63$$
$$x - y = 19.$$

This system can be solved by the addition method. We have

$$\begin{array}{r} x + y = 63 \\ x - y = 19 \\ \hline 2x \quad\;\; = 82. \end{array}$$

From this last equation, $x = 41$. Substitute 41 for x in the first equation and check that $y = 22$. The numbers required in the problem are 41 and 22.

Example 2 Admission prices at a football game were $1.25 for adults and $.50 for children. The total receipts from the game were $530.75. Tickets were sold to 454 people. How many adults and how many children attended the game?

Let a represent the number of adult tickets that were sold, and let c represent the number of child's tickets. The information given in the problem is summarized in the following table.

Kind of ticket	Number sold	Cost of each (in dollars)	Receipts (in dollars)
Adult	a	1.25	$1.25a$
Child	c	.50	$.50c$

The total number of tickets sold was 454, so that

$$a + c = 454.$$

The receipts from the sale of a adult tickets at \$1.25 each are \$1.25a, while the receipts from the sale of c children's tickets at \$.50 each are \$.50c. Since the total receipts were \$530.75,

$$1.25a + .50c = 530.75.$$

We have used the information in the problem to set up the system of equations

$$a + c = 454 \qquad (1)$$

$$1.25a + .50c = 530.75. \qquad (2)$$

Equation (2) can be simplified if we multiply both sides by 100 to clear the decimals.

$$100(1.25a + .50c) = 100(530.75)$$

$$125a + 50c = 53075. \qquad (3)$$

To solve the system of equations, multiply equation (1) on both sides by -50, and then add to equation (3).

$$-50a - 50c = -50(454)$$

$$-50a - 50c = -22700.$$

Now we add.

$$\begin{array}{r} -50a - 50c = -22700 \\ 125a + 50c = 53075 \\ \hline 75a \qquad\quad = 30375. \end{array}$$

From the equation $75a = 30375$, we get $a = 405$. We know that $a + c = 454$; therefore, $c = 49$. (Check this.) There were 405 adults and 49 children at the game.

Example 3 A pharmacist needs 100 gallons of 50% alcohol solution. She has on hand 30% alcohol solution and 80% alcohol solution, which she can mix. How many gallons of each will be required to make the 100 gallons of 50% alcohol solution?

Let x represent the number of gallons of 30% alcohol needed, and let y represent the number of gallons of 80% alcohol. The information of the problem can be summarized in the following table.

Gallons of solution	Percent	Gallons of pure alcohol
x	30	$.30x$
y	80	$.80y$
100	50	$.50(100)$

She will have $.30x$ gallons of alcohol from the x gallons of 30% solution and $.80y$ gallons of alcohol from the y gallons of 80% solution. The total is $.30x + .80y$ gallons of pure alcohol. In the mixture, she wants 100 gallons of 50% solution. This 100 gallons would contain $.50(100) = 50$ gallons of pure alcohol. Since the amounts of pure alcohol must be equal,

$$.30x + .80y = 50.$$

We also know that the total number of gallons is 100, or

$$x + y = 100.$$

These two equations give the system

$$\begin{aligned} .30x + .80y &= 50 \\ x + y &= 100. \end{aligned}$$

Let us solve this system by the substitution method. From the second equation of the system, we have $x = 100 - y$. If we substitute $100 - y$ for x in the first equation, we get

$$\begin{aligned} .30(100 - y) + .80y &= 50 \qquad \text{Let } x = 100 - y \\ 30 - .30y + .80y &= 50 \\ .50y &= 20 \\ y &= 40. \end{aligned}$$

Since $x + y = 100$, then $x = 60$. The pharmacist should use 60 gallons of the 30% solution, and 40 gallons of the 80% solution.

Example 4 Two cars start from positions 400 miles apart and travel toward each other. They meet after four hours. Find the average speed of each car if one car travels 20 miles per hour faster than the other.

We need the formula that relates distance, rate, and time. From the list of formulas inside the back cover, this formula is $d = rt$. Let x be the average speed of the first car, and y the average speed of the second car. Since each car travels for four hours, t for each car is 4.

	r	t	d
First car:	x	4	$4x$
Second car:	y	4	$4y$

Since the total distance traveled by both cars is 400 miles,

$$4x + 4y = 400.$$

One car traveled 20 miles per hour faster than the other. Assume that the first car was faster. Then

$$x = 20 + y.$$

We now have the system of equations

$$\begin{aligned} 4x + 4y &= 400 \\ x &= 20 + y. \end{aligned}$$

This system can be solved by substitution. Replace x with $20 + y$ in the first equation of the system.

$$\begin{aligned} 4(20 + y) + 4y &= 400 \\ 80 + 4y + 4y &= 400 \\ 80 + 8y &= 400 \\ 8y &= 320 \\ y &= 40. \end{aligned}$$

Since $x = 20 + y$, we have $x = 60$. Thus, the speeds of the two cars were 40 miles per hour and 60 miles per hour.

EXERCISES

In Exercises 1–22, write a system of equations for each problem. Then solve the system. Formulas are inside the back cover.

1. The sum of two numbers is 52, and their difference is 34. Find the numbers.

2. Find two numbers whose sum is 56 and whose difference is 18.

3. A certain number is three times as large as a second number. Their sum is 96. What are the two numbers?

4. One number is five times as large as another. The difference of the numbers is 48. Find the numbers.

5. A rectangle is twice as long as it is wide. Its perimeter is 60 inches. Find the dimensions of the rectangle.

6. The perimeter of a triangle is 21 inches. If two sides are of equal length, and the third side is 3 inches longer than one of the equal sides, find the lengths of the three sides.

7. The cashier at the Mustang Joy Ranch has some $10 bills and some $20 bills. The total value of the money is $1480. If there is a total of 85 bills, how many of each type are there?

8. A bank teller has 154 bills of $1 and $5 denominations. How many of each type of bill does he have if the total value of the money is $466?

9. A club secretary bought 8¢ and 10¢ pieces of candy to give to the members. She spent a total of $15.52. If she bought 170 pieces of candy, how many of each kind did she buy?

10. There were 311 tickets sold for a basketball game, some for students and some for non-students. Student tickets cost 25¢ each and non-student tickets cost 75¢ each. The total receipts were $108.75. How many of each type of ticket were sold?

11. Ms. Sullivan has $10,000 to invest, part at 5% and part at 7%. She wants the income from simple interest on the two investments to total $550 yearly. How much should she invest at each rate?

12. Mr. Emerson has twice as much money invested at 7% as he has at 8%. If his yearly income from investments is $440, how much does he have invested at each rate?

13. A 90% antifreeze solution is to be mixed with a 75% solution to make 20 liters of a 78% solution. How many liters of 90% and 75% solutions should be used?

14. A grocer wishes to blend candy selling for 60¢ a pound with candy selling for 90¢ a pound to get a mixture which will be sold for 70¢ a pound. How many pounds of the 60¢ and the 90¢ candy should be used to get 30 pounds of the mixture?

15. How many barrels of olives worth $40 per barrel must be mixed with olives worth $60 per barrel to get 50 barrels of a mixture worth $48 per barrel?

16. A glue merchant wishes to mix some glue worth $70 per barrel with some glue worth $90 per barrel, to get 80 barrels of a mixture worth $77.50 per barrel. How many barrels of each type should be used?

17. If a plane can travel 400 miles per hour into the wind and 540 miles per hour with the wind, find the speed of the wind, and also find the speed of the plane in still air.

18. It takes a boat $1\frac{1}{2}$ hours to go 12 miles downstream, and 6 hours to return. Find the speed of the current and the speed of the boat in still water.

19. At the beginning of a walk for charity, John and Harriet are 30 miles apart. If they leave at the same time and walk in the same direction, John overtakes Harriet in 60 hours. If they walk toward each other, they meet in 5 hours. What are their speeds?

20. Mr. Anderson left Farmersville in a plane at noon to travel to Exeter. Mr. Bentley left Exeter in his automobile at 2 PM to travel to Farmersville. It is 400 miles from Exeter to Farmersville. If the sum of their speeds is 120 miles per hour, and if they met at 4 PM, find the speed of each.

21. The Smith family is coming to visit, and no one knows how many children they have. Janet, one of the girls, says she has as many brothers as sisters; her brother Steve says he has twice as many sisters as brothers. How many boys and how many girls are in the family?

22. In the Lopez family, the number of boys is one more than half the number of girls. One of the Lopez boys, Rico, says that he has one more sister than brothers. How many boys and girls are in the family?

7.6 GRAPHING SYSTEMS OF LINEAR INEQUALITIES

The first example of this section reviews the method of graphing a linear inequality (first discussed in Section 6.4). After this example, we look at ways to graph *systems* of linear inequalities.

Example 1 Graph $x + 3y > 12$.

First, graph the straight line $x + 3y = 12$. Because of the $>$ symbol, make the line dashed, to show that the points of the line itself do not belong to the graph. Choose the point (0, 0) for a test—substitute 0 for x and 0 for y in the original inequality.

$$x + 3y > 12$$
$$0 + 3(0) > 12$$
$$0 > 12. \qquad \text{False}$$

We shade the region on the side of the line that does not include (0, 0), as in Figure 7.5.

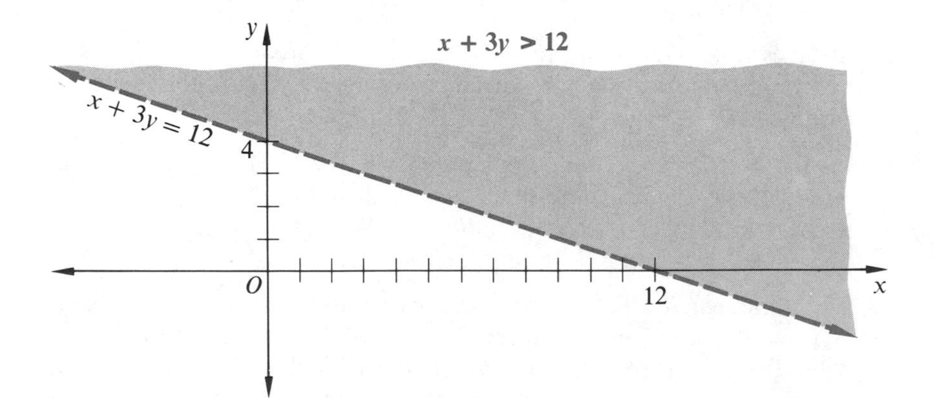

Figure 7-5

Exactly the same procedure can be used to graph the solutions of systems of linear inequalities, as the next example shows.

Example 2 Graph the solution of the system

$$3x + 2y \leq 6$$
$$2x - 5y \geq 10.$$

First, graph the inequality $3x + 2y \leq 6$, using the steps described above. Then, on the same axes, graph the second inequality, $2x - 5y \geq 10$. The solution of the system is given by the overlap of the regions of the two graphs. This solution is the darkest shaded region in Figure 7.6, and includes portions of the two boundary lines.

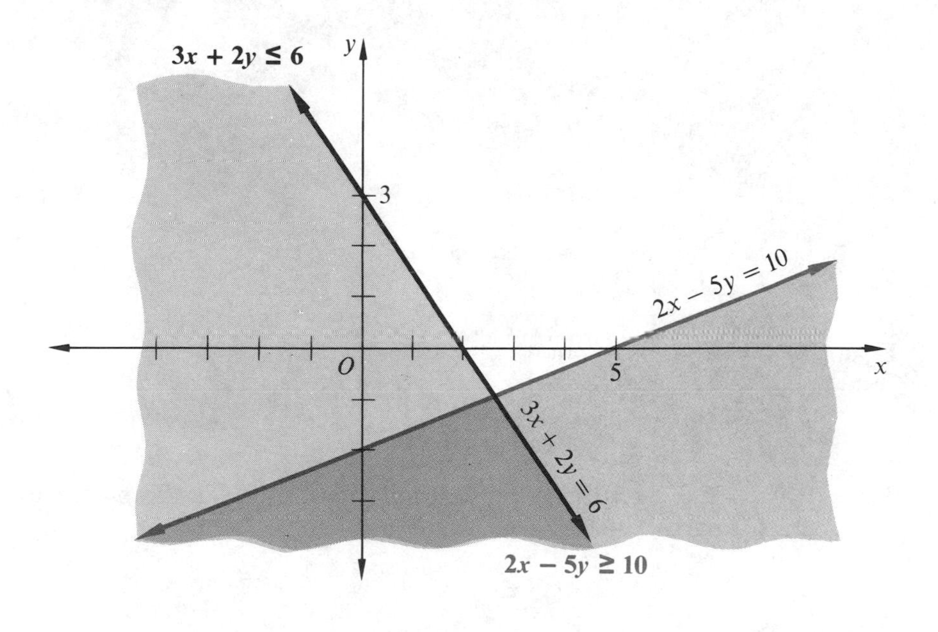

Figure 7-6

Example 3 Graph the solution of the system

$$x - y > 5$$
$$2x + y < 2.$$

Figure 7.7 shows the graphs of both $x - y > 5$ and $2x + y < 2$. Dashed lines show that the graphs of the inequalities do not include their boundary lines. The solution of the system is the darkest shaded region (see over).

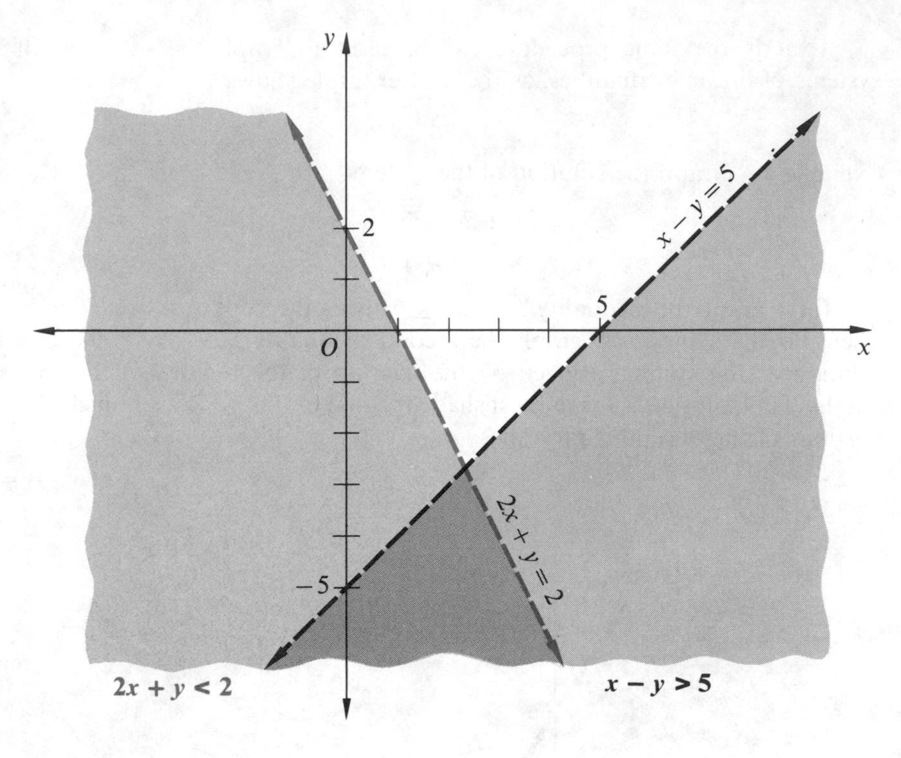

Figure 7-7

Example 4 Graph the solution of the system

$$4x - 3y \leq 8$$
$$x \geq 2.$$

Recall that $x = 2$ is a vertical line through the point (2, 0). The graph of the solution is the darkest shaded region in Figure 7.8.

EXERCISES

In Exercises 1–24, graph the solutions of each system of linear inequalities.

1. $x + y \leq 6$
$x - y \leq 1$

2. $x + y \geq 2$
$x - y \leq 3$

3. $2x - 3y \leq 6$
$x + y + 1 \geq 0$

4. $4x + 5y \leq 20$
$x - y \leq 3$

5. $x - 2y > 6$
$2x + y > 4$

6. $3x + y < 4$
$x + 2y > 2$

7. $x + 4y \leq 8$
$2x - y \leq 4$

8. $3x + y \leq 6$
$2x - y \leq 8$

9. $x - 4y \leq 3$
$x \geq 2y$

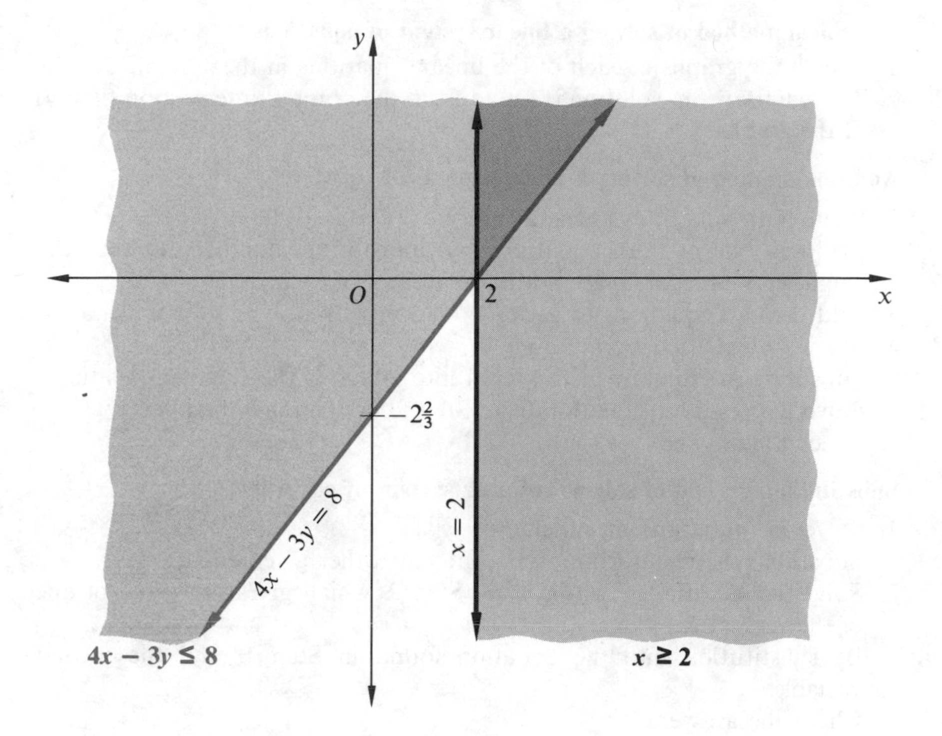

Figure 7-8

10. $2x + 3y \leq 6$
$x - y \geq 5$

11. $x + 2y \leq 4$
$x + 1 \geq y$

12. $x < 2y + 3$
$x + y > 0$

13. $y \leq 2x - 5$
$x - 3y \leq 2$

14. $4x + 3y \leq 6$
$x - 2y \geq 4$

15. $x - 3y \leq 6$
$x \geq -1$

16. $2x + 5y \geq 20$
$x \leq 4$

17. $3x - 2y \geq 9$
$y \leq 3$

18. $4x + 5y < 8$
$y > -2$

19. $2x + 3y < 6$
$4x + 6y > 18$

20. $3x - y \leq 4$
$-6x + 2y \leq -10$

21. $x \geq 2$
$y \leq 3$

22. $x \geq -1$
$y \geq 4$

23. $x \leq 0$
$y \leq 0$

24. $x > 0$
$y > 0$

CHAPTER 7 SUMMARY

Key Words

System of linear equations
Simultaneous solution
System of linear inequalities
Graphical method
Addition method (elimination)
Substitution method

Graphical method of solving a linear system of equations

1. Sketch the graphs of each of the linear equations in the system.
2. The simultaneous solution is found from the point of intersection (if any) of the graphs.

Addition method of solving a linear system of equations

1. Write both equations in the form $ax + by = c$.
2. Multiply one or both equations by appropriate numbers so that the coefficients of x (or y) are additive inverses of each other.
3. Add the two equations to get an equation with only one variable.
4. Solve the equation from Step 3.
5. Substitute the solution from Step 4 into either of the original equations.
6. Solve the resulting equation from Step 5 for the remaining variable.
7. Check the answer.

Substitution method of solving a linear system of equations

1. Solve one equation for either x or y.
2. Substitute the result from Step 1 into the other given equation.
3. Solve the equation resulting from Step 2, which gives the value of one of the variables.
4. By substitution into the equation found in Step 1, find the second variable.
5. Check the answer.

The simultaneous solution and two special cases

1. A linear system has exactly one solution if either the addition or substitution methods give a result such as $x = 2$ or $y = -3$. The graphs of the equations in such a system cross in exactly one point.

If the system does not have exactly one solution, then one of the special cases applies.

2. If the result is a false statement, such as $0 = 4$, there is no simultaneous solution for the system. The graphs of the equations of the system are parallel lines.

3. If the result is a true statement, such as $0 = 0$, then the system has an infinite number of solutions. In this case, the graphs of the equations of the system are the same line.

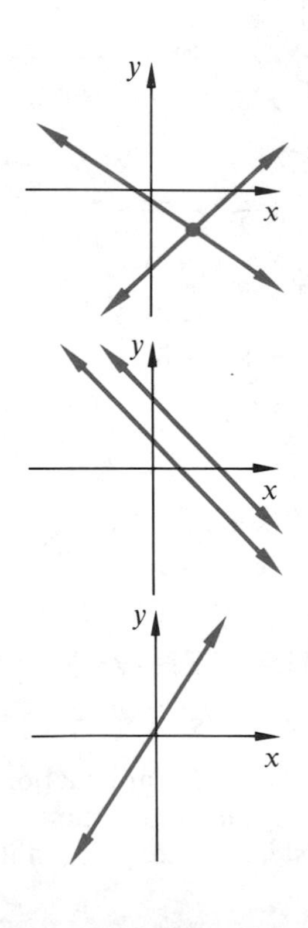

CHAPTER 7 TEST

Solve each of the following systems by the graphical method.

1. $2x + y = 5$
$3x - y = 15$

2. $3x + 2y = 8$
$5x + 4y = 10$

3. $x + 2y = 6$
$2x - y = 7$

Solve each of the following systems by the addition method.

4. $2x - 5y = -13$
$3x + 5y = 43$

5. $4x + 3y = 26$
$5x + 4y = 32$

6. $6x + 5y = -13$
$3x + 2y = -4$

7. $4x + 5y = 8$
$-8x - 10y = -6$

8. $4x - 3y = 6$
$x + 2y = -4$

9. $6x - 5y = 2$
$-2x + 3y = 2$

10. $2x - y = 5$
$4x + 3y = 0$

11. $3x + 2y = 16$
$9x - 3y = -6$

Solve each of the following systems by substitution.

12. $2x + y = 1$
$x = 8 + y$

13. $4x + 3y = 0$
$x = 2 - y$

14. $3x - y = 6$
$y = 3x - 6$

Solve each of the following by any method.

15. The sum of two numbers is 39. If one number is doubled, it equals three less than the other. Find the numbers.

16. The local record shop is having a sale. Some records cost \$2.50 and some cost \$3.75. Joe has exactly \$20 to spend and wants to buy six records. How many can he buy at each price?

17. Two cars leave from the same place and travel in the same direction. One car travels one and one-third times as fast as the other. After three hours, they are 45 miles apart. What was the speed of each car?

Graph the solution of each system of inequalities.

18. $2x + 7y \leq 14$
$x - y \geq 1$

19. $2x - y \leq 6$
$4y + 12 \geq -3x$

20. $3x - 5y < 15$
$y < 2$

8 ROOTS AND RADICALS

8.1 ROOTS

A number is squared by multiplying it by itself.

$$\text{if } a = 7, \text{ then } a^2 = 7 \cdot 7 = 49$$
$$\text{if } a = 9, \text{ then } a^2 = 9 \cdot 9 = 81$$
$$\text{if } a = -5, \text{ then } a^2 = (-5)(-5) = 25.$$

In this chapter we consider the opposite problem.

$$\text{if } a^2 = 49, \text{ then } a = \,?$$
$$\text{if } a^2 = 81, \text{ then } a = \,?$$
$$\text{if } a^2 = 25, \text{ then } a = \,?$$

To answer these questions, we must find a number, when we are given the square of the number. The number that we find is called the **square root** of the given number. Finding (or extracting, or taking) the square root is the inverse operation of squaring a number, just as subtraction is the inverse of addition.

Example 1 One square root of 49 is 7, since $7^2 = 49$. Another square root of 49 is -7, since $(-7)^2 = 49$ also. Thus, 49 has *two* square roots, 7 and -7, one positive and one negative.

Also, there are two square roots of 81, one which is positive, 9, and one which is negative, -9. The numbers 49 and 81 have square roots which are integers. Any number which has integer square roots is called a *perfect square.* The first hundred positive perfect squares are listed inside the front cover.

To write the square root of a number, use the symbol $\sqrt{}$. For example, the positive square root of 121 is 11, written

$$\sqrt{121} = 11.$$

Also, $\sqrt{169} = 13$ and $\sqrt{7396} = 86$.

The symbol $\sqrt{\ }$ is called a **radical sign,** and the number inside the radical sign is called the **radicand.** The entire expression, radical sign and radicand, is a **radical.**

The radical sign, used alone, always represents the *positive* square root (except that $\sqrt{0} = 0$). For example, since the positive square root of 81 is 9, we have $\sqrt{81} = 9$. To write the negative square root of 81, write $-\sqrt{81} = -9$.

Example 2 The square roots of 225 are 15 and -15.

Example 3 $\sqrt{225} = 15$ and $-\sqrt{225} = -15$.

Example 4 **(a)** $-\sqrt{144} = -12$.

(b) $\sqrt{10{,}000} = 100$.

(c) $-\sqrt{10{,}000} = -100$.

(d) $-\sqrt{5041} = -71$.

Summarizing, for any *positive* number a, there are always *two* square roots, one positive, written $\sqrt{a}$, and one negative, written $-\sqrt{a}$. If the number a is a perfect square, then $\sqrt{a}$ is an integer. The square root of 0, written $\sqrt{0}$, is 0. If a is a *negative* number, then there is *no* real number which equals $\sqrt{a}$. For example, $\sqrt{-16}$ is not a real number, since there is no real number whose square is -16. (The square of any real number is non-negative.)

Example 5 **(a)** There are two square roots of 256,

$$\sqrt{256} = 16 \quad \text{and} \quad -\sqrt{256} = -16.$$

(b) There are two square roots of 15,

$$\sqrt{15} \quad \text{and} \quad -\sqrt{15}.$$

Since 15 is not a perfect square, $\sqrt{15}$ cannot be written as an integer.

Example 6 There is no real number square root of -36, so that $\sqrt{-36}$ is not a real number.

A real number which is not rational is called an **irrational number.** In general, it is difficult to prove that a given number is irrational. However, if a is a positive integer that is not a perfect square, then $\sqrt{a}$ is irrational. For example, since 17, 35, 52, and 78 are not perfect squares, then $\sqrt{17}$, $\sqrt{35}$, $\sqrt{52}$, and $\sqrt{78}$ are irrational. Since 81 is a perfect square, $\sqrt{81}$ (or 9) is a rational number.

Not all irrational numbers are roots of integers. For example, π (approximately 3.14159) is an irrational number that is not a square root of any integer.

Example 7 Write rational or irrational for each square root: **(a)** $\sqrt{64}$; **(b)** $\sqrt{90}$; **(c)** $\sqrt{7}$.

Since $\sqrt{64} = 8$, which is a rational number, example (a) is rational. There is no rational number whose square is 90, and no rational number whose square is 7. Examples (b) and (c) represent irrational numbers.

In summary, we know the following facts about square roots.

1. For any positive number a, the number $\sqrt{a}$ represents a positive number with the property that

$$\sqrt{a} \cdot \sqrt{a} = (\sqrt{a})^2 = a.$$

The number $\sqrt{a}$ is the positive square root of a, while $-\sqrt{a}$ is the negative square root of a.

2. If a is a positive perfect square, then $\sqrt{a}$ is an integer. Otherwise, $\sqrt{a}$ is an irrational number.
3. $\sqrt{0} = 0$.
4. If a is a negative number, then $\sqrt{a}$ is not a real number.

Finding the square root of a number is the inverse of squaring a number. In a similar way, there are inverses to finding the cube of a number, or finding the fourth or higher power of a number. These inverses are called finding the **cube root,** written $\sqrt[3]{a}$, the **fourth root,** written $\sqrt[4]{a}$, and so on. We could write $\sqrt[2]{a}$ instead of $\sqrt{a}$, but the simpler symbol $\sqrt{a}$ is customary since the square root is the most commonly used root.

Example 8 Find $\sqrt[3]{8}$.

We need a number such that the cube of the number is 8. Since $2^3 = 8$, we find that $\sqrt[3]{8} = 2$.

Example 9 Find $\sqrt[3]{-8}$.

Since $(-2)^3 = -8$, we have $\sqrt[3]{-8} = -2$. In general, the cube root of a positive number is positive, and the cube root of a negative number is negative. There is only one real number cube root for a given number.

Example 10 **(a)** $\sqrt[4]{16} = 2$ and $-\sqrt[4]{16} = -2$.

(b) $\sqrt[3]{64} = 4$.

(c) $\sqrt[3]{-64} = -4$.

(d) $-\sqrt[5]{32} = -2$ (and $\sqrt[5]{32} = 2$).

EXERCISES

In Exercises 1–20, find all the roots of the numbers. If a root cannot be found, write "not a real number." Use the square root table inside the front cover, as necessary.

1. square roots of 9
2. square roots of 16
3. square roots of 121
4. square roots of 196
5. square roots of 400
6. square roots of 900
7. square roots of 8281
8. square roots of 6241
9. square roots of 15
10. square roots of 21
11. square roots of 72
12. square roots of 68
13. square roots of -25
14. square roots of -49
15. cube roots of 125
16. cube roots of 216
17. fourth roots of 625
18. fourth roots of 10,000
19. fifth roots of 1
20. sixth roots of 1

In Exercises 21–42, find all the roots indicated.

21. $\sqrt{64}$
22. $\sqrt{100}$
23. $-\sqrt{81}$
24. $-\sqrt{121}$
25. $\sqrt{256}$
26. $\sqrt{144}$
27. $\sqrt{625}$
28. $\sqrt{1024}$
29. $-\sqrt{1225}$
30. $-\sqrt{2304}$
31. $-\sqrt{1681}$
32. $-\sqrt{3481}$
33. $\sqrt{-9}$
34. $\sqrt{-16}$
35. $\sqrt[3]{27}$
36. $\sqrt[3]{216}$
37. $\sqrt[3]{-1}$
38. $-\sqrt[3]{1}$
39. $\sqrt[4]{81}$
40. $\sqrt[4]{256}$
41. $\sqrt[4]{1}$
42. $-\sqrt[4]{1}$

In Exercises 43–50, find the square roots of the decimals. Use the square-root table, along with trial and error, to find the roots. (For example, $\sqrt{.0036} = .06$, since $.06^2 = .0036$.)

43. $\sqrt{.25}$
44. $\sqrt{.36}$
45. $\sqrt{1.21}$
46. $\sqrt{1.69}$
47. $\sqrt{42.25}$
48. $\sqrt{65.61}$
49. $\sqrt{.0016}$
50. $\sqrt{.0081}$

8.2 PRODUCTS AND QUOTIENTS OF RADICALS

In this section, we develop rules that can be used to find products and quotients of radicals. To find a rule for products, first look at the following examples.

$$\sqrt{4} \cdot \sqrt{9} = 2 \cdot 3 = 6 \qquad \text{and} \qquad \sqrt{4 \cdot 9} = \sqrt{36} = 6.$$

Then

$$\sqrt{4} \cdot \sqrt{9} = \sqrt{4 \cdot 9}.$$

This example generalizes as the **product rule for radicals.***

If $x \geq 0$ and $y \geq 0$, then $\sqrt{x} \cdot \sqrt{y} = \sqrt{xy}$.

In other words, the product of two radicals is the radical of the product, as long as both radicals are square roots, or both are cube roots, etc.

Example 1 **(a)** $\sqrt{2} \cdot \sqrt{3} = \sqrt{2 \cdot 3} = \sqrt{6}.$

(b) $\sqrt{8} \cdot \sqrt{5} = \sqrt{8 \cdot 5} = \sqrt{40}.$

(c) $\sqrt{7} \cdot \sqrt{a} = \sqrt{7a}.$

The most important use of the product rule is in writing radicals in a simplified form, in which the radicand has no perfect square factor. For example, to simplify $\sqrt{20}$, note that 20 is exactly divisible by 4, which is a perfect square.

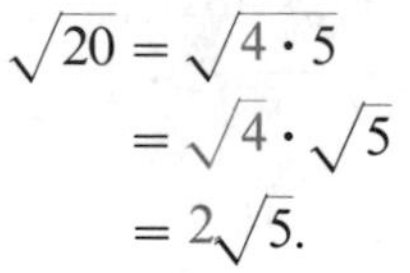

$$\begin{aligned} \sqrt{20} &= \sqrt{4 \cdot 5} \\ &= \sqrt{4} \cdot \sqrt{5} \\ &= 2\sqrt{5}. \end{aligned}$$

Thus $\sqrt{20} = 2\sqrt{5}$. Since 5 is not divisible by a perfect square, $2\sqrt{5}$ is the **simplified form** of $\sqrt{20}$.

Example 2 Simplify $\sqrt{72}$.

Since 72 is divisible by the perfect square 36, then

$$\begin{aligned} \sqrt{72} &= \sqrt{36 \cdot 2} \\ &= \sqrt{36} \cdot \sqrt{2} \\ &= 6\sqrt{2}. \end{aligned}$$

* All the rules for radicals in this chapter are given only for square roots. However, the rules are also valid for cube roots, fourth roots, etc.

Example 3 $\sqrt{300} = \sqrt{100 \cdot 3}$
$= \sqrt{100} \cdot \sqrt{3}$
$= 10\sqrt{3}.$

Example 4 The number 15 is not evenly divisible by a perfect square. Thus, $\sqrt{15}$ cannot be simplified.

Example 5 $\sqrt{25} \cdot \sqrt{75} = 5\sqrt{75}$
$= 5\sqrt{25 \cdot 3}$
$= 5(\sqrt{25} \cdot \sqrt{3})$
$= 5(5\sqrt{3})$
$= 25\sqrt{3}.$

A rule similar to the product rule for radicals can be obtained for quotients of radicals. To obtain this rule, first use the product rule to obtain

$$\frac{\sqrt{x}}{\sqrt{y}} \cdot \frac{\sqrt{x}}{\sqrt{y}} = \frac{\sqrt{x} \cdot \sqrt{x}}{\sqrt{y} \cdot \sqrt{y}} = \frac{(\sqrt{x})^2}{(\sqrt{y})^2} = \frac{x}{y}. \qquad [\text{Recall: } \sqrt{x} \cdot \sqrt{x} = x.]$$

Then check that

$$\sqrt{\frac{x}{y}} \cdot \sqrt{\frac{x}{y}} = \left(\sqrt{\frac{x}{y}}\right)^2 = \frac{x}{y}.$$

Since $\sqrt{x/y}$ and $\sqrt{x}/\sqrt{y}$ both represent the positive square root of x/y, and since there is only one positive square root of x/y, we have

$$\sqrt{\frac{x}{y}} = \frac{\sqrt{x}}{\sqrt{y}}$$

for $x \geq 0$ and $y > 0$. This result is called the **quotient rule for radicals**. In other words: the radical of a quotient is the quotient of the radicals. Do not forget that we still cannot have a denominator of 0.

Example 6 $\sqrt{\frac{25}{9}} = \frac{\sqrt{25}}{\sqrt{9}} = \frac{5}{3}.$

Example 7 $\sqrt{\frac{144}{49}} = \frac{\sqrt{144}}{\sqrt{49}} = \frac{12}{7}.$

Example 8 $\sqrt{\frac{3}{4}} = \frac{\sqrt{3}}{\sqrt{4}} = \frac{\sqrt{3}}{2}.$

Since $\sqrt{3}$ cannot be written in a simpler form, the simplified form of $\sqrt{3/4}$ is $\sqrt{3}/2$.

Example 9 $\dfrac{27\sqrt{15}}{9\sqrt{3}} = \dfrac{27}{9} \cdot \dfrac{\sqrt{15}}{\sqrt{3}} = 3\sqrt{\dfrac{15}{3}} = 3\sqrt{5}.$

Some problems require both the product and the quotient rules, as the next example shows.

Example 10 Simplify $\sqrt{\dfrac{3}{8}} \cdot \sqrt{\dfrac{27}{32}}$.

Use the product rule and the quotient rule.

$$\sqrt{\frac{3}{8}} \cdot \sqrt{\frac{27}{32}} = \frac{\sqrt{3}}{\sqrt{8}} \cdot \frac{\sqrt{27}}{\sqrt{32}} = \frac{\sqrt{3} \cdot \sqrt{27}}{\sqrt{8} \cdot \sqrt{32}} = \frac{\sqrt{3 \cdot 27}}{\sqrt{8 \cdot 32}} = \frac{\sqrt{81}}{\sqrt{256}} = \frac{9}{16}.$$

The properties of this section are also valid when variables appear under the radical sign, as long as it is assumed that all the variables represent only positive numbers.

Example 11 **(a)** $\sqrt{25m^4} = \sqrt{25} \cdot \sqrt{m^4} = 5m^2.$

(b) $\sqrt{64p^{10}} = 8p^5.$

(c) $\sqrt{r^9} = \sqrt{r^8 \cdot r} = \sqrt{r^8} \cdot \sqrt{r} = r^4\sqrt{r}.$

(d) $\sqrt{\dfrac{5}{x^2}} = \dfrac{\sqrt{5}}{\sqrt{x^2}} = \dfrac{\sqrt{5}}{x}.$

(e) $\sqrt[3]{a^{12}} = a^4.$

EXERCISES

In Exercises 1–36, use the product rule to simplify each expression.

1. $\sqrt{27}$

2. $\sqrt{45}$

3. $\sqrt{28}$

4. $\sqrt{40}$

5. $\sqrt{18}$

6. $\sqrt{75}$

7. $\sqrt{48}$

8. $\sqrt{80}$

9. $\sqrt{125}$

10. $\sqrt{150}$

11. $\sqrt{700}$

12. $\sqrt{1100}$

13. $\sqrt{8} \cdot \sqrt{2}$

14. $\sqrt{27} \cdot \sqrt{3}$

15. $\sqrt{6} \cdot \sqrt{6}$

16. $\sqrt{11} \cdot \sqrt{11}$

17. $\sqrt{21} \cdot \sqrt{21}$

18. $\sqrt{17} \cdot \sqrt{17}$

19. $\sqrt{3} \cdot \sqrt{7}$

20. $\sqrt{2} \cdot \sqrt{5}$

21. $\sqrt{9} \cdot \sqrt{75}$

22. $\sqrt{16} \cdot \sqrt{8}$

23. $\sqrt{80} \cdot \sqrt{20}$

24. $\sqrt{200} \cdot \sqrt{2}$

25. $\sqrt{27} \cdot \sqrt{48}$

26. $\sqrt{75} \cdot \sqrt{27}$

27. $\sqrt{50} \cdot \sqrt{72}$

28. $\sqrt{98} \cdot \sqrt{8}$

29. $\sqrt{7} \cdot \sqrt{21}$

30. $\sqrt{12} \cdot \sqrt{48}$

31. $\sqrt{15} \cdot \sqrt{45}$

32. $\sqrt{20} \cdot \sqrt{45}$

33. $\sqrt{80} \cdot \sqrt{15}$

34. $\sqrt{60} \cdot \sqrt{12}$

35. $\sqrt{50} \cdot \sqrt{20}$

36. $\sqrt{72} \cdot \sqrt{12}$

In Exercises 37–56, use the quotient rule and the product rule, as necessary, to simplify each expression.

37. $\sqrt{\frac{100}{9}}$

38. $\sqrt{\frac{225}{16}}$

39. $\sqrt{\frac{36}{49}}$

40. $\sqrt{\frac{256}{9}}$

41. $\sqrt{\frac{5}{16}}$

42. $\sqrt{\frac{11}{25}}$

43. $\sqrt{\frac{30}{49}}$

44. $\sqrt{\frac{10}{121}}$

45. $\sqrt{\frac{1}{5}} \cdot \sqrt{\frac{4}{5}}$

46. $\sqrt{\frac{2}{3}} \cdot \sqrt{\frac{2}{27}}$

47. $\sqrt{\frac{2}{5}} \cdot \sqrt{\frac{8}{125}}$

48. $\sqrt{\frac{3}{8}} \cdot \sqrt{\frac{3}{2}}$

49. $\frac{\sqrt{75}}{\sqrt{3}}$

50. $\frac{\sqrt{200}}{\sqrt{2}}$

51. $\frac{\sqrt{48}}{\sqrt{3}}$

52. $\frac{\sqrt{72}}{\sqrt{8}}$

53. $\frac{15\sqrt{10}}{5\sqrt{2}}$

54. $\frac{18\sqrt{20}}{2\sqrt{10}}$

55. $\frac{25\sqrt{50}}{5\sqrt{5}}$

56. $\frac{26\sqrt{10}}{13\sqrt{5}}$

Simplify each expression in Exercises 57–72. Assume that all variables represent non-negative numbers.

57. $\sqrt{y} \cdot \sqrt{y}$

58. $\sqrt{m} \cdot \sqrt{m}$

59. $\sqrt{x} \cdot \sqrt{z}$

60. $\sqrt{p} \cdot \sqrt{q}$

61. $\sqrt{x^2}$

62. $\sqrt{y^2}$

63. $\sqrt{x^4}$

64. $\sqrt{y^4}$

65. $\sqrt{x^2 y^4}$

66. $\sqrt{x^4 y^8}$

67. $\sqrt{x^3}$

68. $\sqrt{y^3}$

69. $\sqrt[3]{x^6}$

70. $\sqrt[3]{y^9}$

71. $\sqrt[3]{a^3 b^6}$

72. $\sqrt[3]{m^9 n^{15}}$

8.3 ADDITION AND SUBTRACTION OF RADICALS

In the preceding section we discussed the product and quotient rules for radicals: for $x \geq 0$ and $y \geq 0$,

$$\sqrt{x} \cdot \sqrt{y} = \sqrt{xy} \quad \text{and} \quad \frac{\sqrt{x}}{\sqrt{y}} = \sqrt{\frac{x}{y}} \quad (y \neq 0).$$

In this section we look at addition and subtraction of radicals. It would be helpful if addition of radicals obeyed a rule similar to that for multiplication: that is, we would like it to be true that $\sqrt{x} + \sqrt{y} = \sqrt{x + y}$ for all positive real numbers x and y. To see if this rule is true, let us substitute some numerical values for x and y. If we let $x = 4$ and $y = 9$, we see that

$$\sqrt{4} + \sqrt{9} = 2 + 3 = 5,$$

while

$$\sqrt{4 + 9} = \sqrt{13}.$$

Since $\sqrt{4} + \sqrt{9} \neq \sqrt{4 + 9}$, it is *not* generally true that $\sqrt{x} + \sqrt{y}$ and $\sqrt{x + y}$ are equal. Because of this, it is not always possible to simplify a sum or difference of two radicals. For example, $\sqrt{3} + \sqrt{7}$ cannot be further simplified, nor can $\sqrt{19} - \sqrt{3}$.

However, if the radicals in a sum or difference have the same radicands, then the distributive property can be used to simplify the (radical) expression.

Example 1 **(a)** $3\sqrt{6} + 5\sqrt{6} = (3 + 5)\sqrt{6} = 8\sqrt{6}$.

(b) $5\sqrt{10} - 7\sqrt{10} = (5 - 7)\sqrt{10} = -2\sqrt{10}$.

(c) $\sqrt{5} + \sqrt{5} = 1\sqrt{5} + 1\sqrt{5} = (1 + 1)\sqrt{5} = 2\sqrt{5}$.

(d) $\sqrt{7} + 2\sqrt{7} = 1\sqrt{7} + 2\sqrt{7} = 3\sqrt{7}$.

Sometimes it is necessary first to simplify each radical expression in a sum or difference. Then simplify further if possible.

Example 2
$$\begin{aligned} 3\sqrt{2} + \sqrt{8} &= 3\sqrt{2} + \sqrt{4 \cdot 2} \\ &= 3\sqrt{2} + \sqrt{4} \cdot \sqrt{2} \\ &= 3\sqrt{2} + 2\sqrt{2} \\ &= 5\sqrt{2}. \end{aligned}$$

Example 3
$$\begin{aligned} \sqrt{18} - \sqrt{27} &= \sqrt{9 \cdot 2} - \sqrt{9 \cdot 3} \\ &= \sqrt{9} \cdot \sqrt{2} - \sqrt{9} \cdot \sqrt{3} \\ &= 3\sqrt{2} - 3\sqrt{3}. \end{aligned}$$

Since $\sqrt{2}$ and $\sqrt{3}$ are unlike radicals, this difference cannot be further simplified.

Example 4 $2\sqrt{12} + 3\sqrt{75} = 2(\sqrt{4} \cdot \sqrt{3}) + 3(\sqrt{25} \cdot \sqrt{3})$
$= 2(2\sqrt{3}) + 3(5\sqrt{3})$
$= 4\sqrt{3} + 15\sqrt{3}$
$= 19\sqrt{3}.$

Example 5 $\sqrt{5} \cdot \sqrt{15} + 4\sqrt{3} = \sqrt{5 \cdot 15} + 4\sqrt{3}$
$= \sqrt{75} + 4\sqrt{3}$
$= \sqrt{25 \cdot 3} + 4\sqrt{3}$
$= 5\sqrt{3} + 4\sqrt{3}$
$= 9\sqrt{3}.$

Example 6 $\sqrt{12k} + \sqrt{27k} = \sqrt{4 \cdot 3k} + \sqrt{9 \cdot 3k}$
$= \sqrt{4} \cdot \sqrt{3k} + \sqrt{9} \cdot \sqrt{3k}$
$= 2\sqrt{3k} + 3\sqrt{3k}$
$= 5\sqrt{3k}.$

In summary, a sum or difference of radicals can be simplified only if the radicals are *like* radicals.

EXERCISES

In Exercises 1–40, simplify and combine terms wherever possible.

1. $2\sqrt{3} + 5\sqrt{3}$

2. $6\sqrt{5} + 8\sqrt{5}$

3. $4\sqrt{7} - 9\sqrt{7}$

4. $6\sqrt{2} - 8\sqrt{2}$

5. $\sqrt{6} + \sqrt{6}$

6. $\sqrt{11} + \sqrt{11}$

7. $\sqrt{17} + 2\sqrt{17}$

8. $3\sqrt{19} + \sqrt{19}$

9. $5\sqrt{7} - \sqrt{7}$

10. $12\sqrt{14} - \sqrt{14}$

11. $5\sqrt{8} + \sqrt{8}$

12. $3\sqrt{27} - \sqrt{27}$

13. $\sqrt{45} + 2\sqrt{20}$

14. $\sqrt{24} + 5\sqrt{54}$

15. $3\sqrt{18} + \sqrt{8}$

16. $2\sqrt{27} - \sqrt{3}$

17. $-\sqrt{12} + \sqrt{75}$

18. $2\sqrt{27} - \sqrt{300}$

19. $5\sqrt{72} - 2\sqrt{50}$

20. $6\sqrt{18} - 4\sqrt{32}$

21. $-5\sqrt{32} + \sqrt{98}$

22. $4\sqrt{75} + 3\sqrt{12}$

23. $5\sqrt{7} - 2\sqrt{28} + 6\sqrt{63}$

24. $3\sqrt{11} + 5\sqrt{44} - 3\sqrt{99}$

25. $6\sqrt{5} + 3\sqrt{20} - 8\sqrt{45}$

26. $7\sqrt{3} + 2\sqrt{12} - 5\sqrt{27}$

27. $6\sqrt{2} + 5\sqrt{27} - 4\sqrt{12}$

28. $9\sqrt{24} - 2\sqrt{54} + 3\sqrt{20}$

29. $2\sqrt{8} - 5\sqrt{32} + 2\sqrt{48}$

30. $5\sqrt{72} - 3\sqrt{48} - 4\sqrt{128}$

31. $4\sqrt{50} + 3\sqrt{12} + 5\sqrt{45}$

32. $6\sqrt{18} + 2\sqrt{48} - 6\sqrt{28}$

33. $\frac{1}{4}\sqrt{288} - \frac{1}{6}\sqrt{72}$

34. $\frac{2}{3}\sqrt{27} - \frac{3}{4}\sqrt{48}$

35. $\frac{3}{5}\sqrt{75} - \frac{2}{3}\sqrt{45}$

36. $\frac{5}{8}\sqrt{128} - \frac{3}{4}\sqrt{160}$

37. $\sqrt{6} \cdot \sqrt{2} + 3\sqrt{3}$

38. $4\sqrt{15} \cdot \sqrt{3} - 2\sqrt{5}$

39. $\sqrt{3} \cdot \sqrt{7} + 2\sqrt{21} - \sqrt{7}$

40. $\sqrt{13} \cdot \sqrt{2} + 3\sqrt{26}$

Simplify each expression in Exercises 41–50. Assume that all variables represent non-negative real numbers.

41. $\sqrt{9x} + \sqrt{49x} - \sqrt{16x}$

42. $\sqrt{4a} - \sqrt{16a} + \sqrt{9a}$

43. $\sqrt{4a} + 6\sqrt{a} + \sqrt{25a}$

44. $\sqrt{6x^2} + x\sqrt{54}$

45. $\sqrt{75x^2} + x\sqrt{300}$

46. $\sqrt{20y^2} - 3y\sqrt{5}$

47. $3\sqrt{8x^2} - 4x\sqrt{2}$

48. $6r\sqrt{27r^2s} + 3r^2\sqrt{3s}$

49. $\sqrt{x^2y^3w} - xy\sqrt{yw}$

50. $2\sqrt{3m^2n} + 5m\sqrt{3n}$

8.4 RATIONALIZING THE DENOMINATOR

In Section 8.6 we look at ways of finding decimal approximations of radicals. It is shown in that section that it is easier to find these decimal approximations if denominators contain no radicals. For example, the quotient $\sqrt{3}/\sqrt{2}$ would be numerically easier to work with if the denominator contained no radicals.

To eliminate the radical in the denominator, multiply numerator and denominator by $\sqrt{2}$. We multiply by $\sqrt{2}$ since we wish to make the denominator into a rational number, and we know that $\sqrt{2} \cdot \sqrt{2} = 2$, a rational number. Multiplying both numerator and denominator by $\sqrt{2}$ gives

$$\frac{\sqrt{3}}{} = \frac{\sqrt{3} \cdot \sqrt{2}}{\sqrt{2} \cdot \sqrt{2}} = \frac{\sqrt{6}}{}.$$

This process of changing the denominator from a radical (irrational number) to a rational number is called **rationalizing the denominator**. Note that the value of the number is not changed, only the form of the number is changed.

Example 1 Rationalize the denominator in the quotient $9/\sqrt{6}$.

Multiply both numerator and denominator by $\sqrt{6}$.

$$\frac{9}{\sqrt{6}} = \frac{9 \cdot \sqrt{6}}{\sqrt{6} \cdot \sqrt{6}} = \frac{9\sqrt{6}}{6} = \frac{3\sqrt{6}}{2}.$$

Example 2 Rationalize the denominator in the quotient $12/\sqrt{8}$.

We could rationalize the denominator here by multiplying by $\sqrt{8}$. However, we can eliminate the radical more readily if we multiply by $\sqrt{2}$. This will give a denominator of $\sqrt{8} \cdot \sqrt{2} = \sqrt{16} = 4$, a rational number.

$$\begin{aligned} \frac{12}{\sqrt{8}} &= \frac{12\sqrt{2}}{\sqrt{8} \cdot \sqrt{2}} \\ &= \frac{12\sqrt{2}}{\sqrt{16}} \\ &= \frac{12\sqrt{2}}{4} \\ &= 3\sqrt{2}. \end{aligned}$$

Example 3 Simplify $\sqrt{\dfrac{27}{5}}$ by rationalizing the denominator.

First use the quotient rule for radicals, and then multiply both numerator and denominator by $\sqrt{5}$.

$$\begin{aligned} \sqrt{\frac{27}{5}} &= \frac{\sqrt{27}}{\sqrt{5}} \\ &= \frac{\sqrt{27} \cdot \sqrt{5}}{\sqrt{5} \cdot \sqrt{5}} \\ &= \frac{(\sqrt{9} \cdot \sqrt{3})\sqrt{5}}{5} \\ &= \frac{3(\sqrt{3} \cdot \sqrt{5})}{5} \\ &= \frac{3\sqrt{15}}{5}. \end{aligned}$$

Example 4 Rationalize the denominator in the quotient $\sqrt{4x}/\sqrt{y}$. Assume $x \geq 0$ and $y > 0$.

Multiply numerator and denominator by $\sqrt{y}$.

$$\frac{\sqrt{4x}}{\sqrt{y}} = \frac{\sqrt{4x} \cdot \sqrt{y}}{\sqrt{y} \cdot \sqrt{y}}$$

$$= \frac{\sqrt{4xy}}{y}$$

$$= \frac{\sqrt{4} \cdot \sqrt{xy}}{y}$$

$$= \frac{2\sqrt{xy}}{y}.$$

Example 5 Rationalize the denominator in $\sqrt[3]{\frac{1}{9}}$.

By the quotient rule, we get

$$\sqrt[3]{\frac{1}{9}} = \frac{\sqrt[3]{1}}{\sqrt[3]{9}} = \frac{1}{\sqrt[3]{9}}.$$

We can eliminate the radical sign in the denominator if we can replace the 9 by a number which is a perfect cube. The number 27 is a perfect cube ($3^3 = 27$). Since $9 \cdot 3 = 27$, we can rationalize the denominator by multiplying both numerator and denominator by $\sqrt[3]{3}$.

$$\frac{1}{\sqrt[3]{9}} = \frac{1 \cdot \sqrt[3]{3}}{\sqrt[3]{9} \cdot \sqrt[3]{3}}$$

$$= \frac{\sqrt[3]{3}}{\sqrt[3]{27}}$$

$$= \frac{\sqrt[3]{3}}{3}.$$

EXERCISES

In Exercises 1–40, perform the indicated operations. Write all answers in simplest form. Rationalize all denominators.

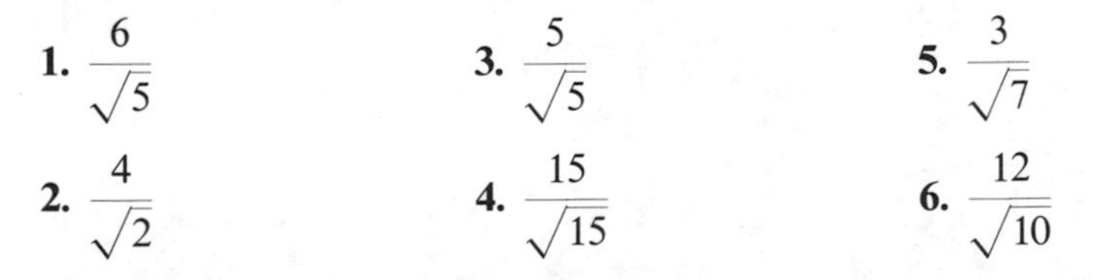

1. $\frac{6}{\sqrt{5}}$

2. $\frac{4}{\sqrt{2}}$

3. $\frac{5}{\sqrt{5}}$

4. $\frac{15}{\sqrt{15}}$

5. $\frac{3}{\sqrt{7}}$

6. $\frac{12}{\sqrt{10}}$

7. $\frac{8\sqrt{3}}{\sqrt{5}}$

8. $\frac{9\sqrt{6}}{\sqrt{5}}$

9. $\frac{12\sqrt{10}}{8\sqrt{3}}$

10. $\frac{9\sqrt{15}}{6\sqrt{2}}$

11. $\frac{8}{\sqrt{27}}$

12. $\frac{12}{\sqrt{18}}$

13. $\frac{3}{\sqrt{50}}$

14. $\frac{5}{\sqrt{75}}$

15. $\frac{12}{\sqrt{72}}$

16. $\frac{21}{\sqrt{45}}$

17. $\frac{9}{\sqrt{32}}$

18. $\frac{50}{\sqrt{125}}$

19. $\frac{\sqrt{8}}{\sqrt{2}}$

20. $\frac{\sqrt{27}}{\sqrt{3}}$

21. $\frac{\sqrt{10}}{\sqrt{5}}$

22. $\frac{\sqrt{6}}{\sqrt{3}}$

23. $\frac{\sqrt{40}}{\sqrt{3}}$

24. $\frac{\sqrt{5}}{\sqrt{8}}$

25. $\sqrt{\frac{1}{2}}$

26. $\sqrt{\frac{1}{8}}$

27. $\sqrt{\frac{10}{7}}$

28. $\sqrt{\frac{2}{3}}$

29. $\sqrt{\frac{9}{5}}$

30. $\sqrt{\frac{16}{7}}$

31. $\sqrt{\frac{7}{5}} \cdot \sqrt{10}$

32. $\sqrt{\frac{1}{3}} \cdot \sqrt{3}$

33. $\sqrt{\frac{3}{4}} \cdot \sqrt{\frac{1}{5}}$

34. $\sqrt{\frac{1}{10}} \cdot \sqrt{\frac{10}{3}}$

35. $\sqrt[3]{\frac{1}{4}}$

36. $\sqrt[3]{\frac{1}{32}}$

37. $\sqrt[3]{\frac{3}{2}}$

38. $\sqrt[3]{\frac{2}{5}}$

39. $\sqrt[3]{\frac{4}{9}}$

40. $\sqrt[3]{\frac{3}{4}}$

In Exercises 41–50, perform the indicated operations. Write all answers in simplest form. Rationalize all denominators. Assume that all variables represent positive real numbers.

41. $\sqrt{\frac{5}{x}}$

42. $\sqrt{\frac{6}{p}}$

43. $\sqrt{\frac{4r^3s}{r}}$

44. $\sqrt{\frac{6p^3}{3m}}$

45. $\sqrt{\frac{72a^3b}{6}}$

46. $\sqrt{\frac{x^2}{4y}}$

47. $\sqrt{\frac{2x^2z^4}{3y}}$

48. $\frac{\sqrt{18m^2n^3p}}{\sqrt{3mp}}$

49. $\sqrt[3]{\frac{1}{r^2}}$

50. $\sqrt[3]{\frac{3x^2}{2}}$

8.5 SIMPLIFYING RADICAL EXPRESSIONS

It may not be always clear what is the "simplest" form of a radical. In this book, a radical expression is simplified when the following five rules are satisfied.

1. If a radical represents a rational number, then that rational number should be used in place of the radical.

For example, $\sqrt{49}$ is simplified by writing 7; $\sqrt{64}$ as 8; $\sqrt{169/9}$ as 13/3.

2. If a radical expression contains products of radicals, the product rule for radicals, $\sqrt{x} \cdot \sqrt{y} = \sqrt{xy}$, should be used to get a single radical.

For example, $\sqrt{3} \cdot \sqrt{2}$ is simplified to $\sqrt{6}$; $\sqrt{5} \cdot \sqrt{x}$ to $\sqrt{5x}$.

3. If a radicand has a factor that is a perfect square, the radical should be expressed as the product of the positive square root of the perfect square and the remaining radical factor.

For example, $\sqrt{20}$ is simplified to $\sqrt{20} = \sqrt{4 \cdot 5} = \sqrt{4} \cdot \sqrt{5} = 2\sqrt{5}$; $\sqrt{75}$ as $5\sqrt{3}$.

4. If a radical expression contains sums or differences of radicals, the distributive property should be used to combine terms, if possible.

For example, $3\sqrt{2} + 4\sqrt{2}$ is combined as $7\sqrt{2}$; but $3\sqrt{2} + 4\sqrt{3}$ cannot be further combined.

5. Any radicals in the denominator should be changed to rational numbers.

For example, $5/\sqrt{3}$ is rationalized as $5/\sqrt{3} = 5\sqrt{3}/3$.

Example 1 Simplify $\sqrt{16} + \sqrt{9}$.

We have $\sqrt{16} + \sqrt{9} = 4 + 3 = 7$.

Example 2 Simplify $5\sqrt{2} + 2\sqrt{18}$.

First, simplify $\sqrt{18}$.

$$\begin{aligned} 5\sqrt{2} + 2\sqrt{18} &= 5\sqrt{2} + 2(\sqrt{9} \cdot \sqrt{2}) \\ &= 5\sqrt{2} + 2(3\sqrt{2}) \\ &= 5\sqrt{2} + 6\sqrt{2} \\ &= 11\sqrt{2}. \end{aligned}$$

Example 3 Simplify $\sqrt{5}(\sqrt{8} - \sqrt{32})$.

Using the distributive property, we have

$$\begin{aligned} \sqrt{5}(\sqrt{8} - \sqrt{32}) &= \sqrt{5}\cdot\sqrt{8} - \sqrt{5}\cdot\sqrt{32} \\ &= \sqrt{40} - \sqrt{160} \\ &= \sqrt{4}\cdot\sqrt{10} - \sqrt{16}\cdot\sqrt{10} \\ &= 2\sqrt{10} - 4\sqrt{10} \\ &= -2\sqrt{10}. \end{aligned}$$

Example 4 Simplify the product $(\sqrt{3} + 2\sqrt{5})(\sqrt{3} - 4\sqrt{5})$.

The product of these sums of radicals can be found in much the same way that we found the product of binomials in Chapter 3. The pattern of multiplication is the same.

$$\begin{aligned} (\sqrt{3} + 2\sqrt{5})(\sqrt{3} - 4\sqrt{5}) &= \sqrt{3}\cdot\sqrt{3} + \sqrt{3}(-4\sqrt{5}) + 2\sqrt{5}\cdot\sqrt{3} \\ &\quad + 2\sqrt{5}(-4\sqrt{5}) \\ &= 3 - 4\sqrt{15} + 2\sqrt{15} - 8\cdot 5 \\ &= 3 - 2\sqrt{15} - 40 \\ &= -37 - 2\sqrt{15}. \end{aligned}$$

Example 5 Simplify the product $(4 - \sqrt{3})(4 + \sqrt{3})$.

Following the pattern of binomial multiplication,

$$\begin{aligned} (4 - \sqrt{3})(4 + \sqrt{3}) &= 4\cdot 4 + 4\sqrt{3} - 4\sqrt{3} - \sqrt{3}\cdot\sqrt{3} \\ &= 16 - 3 \\ &= 13. \end{aligned}$$

Example 6 Simplify $(\sqrt{12} - \sqrt{6})(\sqrt{12} + \sqrt{6})$.

$$\begin{aligned} (\sqrt{12} - \sqrt{6})(\sqrt{12} + \sqrt{6}) &= \sqrt{12}\cdot\sqrt{12} + \sqrt{12}\cdot\sqrt{6} \\ &\quad - \sqrt{12}\cdot\sqrt{6} - \sqrt{6}\cdot\sqrt{6} \\ &= 12 - 6 \\ &= 6. \end{aligned}$$

The statement $(\sqrt{12} - \sqrt{6})(\sqrt{12} + \sqrt{6}) = 12 - 6 = 6$ is an example of the fact that $(a + b)(a - b) = a^2 - b^2$.

How can we rationalize the denominator of a radical expression such as the following?

$$\frac{2}{4 - \sqrt{3}}$$

We saw in Example 5 above that if we multiply this denominator, $4 - \sqrt{3}$, by the radical $4 + \sqrt{3}$, then the product, $(4 - \sqrt{3})(4 + \sqrt{3})$, is the rational number 13. Multiplying numerator and denominator by $4 + \sqrt{3}$ gives the following result:

$$\frac{2}{4 - \sqrt{3}} = \frac{2(4 + \sqrt{3})}{(4 - \sqrt{3})(4 + \sqrt{3})}$$
$$= \frac{2(4 + \sqrt{3})}{13}.$$

The denominator has now been rationalized—it contains no radical signs.

Example 6 Rationalize the denominator in the quotient

$$\frac{4}{3 + \sqrt{5}}.$$

To eliminate the radical in the denominator, multiply numerator and denominator by $3 - \sqrt{5}$.

$$\frac{4}{3 + \sqrt{5}} = \frac{4(3 - \sqrt{5})}{(3 + \sqrt{5})(3 - \sqrt{5})}$$
$$= \frac{4(3 - \sqrt{5})}{9 + 3\sqrt{5} - 3\sqrt{5} - 5}$$
$$= \frac{4(3 - \sqrt{5})}{9 - 5}$$
$$= \frac{4(3 - \sqrt{5})}{4}$$
$$= 3 - \sqrt{5}.$$

The two expressions $3 + \sqrt{5}$ and $3 - \sqrt{5}$ are called **conjugates** of each other.

Example 7 Simplify $\dfrac{6 + \sqrt{2}}{\sqrt{2} - 5}$.

Multiply numerator and denominator by $\sqrt{2} + 5$:

$$\begin{aligned}\frac{6+\sqrt{2}}{\sqrt{2}-5} &= \frac{(6+\sqrt{2})(\sqrt{2}+5)}{(\sqrt{2}-5)(\sqrt{2}+5)} \\ &= \frac{6\sqrt{2}+30+\sqrt{2}\cdot\sqrt{2}+5\sqrt{2}}{\sqrt{2}\cdot\sqrt{2}+5\sqrt{2}-5\sqrt{2}-25} \\ &= \frac{6\sqrt{2}+30+2+5\sqrt{2}}{2-25} \\ &= \frac{11\sqrt{2}+32}{-23} \\ &= -\frac{11\sqrt{2}+32}{23}.\end{aligned}$$

EXERCISES

Simplify each expression in Exercises 1–32. Use the five rules given in the text.

1. $3\sqrt{5} + 8\sqrt{45}$
2. $6\sqrt{2} + 4\sqrt{18}$
3. $9\sqrt{50} - 4\sqrt{72}$
4. $3\sqrt{80} - 5\sqrt{45}$
5. $\sqrt{2}(\sqrt{8} - \sqrt{32})$
6. $\sqrt{3}(\sqrt{27} - \sqrt{3})$
7. $\sqrt{5}(\sqrt{3} + \sqrt{7})$
8. $\sqrt{7}(\sqrt{10} - \sqrt{3})$
9. $2\sqrt{5}(\sqrt{2} + \sqrt{5})$
10. $3\sqrt{7}(2\sqrt{7} - 4\sqrt{5})$
11. $-\sqrt{14}\cdot\sqrt{2} - \sqrt{28}$
12. $\sqrt{6}\cdot\sqrt{3} - 2\sqrt{50}$
13. $(2\sqrt{6} + 3)(3\sqrt{6} - 5)$
14. $(4\sqrt{5} - 2)(2\sqrt{5} + 3)$
15. $(5\sqrt{7} - 2\sqrt{3})(3\sqrt{7} + 3\sqrt{3})$
16. $(2\sqrt{10} + 5\sqrt{2})(3\sqrt{10} - 4\sqrt{2})$
17. $(3\sqrt{2} + 4)(3\sqrt{2} + 4)$
18. $(4\sqrt{5} - 1)(4\sqrt{5} - 1)$
19. $(2\sqrt{7} - 3)^2$
20. $(3\sqrt{5} + 5)^2$
21. $(3 - \sqrt{2})(3 + \sqrt{2})$
22. $(7 - \sqrt{5})(7 + \sqrt{5})$
23. $(2 + \sqrt{8})(2 - \sqrt{8})$
24. $(3 + \sqrt{11})(3 - \sqrt{11})$
25. $(\sqrt{6} - \sqrt{5})(\sqrt{6} + \sqrt{5})$
26. $(\sqrt{11} + \sqrt{10})(\sqrt{11} - \sqrt{10})$
27. $(\sqrt{18} + \sqrt{2})(\sqrt{18} - \sqrt{2})$
28. $(\sqrt{21} - \sqrt{5})(\sqrt{21} + \sqrt{5})$
29. $(\sqrt{80} - \sqrt{60})(\sqrt{80} + \sqrt{60})$
30. $(\sqrt{92} + \sqrt{72})(\sqrt{92} - \sqrt{72})$
31. $(2\sqrt{5} + 3\sqrt{2})(2\sqrt{5} - 3\sqrt{2})$
32. $(5\sqrt{2} + 6\sqrt{3})(5\sqrt{2} - 6\sqrt{3})$

Rationalize the denominators in Exercises 33–50.

33. $\dfrac{1}{3+\sqrt{2}}$

34. $\dfrac{1}{4-\sqrt{3}}$

35. $\dfrac{5}{2+\sqrt{5}}$

36. $\dfrac{6}{3+\sqrt{7}}$

37. $\dfrac{7}{2-\sqrt{11}}$

38. $\dfrac{38}{5-\sqrt{6}}$

39. $\dfrac{\sqrt{18}}{1+\sqrt{2}}$

40. $\dfrac{\sqrt{27}}{2-\sqrt{3}}$

41. $\dfrac{\sqrt{45}}{1-\sqrt{5}}$

42. $\dfrac{\sqrt{12}}{2+\sqrt{3}}$

43. $\dfrac{\sqrt{48}}{\sqrt{3}+1}$

44. $\dfrac{\sqrt{18}}{\sqrt{2}-1}$

45. $\dfrac{\sqrt{3}}{\sqrt{27}+1}$

46. $\dfrac{\sqrt{12}}{2-\sqrt{2}}$

47. $\dfrac{\sqrt{6}-1}{\sqrt{11}+\sqrt{3}}$

48. $\dfrac{2+\sqrt{18}}{\sqrt{2}+\sqrt{3}}$

49. $\dfrac{\sqrt{6}+1}{\sqrt{12}-3}$

50. $\dfrac{8+\sqrt{5}}{\sqrt{5}-8}$

8.6 DECIMAL APPROXIMATIONS OF RADICALS

Any rational number can be expressed as a decimal by dividing the numerator by the denominator. For example,

$$\frac{6}{5} = 1.2 \qquad \text{and} \qquad \frac{3}{4} = .75.$$

Not all such divisions terminate. For example

$$\frac{2}{3} = .66666666\ldots \qquad \text{and} \qquad \frac{3}{11} = .2727272727\ldots.$$

These two rational numbers have decimal values which are strings of digits that repeat without stopping. By division in this way, every rational number can be expressed as either a terminating or a repeating decimal.

Irrational numbers, on the other hand, have non-repeating, non-terminating decimal representations. Thus, it is not as easy to find decimal values for irrational numbers such as square roots. While formulas from calculus are often necessary to calculate decimal values for roots, we have the convenience of using the square root table, inside the front cover.

This table gives decimal values of square roots of all integers 1–100 to the nearest thousandth. Integers are listed in the first column, headed n. Square roots are given to 3 decimal places in the column headed $\sqrt{n}$. For example,

$$\sqrt{2} = 1.414$$
$$\sqrt{5} = 2.236.$$

(If more accuracy is needed, it is necessary to consult a larger table. The *CRC Standard Mathematical Tables* is a book found in most college bookstores or libraries. Square roots are also given to 8 or 10 places by several types of modern calculators.)

The square root table gives approximations for many square roots. For example, 1.732 is listed for $\sqrt{3}$. In fact, $(\sqrt{3})^2 = 3$, while $(1.732)^2 = 2.999824$. However, to the nearest thousandth, 1.732 is the best possible approximation to $\sqrt{3}$. Some pocket calculators give more decimal places of accuracy, for example,

$$\sqrt{3} = 1.7320508.$$

To show that 1.732 is only an approximation for $\sqrt{3}$, we use the symbol $\approx$ (read "is approximately equal to") and write

$$\sqrt{3} \approx 1.732$$

or

$$\sqrt{3} \approx 1.7320508.$$

The table also contains a column headed $\sqrt{10n}$, which can be used to find the square root of those integers from 100 through 1000 that end in 0. For example, to find $\sqrt{430}$, look down the left-hand column for the number 43. The column headed $\sqrt{10n}$ gives the value of $\sqrt{10 \cdot 43}$ as 20.735.

In this way you can find 3-place approximations for square roots like $\sqrt{180}$, $\sqrt{400}$, $\sqrt{750}$, and $\sqrt{1000}$.

$$\sqrt{180} \approx 13.416$$
$$\sqrt{400} = 20.000$$
$$\sqrt{750} \approx 27.386$$
$$\sqrt{1000} \approx 31.623$$

Approximate square roots of larger and smaller numbers can be worked out using the table and the rules for radicals. The following examples demonstrate this method.

Example 1 $\sqrt{850{,}000} = \sqrt{85 \cdot 10{,}000} = \sqrt{85} \cdot \sqrt{10{,}000}$
$\approx 9.220(100) = 922.0.$

Example 2 $\sqrt{16{,}000} = \sqrt{160 \cdot 100} = \sqrt{160} \cdot \sqrt{100}$
$\approx 12.649(10) = 126.49.$

Example 3 $\sqrt{.03} = \sqrt{\dfrac{3}{100}} = \dfrac{\sqrt{3}}{\sqrt{100}}$

$$\approx \frac{1.732}{10} = .1732.$$

Example 4 $\sqrt{.005} = \sqrt{\dfrac{50}{10{,}000}} = \dfrac{\sqrt{50}}{\sqrt{10{,}000}}$

$$\approx \frac{7.071}{100} = .07071.$$

In Examples 3 and 4, the first step was to write the denominator so that it contains an even number of zeros. Thus the square root is a rational number.

The next example shows why we learned to rationalize the denominator.

Example 5 Using the table, find a decimal approximation for $9/\sqrt{3}$.

The table shows that $\sqrt{3} \approx 1.732$. Thus,

$$\frac{9}{\sqrt{3}} \approx \frac{9}{1.732}.$$

It would be inconvenient to divide 1.732 into 9. The problem would be easier if we first rationalized the denominator. Multiply numerator and denominator by $\sqrt{3}$.

$$\frac{9}{\sqrt{3}} = \frac{9\sqrt{3}}{\sqrt{3} \cdot \sqrt{3}} = \frac{9\sqrt{3}}{3} = 3\sqrt{3}.$$

Now substitute the decimal approximation of 1.732 for $\sqrt{3}$.

$$3\sqrt{3} \approx 3(1.732) = 5.196.$$

Example 6 Find a decimal approximation for

$$\frac{14}{3 - \sqrt{2}}.$$

To rationalize the denominator, multiply numerator and denominator by $3 + \sqrt{2}$, the conjugate of $3 - \sqrt{2}$.

$$\frac{14}{3 - \sqrt{2}} = \frac{14(3 + \sqrt{2})}{(3 - \sqrt{2})(3 + \sqrt{2})}$$

$$= \frac{14(3 + \sqrt{2})}{9 - 2}$$

$$= \frac{14(3 + \sqrt{2})}{7}$$

$$= 2(3 + \sqrt{2}).$$

Now use the square root table: $\sqrt{2} \approx 1.414$. Thus,

$$\begin{aligned} 2(3 + \sqrt{2}) &\approx 2(3 + 1.414) \\ &= 2(4.414) \\ &= 8.828. \end{aligned}$$

EXERCISES

In Exercises 1–32, find a decimal approximation for each square root.

1. $\sqrt{13}$
2. $\sqrt{42}$
3. $\sqrt{51}$
4. $\sqrt{89}$
5. $\sqrt{150}$
6. $\sqrt{210}$
7. $\sqrt{580}$
8. $\sqrt{650}$
9. $\sqrt{750}$
10. $\sqrt{990}$
11. $\sqrt{5800}$
12. $\sqrt{9600}$
13. $\sqrt{1400}$
14. $\sqrt{2700}$
15. $\sqrt{380{,}000}$
16. $\sqrt{920{,}000}$
17. $\sqrt{53{,}000{,}000}$
18. $\sqrt{71{,}000{,}000}$
19. $\sqrt{14{,}000}$
20. $\sqrt{32{,}000}$
21. $\sqrt{79{,}000}$
22. $\sqrt{84{,}000}$
23. $\sqrt{.08}$
24. $\sqrt{.05}$
25. $\sqrt{.62}$
26. $\sqrt{.38}$
27. $\sqrt{.0054}$
28. $\sqrt{.0091}$
29. $\sqrt{.002}$
30. $\sqrt{.005}$
31. $\sqrt{.016}$
32. $\sqrt{.098}$

In Exercises 33–50, first rationalize the denominator of the given expression. Then find a decimal approximation for the given expression.

33. $\dfrac{14}{\sqrt{7}}$
34. $\dfrac{10}{\sqrt{5}}$
35. $\dfrac{6}{\sqrt{3}}$
36. $\dfrac{8}{\sqrt{2}}$
37. $\dfrac{2}{\sqrt{3}}$
38. $\dfrac{5}{\sqrt{2}}$
39. $\dfrac{8}{\sqrt{6}}$
40. $\dfrac{10}{\sqrt{3}}$
41. $\dfrac{3 + \sqrt{5}}{\sqrt{5}}$
42. $\dfrac{2 - \sqrt{6}}{\sqrt{6}}$
43. $\dfrac{2}{1 + \sqrt{5}}$
44. $\dfrac{8}{2 - \sqrt{6}}$
45. $\dfrac{3}{1 + \sqrt{2}}$
46. $\dfrac{5}{2 - \sqrt{3}}$
47. $\dfrac{2}{3 + \sqrt{5}}$
48. $\dfrac{3}{2 + \sqrt{6}}$
49. $\dfrac{5\sqrt{2}}{3 + \sqrt{8}}$
50. $\dfrac{6\sqrt{3}}{4 - \sqrt{27}}$

8.7 EQUATIONS WITH RADICALS

How can we solve an equation involving radicals, such as

$$\sqrt{x+1} = 3?$$

The addition and multiplication properties of equality will not help us get started; we need another property.

Note that if $a = b$, then $a^2 = b^2$. For example, if $y = 4$, then we can square both sides of the equation and get

$$y^2 = 4^2, \qquad \text{or} \qquad y^2 = 16.$$

However, the equation $y = 4$ has only one solution, the number 4. The equation we got by squaring both sides, $y^2 = 16$, has *two* solutions, the numbers 4 and -4.

As shown by this example, squaring both sides of an equation can lead to a new equation with more solutions than the original equation. Because of this possibility, it is necessary to check all proposed solutions in the *original* equation.

In summary, we have the **squaring property of equality.**

If both sides of a given equation are squared, all solutions of the original equation are also solutions of the squared equation.

Example 1 Solve the equation $\sqrt{x+1} = 3$.

Use the squaring property of equality, and square both sides of the equation.

$$(\sqrt{x+1})^2 = 3^2$$
$$x + 1 = 9$$
$$x = 8.$$

Now check this answer in the original equation.

$$\sqrt{x+1} = 3$$
$$\sqrt{8+1} = 3$$
$$\sqrt{9} = 3$$
$$3 = 3. \qquad \text{True}$$

Since this statement is true, the solution of $\sqrt{x+1} = 3$ is the number 8.

Example 2 Solve $3\sqrt{x} = \sqrt{x+8}$.
Squaring both sides gives

$$\begin{aligned}(3\sqrt{x})^2 &= (\sqrt{x+8})^2\\ 3^2(\sqrt{x})^2 &= (\sqrt{x+8})^2\\ 9x &= x+8\\ 8x &= 8\\ x &= 1.\end{aligned}$$

Check this proposed solution.

$$\begin{aligned}3\sqrt{x} &= \sqrt{x+8}\\ 3\sqrt{1} &= \sqrt{1+8}\\ 3(1) &= \sqrt{9}\\ 3 &= 3. \qquad \text{True}\end{aligned}$$

The solution of $3\sqrt{x} = \sqrt{x+8}$ is the number 1.

Example 3 Solve the equation $\sqrt{x} = -3$.
Square both sides.

$$\begin{aligned}(\sqrt{x})^2 &= (-3)^2\\ x &= 9.\end{aligned}$$

Check this proposed answer in the original equation.

$$\begin{aligned}\sqrt{x} &= -3\\ \sqrt{9} &= -3\\ 3 &= -3. \qquad \text{False}\end{aligned}$$

Since the statement $3 = -3$ is false, the number 9 is not a solution of the given equation. Thus, $\sqrt{x} = -3$ has no real number solutions at all.

Example 4 Solve the equation $\sqrt{2x-3} = x - 3$.
To square both sides, first recall that

$$(x-3)^2 = (x-3)(x-3) = x^2 - 6x + 9.$$

Squaring both sides,

$$(\sqrt{2x-3})^2 = (x-3)^2,$$

leads to

$$2x - 3 = x^2 - 6x + 9.$$

Add $-2x$ and 3 to both sides.

$$0 = x^2 - 8x + 12.$$

The quadratic equation $0 = x^2 - 8x + 12$ can be solved by factoring.

$$0 = (x - 6)(x - 2)$$

$$x - 6 = 0 \quad \text{or} \quad x - 2 \neq 0$$

$$x = 6 \quad \text{or} \quad x = 2.$$

Check both these proposed answers in the original equation.

If $x = 6$,

$$\sqrt{2x - 3} = \sqrt{2 \cdot 6 - 3} = \sqrt{12 - 3} = \sqrt{9} = 3$$

$$x - 3 = 6 - 3 = 3.$$

Since both sides of the equation lead to the same answer, 6 is a solution.

If $x = 2$,

$$\sqrt{2x - 3} = \sqrt{2 \cdot 2 - 3} = \sqrt{4 - 3} = \sqrt{1} = 1$$

$$x - 3 = 2 - 3 = -1.$$

Here the two sides of the equation give different results, so 2 is not a solution. The only solution for the given equation is the number 6.

Example 5 Solve the equation $\sqrt{x} + 1 = 2x$.

Square both sides:

$$(\sqrt{x} + 1)^2 = (2x)^2$$

$$x + 2\sqrt{x} + 1 = 4x^2.$$

Squaring both sides of the given equation produced an equation still containing a radical. It would be better to rewrite the given equation so that the radical is alone on one side of the equals sign. To do this, add -1 to both sides.

$$\sqrt{x} = 2x - 1.$$

Now square both sides.

$$(\sqrt{x})^2 = (2x - 1)^2$$

$$x = 4x^2 - 4x + 1.$$

Add $-x$ to both sides.

$$0 = 4x^2 - 5x + 1.$$

This equation is a quadratic equation, which can be solved by factoring.

$$0 = (4x - 1)(x - 1)$$

$$4x - 1 = 0 \quad \text{or} \quad x - 1 = 0$$

$$x = \frac{1}{4} \quad \text{or} \quad x = 1.$$

Both these proposed solutions must be checked in the original equation.

If $x = \frac{1}{4}$,

$$\sqrt{x} + 1 = 2x$$

$$\sqrt{\frac{1}{4}} + 1 = 2\left(\frac{1}{4}\right)$$

$$\frac{1}{2} + 1 = \frac{1}{2}$$

$$\frac{3}{2} = \frac{1}{2}. \qquad \text{False}$$

If $x = 1$,

$$\sqrt{x} + 1 = 2x$$

$$\sqrt{1} + 1 = 2(1)$$

$$1 + 1 = 2$$

$$2 = 2. \qquad \text{True}$$

For $x = \frac{1}{4}$, we obtained a false statement, while $x = 1$ led to a true statement. Therefore, the only solution to the original equation is the number 1.

EXERCISES

In Exercises 1–22, find all solutions for each equation.

1. $\sqrt{x} = 2$

2. $\sqrt{x} = 5$

3. $\sqrt{x + 3} = 2$

4. $\sqrt{x + 1} = 5$

5. $\sqrt{x - 3} = 2$

6. $\sqrt{x + 5} = 4$

7. $\sqrt{x + 8} = 1$

8. $\sqrt{x + 10} = 2$

9. $\sqrt{x + 5} = 0$

10. $\sqrt{x - 4} = 0$

11. $\sqrt{x + 5} = -2$

12. $\sqrt{x - 3} = -2$

13. $\sqrt{5x - 9} = 2\sqrt{x}$

14. $\sqrt{3x + 4} = 2\sqrt{x}$

15. $3\sqrt{x} = \sqrt{8x + 16}$

16. $2\sqrt{x} = \sqrt{3x + 9}$

17. $\sqrt{5x - 5} = \sqrt{4x + 1}$

18. $\sqrt{2x + 2} = \sqrt{3x - 5}$

19. $\sqrt{x + 2} = \sqrt{2x - 5}$

20. $\sqrt{3x + 3} = \sqrt{5x - 1}$

21. $\sqrt{2x + 9} = \sqrt{x + 5}$

22. $\sqrt{6x + 22} = \sqrt{2x + 10}$

In Exercises 23–36, find all solutions for each equation. Remember that

$$(a + b)^2 = a^2 + 2ab + b^2 \qquad \text{and} \qquad (\sqrt{a})^2 = a.$$

23. $\sqrt{2x + 1} = x - 7$

24. $\sqrt{5x + 1} = x + 1$

25. $\sqrt{3x + 10} = 2x - 5$

26. $\sqrt{4x + 13} = 2x - 1$

27. $\sqrt{x + 1} - 1 = x$

28. $\sqrt{3x + 3} + 5 = x$

29. $\sqrt{4x+5} - 2 = 2x - 7$

30. $\sqrt{6x+7} - 1 = x + 1$

31. $3\sqrt{x+13} = x + 9$

32. $2\sqrt{x+7} = x - 1$

33. $\sqrt{4x} - x + 3 = 0$

34. $\sqrt{2x} - x + 4 = 0$

35. $\sqrt{x} = \sqrt{x-5} + 1$
(*Hint:* Square both sides of the equation twice.)

36. $\sqrt{2x} = \sqrt{x+7} - 1$

Solve the word problems of Exercises 37–41.

37. The square root of the sum of a number and 4 is 5. Find the number.

38. A certain number is the same as the square root of the product of 3 and the number. Find the number.

39. Three times the square root of two equals the square root of the sum of some number and ten. Find the number.

40. The negative square root of a number equals that number decreased by two. Find the number.

41. To estimate the speed at which a car was traveling at the time of an accident, police sometimes use the following procedure: A police officer drives the car involved in the accident under conditions similar to those when the accident took place, and skids to a stop. If the car is driven at 30 miles per hour, then the speed at the time of the accident is given by

$$s = \sqrt{\frac{900a}{p}}$$

where a is the length of the skid marks left at the time of the accident, and p is the length of the skid marks in the police test. Find s if
(a) $a = 900$ feet and $p = 100$ feet
(b) $a = 400$ feet and $p = 100$ feet
(c) $a = 80$ feet and $p = 20$ feet.

CHAPTER 8 SUMMARY

Key Words

Root	Radical	Rationalize the denominator
Square root	Radical sign	Decimal approximation
Cube root	Radicand	Conjugates
Fourth root	Simplified form	Irrational number

Definitions

If $a > 0$, then $\sqrt{a}$ is the positive number such that

$$\sqrt{a} \cdot \sqrt{a} = a,$$

and $-\sqrt{a}$ is the negative number such that

$$(-\sqrt{a})(-\sqrt{a}) = a.$$

Also, $\sqrt{0} = 0$.

For all non-negative numbers a and b,

$$\sqrt{a} \cdot \sqrt{b} = \sqrt{ab}; \qquad \text{Product Rule}$$

and, if $b \neq 0$,

$$\frac{\sqrt{a}}{\sqrt{b}} = \sqrt{\frac{a}{b}}. \qquad \text{Quotient Rule}$$

Simplifying Radical Expressions

1. If a radical represents a rational number, use the rational number in place of the radical.
2. If a radical expression contains products of radicals, use the product rule to get a single radical.
3. If a radicand has a perfect square factor, express the radical as a product of the square root and a radical.
4. Combine terms for sums and differences of radicals if possible.
5. Rationalize the denominator.

Squaring Property of Equality

If both sides of a given equation are squared, all solutions of the original equation are also solutions of the squared equation.

CHAPTER 8 TEST

Find the following roots.

1. square root(s) of 100
2. square root(s) of 625
3. cube root(s) of -27
4. fourth root(s) of 16
5. fourth root(s) of -81

Simplify each of the following.

6. $\sqrt{64}$
7. $\sqrt{196}$
8. $\sqrt{50}$
9. $\sqrt{128}$
10. $\sqrt{75}$
11. $\sqrt{8} + 2\sqrt{18}$
12. $\sqrt{6} + 5\sqrt{24}$
13. $3\sqrt{27x} - 4\sqrt{48x}$
14. $\sqrt{32x^2y^3}$
15. $(\sqrt{2} - \sqrt{3})(\sqrt{2} + \sqrt{3})$
16. $(\sqrt{5} + \sqrt{6})^2$
17. $\dfrac{4}{\sqrt{3}}$
18. $\dfrac{3\sqrt{2}}{\sqrt{6}}$
19. $\dfrac{2}{3 - \sqrt{7}}$
20. $\dfrac{-2}{4 - \sqrt{3}}$

Find decimal approximations for each of the following.

21. $\sqrt{390}$
22. $\sqrt{7400}$
23. $\sqrt{95{,}000}$
24. $\dfrac{6}{\sqrt{5}}$
25. $\dfrac{4}{2 - \sqrt{3}}$

Solve each of the following equations.

26. $\sqrt{x + 2} = 5$
27. $\sqrt{x - 1} = 6$
28. $\sqrt{2x + 8} = 2\sqrt{x}$
29. $\sqrt{2x + 11} = \sqrt{x + 6}$
30. $\sqrt{x + 2} - x = 2$

9 QUADRATIC EQUATIONS

9.1 SOLVING QUADRATIC EQUATIONS BY THE SQUARE ROOT METHOD

In Chapter 4 we solved quadratic equations (second-degree equations) by factoring. However, not all quadratic equations can be solved by factoring. Another method can be used to solve quadratic equations like

$$(x - 3)^2 = 16,$$

where the square of a binomial is equal to some number. We can use some of the ideas of Chapter 8 about square roots to solve such equations.

If two positive numbers are equal, then they must have the same square roots or square roots which are negatives of each other. That is, for positive real numbers a and b,

$$\text{if } a = b, \text{ then } \sqrt{a} = \sqrt{b} \text{ or } \sqrt{a} = -\sqrt{b}.$$

Example 1 **(a)** If $x^2 = 16$, then $x = 4$ or $x = -4$.
(b) If $p^2 = 9$, then $p = 3$ or $p = -3$.
(c) If $z^2 = 5$, then $z = \sqrt{5}$ or $z = -\sqrt{5}$.
(d) If $m^2 = 8$, then $m = \sqrt{8}$ or $m = -\sqrt{8}$. Since $\sqrt{8} = 2\sqrt{2}$, $m = 2\sqrt{2}$ or $m = -2\sqrt{2}$.
(e) If $y^2 = -4$, then y is not a real number.

Getting back to the equation $(x - 3)^2 = 16$, we can solve it in the same way we solved the equations in Example 1.

$$\text{If } (x - 3)^2 = 16, \text{ then } x - 3 = 4 \text{ or } x - 3 = -4.$$

From the last two equations, we have

$$x = 7 \quad \text{or} \quad x = -1.$$

Check both numbers in the original equation.

$$(7 - 3)^2 = 4^2 = 16 \quad \text{and} \quad (-1 - 3)^2 = (-4)^2 = 16$$

Both 7 and -1 are solutions.

Example 2 Solve the equation $(x-1)^2 = 6$.
Take the square root on both sides.

$$x - 1 = \sqrt{6} \qquad \text{or} \qquad x - 1 = -\sqrt{6}$$
$$x = 1 + \sqrt{6} \qquad\qquad x = 1 - \sqrt{6}$$

Check: $(1 + \sqrt{6} - 1)^2 = (\sqrt{6})^2 = 6;$
$(1 - \sqrt{6} - 1)^2 = (-\sqrt{6})^2 = 6.$

The solutions are $1 + \sqrt{6}$ and $1 - \sqrt{6}$.

Example 3 Solve the equation $(3r-2)^2 = 27$.
Taking square roots gives

$$3r - 2 = \sqrt{27} \qquad \text{or} \qquad 3r - 2 = -\sqrt{27}.$$

Now simplify the radical: $\sqrt{27} = \sqrt{9 \cdot 3} = \sqrt{9} \cdot \sqrt{3} = 3\sqrt{3}$, so

$$3r - 2 = 3\sqrt{3} \qquad \text{or} \qquad 3r - 2 = -3\sqrt{3}$$
$$3r = 2 + 3\sqrt{3} \qquad\qquad 3r = 2 - 3\sqrt{3}$$
$$r = \frac{2 + 3\sqrt{3}}{3} \qquad\qquad r = \frac{2 - 3\sqrt{3}}{3}.$$

The solutions are

$$\frac{2 + 3\sqrt{3}}{3} \quad \text{and} \quad \frac{2 - 3\sqrt{3}}{3}.$$

Example 4 Solve $(x+3)^2 = -9$.
The square root of -9 is not a real number. Hence there is no real number solution.

EXERCISES

Solve each of the following by factoring. See Example 1 on page 138.

1. $x^2 - 5x + 6 = 0$
2. $x^2 + x - 12 = 0$
3. $x^2 + 12x + 20 = 0$
4. $x^2 - 8x + 15 = 0$
5. $2x^2 + 5x - 3 = 0$
6. $3m^2 + m - 2 = 0$
7. $2a^2 + 5a + 3 = 0$
8. $12r^2 + 8r - 15 = 0$
9. $2z^2 + z - 15 = 0$
10. $6x^2 - 7x - 5 = 0$

In Exercises 11–30, solve each equation by taking the square root of both sides. Express all square roots in simplest form.

11. $(x - 2)^2 = 16$

12. $(r + 4)^2 = 25$

13. $(a + 4)^2 = 10$

14. $(r - 3)^2 = 15$

15. $(x - 1)^2 = 32$

16. $(y + 5)^2 = 28$

17. $(2m - 1)^2 = 9$

18. $(3y - 7)^2 = 4$

19. $(3z + 5)^2 = 9$

20. $(2y - 7)^2 = 49$

21. $(6m - 2)^2 = 121$

22. $(7m - 10)^2 = 144$

23. $(2a - 5)^2 = 30$

24. $(2y + 3)^2 = 45$

25. $(3p - 1)^2 = 18$

26. $(5r - 6)^2 = 75$

27. $(2k - 5)^2 = 98$

28. $(4x - 1)^2 = 48$

29. $(3m + 4)^2 = 8$

30. $(5y - 3)^2 = 50$

31. One expert at marksmanship can hold a silver dollar at forehead level, drop it, draw his gun, and shoot the coin as it passes waist level. The distance traveled by a falling object is given by $d = 16t^2$, where d is the distance the object falls in t seconds. If the coin falls about 4 feet, estimate the time that elapses between the dropping of the coin and the shot.

9.2 SOLVING QUADRATIC EQUATIONS BY COMPLETING THE SQUARE

Consider the equation

$$x^2 + 6x + 7 = 0.$$

In the preceding section we learned to solve equations of the type

$$(x + 3)^2 = 2.$$

If we can rewrite the equation

$$x^2 + 6x + 7 = 0$$

in a form like

$$(x + 3)^2 = 2,$$

we can solve it by taking the square roots of both sides. As a start, let us add -7 to both sides of the equation to get

$$x^2 + 6x = -7.$$

We want the quantity on the left-hand side of $x^2 + 6x = -7$ to be a perfect square trinomial. Note that $x^2 + 6x + 9$ is a perfect square, since

$$x^2 + 6x + 9 = (x + 3)^2.$$

Hence, if we add 9 to both sides, we will have an equation with a perfect square trinomial on the left-hand side, as desired.

$$x^2 + 6x + 9 = -7 + 9$$
$$(x + 3)^2 = 2$$

Now take the square root of both sides of the equation to complete the solution.

$$x + 3 = \sqrt{2} \qquad \text{or} \qquad x + 3 = -\sqrt{2}$$
$$x = \sqrt{2} - 3 \qquad\qquad x = -\sqrt{2} - 3.$$

The solutions of the original equation are $\sqrt{2} - 3$ and $-\sqrt{2} - 3$. Verify this by checking.

We changed

$$x^2 + 6x + 7 = 0$$

to an equation where one side is a squared binomial,

$$(x + 3)^2 = 2,$$

so that we could solve the equation by taking square roots on both sides. The process of changing the form of the equation in this way is called **completing the square.**

Example 1 Find the solutions of the quadratic equation

$$m^2 - 5m = 2.$$

We need to add a suitable number to both sides of the equation, so that the left side is a perfect square. To decide which number to use, first make sure that the coefficient of the m^2 term is 1, and if it is not, multiply both sides by an appropriate number to make it 1. Next, take half the coefficient of m and square it. Then add this squared number to both sides of the equation. (We do this because the coefficient of the middle term of the perfect square trinomial $a^2 + 2ab + b^2$ is the product ab times 2.)

In the equation above, the coefficient of m is -5, and half of -5 is $-\frac{5}{2}$. If we square $-\frac{5}{2}$, we get $\frac{25}{4}$, which is the number to be added to both sides.

$$m^2 - 5m + \frac{25}{4} = 2 + \frac{25}{4}$$

The trinomial $m^2 - 5m + \frac{25}{4}$ is a perfect square trinomial.

$$m^2 - 5m + \frac{25}{4} = \left(m - \frac{5}{2}\right)^2.$$

Thus

$$\left(m - \frac{5}{2}\right)^2 = 2 + \frac{25}{4} = \frac{33}{4}.$$

Now take square roots.

$$m - \frac{5}{2} = \sqrt{\frac{33}{4}} \qquad \text{or} \qquad m - \frac{5}{2} = -\sqrt{\frac{33}{4}}$$

$$m = \frac{5}{2} + \sqrt{\frac{33}{4}} \qquad\qquad m = \frac{5}{2} - \sqrt{\frac{33}{4}}$$

Simplify the radical.

$$\sqrt{\frac{33}{4}} = \frac{\sqrt{33}}{\sqrt{4}} = \frac{\sqrt{33}}{2}.$$

The solutions are

$$\frac{5}{2} + \frac{\sqrt{33}}{2} \qquad \text{and} \qquad \frac{5}{2} - \frac{\sqrt{33}}{2}.$$

Example 2 Solve the equation $2x^2 - 7x = 9$.

To begin, we need the coefficient 1 for the x^2 term. Multiply both sides of the equation by $\frac{1}{2}$. The result is

$$x^2 - \frac{7}{2}x = \frac{9}{2}.$$

Now, take half the coefficient of x and square it. Half of $-\frac{7}{2}$ is $-\frac{7}{4}$, and $-\frac{7}{4}$ squared is $\frac{49}{16}$. Add $\frac{49}{16}$ to both sides of the equation, and write the left side as a perfect square.

$$x^2 - \frac{7}{2}x + \frac{49}{16} = \frac{9}{2} + \frac{49}{16}$$

$$\left(x - \frac{7}{4}\right)^2 = \frac{121}{16}$$

Take the square root of both sides.

$$x - \frac{7}{4} = \sqrt{\frac{121}{16}} \qquad \text{or} \qquad x - \frac{7}{4} = -\sqrt{\frac{121}{16}}$$

Since $\sqrt{\frac{121}{16}} = \frac{11}{4}$, we can write

$$x - \frac{7}{4} = \frac{11}{4} \qquad \text{or} \qquad x - \frac{7}{4} = \frac{-11}{4}$$

$$x = \frac{18}{4} = \frac{9}{2} \qquad\qquad x = \frac{-4}{4} = -1.$$

The solutions are $\frac{9}{2}$ and -1. Since the solutions are rational numbers, the original equation could have been factored:

$$2x^2 - 7x - 9 = (2x - 9)(x + 1).$$

Example 3 Use the method of completing the square to solve the equation $4p^2 + 8p + 5 = 0$.

First multiply both sides by $\frac{1}{4}$ to get the coefficient 1 for the p^2 term. The result is

$$p^2 + 2p + \frac{5}{4} = 0.$$

Add $-\frac{5}{4}$ to both sides, which gives

$$p^2 + 2p = \frac{-5}{4}.$$

The coefficient of p is 2. Take half of 2, square the result, and add it to both sides. The left-hand side can then be written as a perfect square.

$$p^2 + 2p + 1 = \frac{-5}{4} + 1$$

$$(p + 1)^2 = \frac{-1}{4}$$

At this point, we should take the square root of both sides. However, in the real number system, the square root of $-\frac{1}{4}$ has no meaning. This equation has no solution in the real number system.

Summary: Completing the Square in a Quadratic Equation

1. If the coefficient of the squared term is 1, proceed to step 2. If the coefficient of the squared term is not 1, but some other number a, multiply both sides of the equation by the reciprocal of a, $1/a$. This gives an equation which has 1 as coefficient of the squared term.
2. Make sure all terms with variables are on one side of the equals sign, and all numbers are on the other side.
3. Take half the coefficient of x, and square it. Add the square to both sides of the equation. The side containing the variables can now be written as a perfect square.
4. Take the square root of both sides. The solutions are determined by solving the two resulting equations.

EXERCISES

Find the number that should be added to each of the following to write it as a perfect square trinomial.

1. $x^2 + 2x$

2. $y^2 - 4y$

3. $x^2 + 18x$

4. $m^2 - 3m$

5. $z^2 + 9z$

6. $p^2 + 22p$

7. $x^2 + 14x$

8. $r^2 + 7r$

9. $y^2 + 5y$

10. $q^2 - 8q$

In Exercises 11–36, solve each equation by completing the square.

11. $x^2 + 4x = -3$

12. $y^2 - 4y = 0$

13. $a^2 + 2a = 5$

14. $m^2 + 4m = 12$

15. $z^2 + 6z = -8$

16. $q^2 - 8q = -16$

17. $x^2 - 6x + 1 = 0$

18. $b^2 - 2b - 2 = 0$

19. $c^2 + 3c = 2$

20. $k^2 + 5k - 3 = 0$

21. $2m^2 + 4m = -7$

22. $3y^2 - 9y + 5 = 0$

23. $6q^2 - 8q + 3 = 0$

24. $4y^2 + 4y - 3 = 0$

25. $-x^2 + 6x = 4$

26. $3y^2 - 6y - 2 = 0$

27. $2m^2 - 4m - 5 = 0$

28. $-x^2 + 4 = 2x$

29. $3x^2 - 2x = 1$

30. $-x^2 - 4 = 2x$

31. $m^2 - 4m + 8 = 6m$

32. $2z^2 = 8z + 5 - 4z^2$

33. $3r^2 - 2 = 6r + 3$

34. $4p - 3 = p^2 + 2p$

35. $(x + 1)(x + 3) = 2$

36. $(x - 3)(x + 1) = 1$

9.3 SOLVING QUADRATIC EQUATIONS BY THE QUADRATIC FORMULA

Completing the square can be used to solve any quadratic equation, but the method is not very handy. In this section we will work out a general formula, the *quadratic formula.* The formula gives the solutions for any quadratic equation.

To get the quadratic formula, we start with the general form of a quadratic equation,

$$ax^2 + bx + c = 0, \qquad a \neq 0.$$

The restriction $a \neq 0$ is important to make sure that the equation is in fact quadratic. If $a = 0$, then the equation becomes $0x^2 + bx + c = 0$, or $bx + c = 0$. This is a linear equation, not a quadratic.

We solve the equation $ax^2 + bx + c = 0$ by completing the square. First, we need the coefficient 1 for the x^2 term. To get this, multiply both sides by $1/a$, which gives

$$x^2 + \frac{b}{a}x + \frac{c}{a} = 0.$$

Next, add $-c/a$ to both sides, to get

$$x^2 + \frac{b}{a}x = -\frac{c}{a}.$$

Now complete the square on the left. To do this, take half the coefficient of x, that is, $b/2a$. Square $b/2a$ to get $b^2/4a^2$. Next, add $b^2/4a^2$ to both sides of the equation.

$$x^2 + \frac{b}{a}x + \frac{b^2}{4a^2} = \frac{-c}{a} + \frac{b^2}{4a^2}$$

Rewrite the left-hand side as a perfect square.

$$\left(x + \frac{b}{2a}\right)^2 = \frac{-c}{a} + \frac{b^2}{4a^2}$$

Now simplify the right-hand side of the equation.

$$\begin{aligned}\left(x + \frac{b}{2a}\right)^2 &= \frac{b^2}{4a^2} + \frac{-c}{a}\\ &= \frac{b^2}{4a^2} + \frac{-4ac}{4a^2}\\ &= \frac{b^2 - 4ac}{4a^2}\end{aligned}$$

Take the square root of both sides.

$$x + \frac{b}{2a} = \sqrt{\frac{b^2 - 4ac}{4a^2}} \quad \text{or} \quad x + \frac{b}{2a} = -\sqrt{\frac{b^2 - 4ac}{4a^2}}$$

Simplify the radical.

$$\sqrt{\frac{b^2 - 4ac}{4a^2}} = \frac{\sqrt{b^2 - 4ac}}{\sqrt{4a^2}} = \frac{\sqrt{b^2 - 4ac}}{2a}$$

Now we can write the solutions, as follows.

$$x + \frac{b}{2a} = \frac{\sqrt{b^2 - 4ac}}{2a} \qquad \text{or} \qquad x + \frac{b}{2a} = \frac{-\sqrt{b^2 - 4ac}}{2a}$$

$$x = \frac{-b}{2a} + \frac{\sqrt{b^2 - 4ac}}{2a} \qquad\qquad x = \frac{-b}{2a} - \frac{\sqrt{b^2 - 4ac}}{2a}$$

$$x = \frac{-b + \sqrt{b^2 - 4ac}}{2a} \qquad\qquad x = \frac{-b - \sqrt{b^2 - 4ac}}{2a}$$

Hence, the solutions of the general quadratic equation $ax^2 + bx + c = 0$ $(a \neq 0)$ are

$$\frac{-b + \sqrt{b^2 - 4ac}}{2a} \quad \text{and} \quad \frac{-b - \sqrt{b^2 - 4ac}}{2a}.$$

For convenience, the solutions are often expressed in compact form by using the symbol $\pm$. The result is called the **quadratic formula.**

$$x = \frac{-b \pm \sqrt{b^2 - 4ac}}{2a}$$

Example 1 Use the quadratic formula to solve $2x^2 - 7x - 9 = 0$.

To begin, match the coefficients of the variables with the letter symbols of the general quadratic equation

$$ax^2 + bx + c = 0.$$

Here, $a = 2$, $b = -7$, and $c = -9$. Substitute these numbers into the quadratic formula, and simplify the result.

$$\begin{aligned} x &= \frac{-b \pm \sqrt{b^2 - 4ac}}{2a} \\ &= \frac{-(-7) \pm \sqrt{(-7)^2 - 4(2)(-9)}}{2(2)} \\ &= \frac{7 \pm \sqrt{49 + 72}}{4} \\ &= \frac{7 \pm \sqrt{121}}{4} \end{aligned}$$

Since $\sqrt{121} = 11$,

$$x = \frac{7 \pm 11}{4}.$$

To write the two solutions separately, first take the plus sign:

$$x = \frac{7 + 11}{4} = \frac{18}{4} = \frac{9}{2}.$$

Then take the minus sign:

$$x = \frac{7 - 11}{4} = \frac{-4}{4} = -1.$$

The solutions of $2x^2 - 7x - 9 = 0$ are $\frac{9}{2}$ and -1.

Example 2 Find the solutions of the quadratic equation $x^2 - 2x - 1 = 0$.

The coefficients are $a = 1$, $b = -2$, and $c = -1$. The solution is found by substituting these values into the quadratic formula.

$$x = \frac{-b \pm \sqrt{b^2 - 4ac}}{2a} = \frac{-(-2) \pm \sqrt{(-2)^2 - 4(1)(-1)}}{2(1)}$$

$$= \frac{2 \pm \sqrt{4 + 4}}{2}$$

$$= \frac{2 \pm \sqrt{8}}{2}.$$

Since $\sqrt{8} = \sqrt{4 \cdot 2} = \sqrt{4} \cdot \sqrt{2} = 2\sqrt{2}$, then

$$x = \frac{2 \pm 2\sqrt{2}}{2}.$$

And since $2 \pm 2\sqrt{2}$ factors as $2(1 \pm \sqrt{2})$, then

$$x = \frac{2(1 \pm \sqrt{2})}{2} = 1 \pm \sqrt{2}.$$

Therefore, the two solutions of this equation are

$$1 + \sqrt{2} \qquad \text{and} \qquad 1 - \sqrt{2}.$$

Example 3 Solve the equation $x^2 + 5x + 8 = 0$.

Here $a = 1$, $b = 5$, and $c = 8$. Substitute into the quadratic formula and simplify the result.

$$x = \frac{-5 \pm \sqrt{5^2 - 4(1)(8)}}{2(1)}$$

$$= \frac{-5 \pm \sqrt{25 - 32}}{2}$$

$$= \frac{-5 \pm \sqrt{-7}}{2}$$

The radical $\sqrt{-7}$ is not a real number and so the equation has no real number solutions.

Example 4 Solve the equation

$$\frac{1}{10}t^2 = \frac{2}{5} - \frac{1}{2}t.$$

To eliminate the denominators, multiply both sides of the equation by the common denominator 10.

$$10\left(\frac{1}{10}t^2\right) = 10\left(\frac{2}{5} - \frac{1}{2}t\right)$$

$$t^2 = 4 - 5t$$

Add $-4 + 5t$ to both sides of the equation to get

$$t^2 + 5t - 4 = 0.$$

In this form, we can identify $a = 1$, $b = 5$, and $c = -4$. Use the quadratic formula to complete the solution.

$$t = \frac{-5 \pm \sqrt{25 - 4(1)(-4)}}{2(1)}$$

$$= \frac{-5 \pm \sqrt{25 + 16}}{2}$$

$$= \frac{-5 \pm \sqrt{41}}{2}$$

The solutions are

$$\frac{-5 + \sqrt{41}}{2} \quad \text{and} \quad \frac{-5 - \sqrt{41}}{2}.$$

EXERCISES

For each of the following, identify the letters a, b, and c of the general quadratic equation $ax^2 + bx + c = 0$.

1. $3x^2 + 4x - 8 = 0$

2. $9x^2 + 2x - 3 = 0$

3. $-8x^2 - 2x - 3 = 0$

4. $-2x^2 + 3x - 8 = 0$

5. $2x^2 = 3x - 2$

6. $9x^2 - 2 = 4x$

7. $x^2 = 2$

8. $x^2 - 3 = 0$

9. $3x^2 - 8x = 0$

10. $5x^2 = 2x$

11. $(x - 3)(x + 4) = 0$

12. $(x + 6)^2 = 3$

13. $9(x - 1)(x + 2) = 8$

14. $(3x - 1)(2x + 5) = x(x - 1)$

Solve each of the following equations, using the quadratic formula.

15. $x^2 + 2x - 2 = 0$

16. $6x^2 + 6x + 1 = 0$

17. $x^2 + 4x + 4 = 0$

18. $3x^2 - 5x + 1 = 0$

19. $-2x^2 + 7x - 1 = 0$

20. $2x^2 = 3x - 2$

21. $x^2 + x - 1 = 0$

22. $x^2 - 10x = -20$

23. $x^2 = 13 - 12x$

24. $x^2 = 8x + 9$

25. $2x^2 + 12x + 5 = 0$

26. $x^2 = -19 + 20x$

27. $5x^2 + 4x - 1 = 0$

28. $5x^2 + x - 1 = 0$

29. $2x^2 = 3x + 5$

30. $-2x^2 + 7x = -30$

31. $3x^2 = -x + 6$

32. $x^2 + 6x + 9 = 0$

33. $x^2 - 2x + 1 = 0$

34. $5x^2 + 5x = 0$

35. $x^2 = 20$

36. $2x^2 + 2x + 4 = 4 - 2x$

37. $3x^2 - 4x + 3 = 8x - 1$

38. $\frac{3}{2}x^2 - x = \frac{4}{3}$

39. $\frac{2}{3}x^2 - \frac{4}{9}x - \frac{1}{3} = 0$

40. $-\frac{2}{5}x^2 + \frac{3}{5}x = -1$

9.4 COMPLEX NUMBERS

The real number system does not contain enough numbers to permit us to solve all quadratic equations. The simple equation

$$x^2 + 1 = 0$$

cannot be solved using real numbers, since the solution requires taking the square root of -1 as shown below.

$$x^2 = -1$$

$$x = \sqrt{-1} \qquad \text{or} \qquad x = -\sqrt{-1}.$$

In this section we define a new set of numbers, which are not real numbers, so that every quadratic equation will have a solution. We begin by defining

$$\sqrt{-1} = i.$$

The symbol i represents a number, just as the symbol 3 represents a number. If we square both sides of the equation above, we have

$$i^2 = \sqrt{-1} \cdot \sqrt{-1} = -1.$$

Because of this, we define the square of the number i to be -1, a real number.

If $a \geq 0$, we assume that

$$\sqrt{-a} = \sqrt{a(-1)} = \sqrt{a} \cdot \sqrt{-1} = \sqrt{a}\,i.$$

We can use this result to express square roots of negative numbers as the product of i and some real number (rational or irrational).

Example 1 $\sqrt{-25} = \sqrt{25(-1)} = \sqrt{25} \cdot \sqrt{-1} = 5\sqrt{-1} = 5i.$

Example 2 $\sqrt{-72} = \sqrt{(36)(2)(-1)} = \sqrt{36} \cdot \sqrt{2} \cdot \sqrt{-1} = 6\sqrt{2}i.$

It is important to note that if $a \geq 0$ and $b \geq 0$, then $\sqrt{-a} \cdot \sqrt{-b} \neq \sqrt{(-a)(-b)}$. For example, let $a = 4$ and $b = 9$. Then $\sqrt{-a} = \sqrt{-4} = \sqrt{4} \cdot \sqrt{-1} = 2i$, and $\sqrt{-b} = \sqrt{-9} = \sqrt{9} \cdot \sqrt{-1} = 3i$. Thus the product

$$\sqrt{-4} \cdot \sqrt{-9} = 2i \cdot 3i = 6i^2 = 6(-1) = -6.$$

However, $\sqrt{(-4)(-9)} = \sqrt{36} = 6$. *The rules for multiplication and division of radicals do not apply unless both radicands are positive.* It is always necessary to rewrite the square root of a negative number as a multiple of i *before* performing any operations such as multiplication or division.

Any number of the form $a + bi$ where a and b are any real numbers is called a **complex number**. For example, $2 + 3i$, $-1 + 2i$ and $4 - i$ are complex numbers. Every real number is also a complex number, since a real number can be written as the sum of itself and $0i$. For example, $5 = 5 + 0i$, so that 5 is a complex number. Any non-real complex number is called an **imaginary number**. Thus,

$$2 - 5i, \quad -3i, \quad i, \quad \text{and} \quad 6 + 4i$$

are all imaginary numbers.

The name "imaginary" is unfortunate, since one might think that an imaginary number is non-existent. However, this is not the case. Imaginary numbers are just as "real" as real numbers, and have numerous practical applications. One application is in the theory of electronics.

The set of complex numbers satisfies all the properties that we have studied for real numbers, except that the complex numbers cannot be ordered. That is, given any two complex numbers, there is no way to determine that one is larger than the other. However, the closure, commutative, associative, and distributive properties can be used when simplifying expressions which include complex numbers.

Example 3 Simplify $2\sqrt{-96} + \sqrt{-54}$.

First, we must rewrite $\sqrt{-96}$ and $\sqrt{-54}$ as the product of i and a real number:

$$\sqrt{-96} = \sqrt{16(6)(-1)} = 4\sqrt{6}\,i$$

$$\sqrt{-54} = \sqrt{9(6)(-1)} = 3\sqrt{6}\,i.$$

Substituting into the given expression, we have

$$2\sqrt{-96} + \sqrt{-54} = 2(4\sqrt{6}\,i) + 3\sqrt{6}\,i = 8\sqrt{6}\,i + 3\sqrt{6}\,i = 11\sqrt{6}\,i.$$

We used the distributive property in the last step, as we always do when we combine like radicals.

Example 4 Simplify $\dfrac{-6 + 3\sqrt{2}\,i}{3}$.

We begin by using the distributive property to factor the numerator, so that we can then divide by the denominator, 3.

$$\frac{-6 + 3\sqrt{2}\,i}{3} = \frac{3(-2 + \sqrt{2}\,i)}{3} = -2 + \sqrt{2}\,i.$$

EXERCISES

In Exercises 1–8, answer *always*, *sometimes*, or *never* for each statement.

1. An integer is a complex number.

2. A complex number is an integer.

3. An integer is a rational number.

4. A rational number is a whole number.

5. An imaginary number is a real number.

6. A complex number is a real number.

7. An imaginary number is a complex number.

8. A whole number is a complex number.

Simplify each of the following as much as possible.

9. $\sqrt{-4}$

10. $\sqrt{-9}$

11. $\sqrt{-16}$

12. $\sqrt{-25}$

13. $-\sqrt{-49}$

14. $-\sqrt{-64}$

15. $\sqrt{-12}$

16. $\sqrt{-48}$

17. $\sqrt{-98}$

18. $\sqrt{-8}$

19. $-\sqrt{-27}$

20. $-\sqrt{-54}$

21. $1 + \sqrt{-36}$

22. $3 - \sqrt{-4}$

23. $2 - \sqrt{-8}$

24. $5 + \sqrt{-25}$

25. $\sqrt{-48} - \sqrt{-12}$

26. $\sqrt{-72} + \sqrt{-8}$

27. $\sqrt{-27} + \sqrt{-48}$

28. $\sqrt{-45} + \sqrt{-80}$

29. $\dfrac{2 + 2i}{2}$

30. $\dfrac{4 + 2i}{2}$

31. $\dfrac{3 + 3\sqrt{2}\,i}{3}$

32. $\dfrac{5 + 10\sqrt{3}\,i}{5}$

33. $\dfrac{6 - 12i}{12}$

34. $\dfrac{8 - 4i}{8}$

9.5 THE QUADRATIC EQUATION—COMPLETE SOLUTION

Now that we have studied complex numbers, we can use the quadratic formula to find all solutions of any quadratic equation. This means that we can find imaginary solutions as well as real solutions, in other words, solutions to quadratic equations that are imaginary numbers.

Example 1 Use the quadratic formula to solve $x^2 - 2x + 2 = 0$.

Substitute the values $a = 1, b = -2$, and $c = 2$ in the quadratic formula.

$$x = \frac{-b \pm \sqrt{b^2 - 4ac}}{2a}$$

$$x = \frac{-(-2) \pm \sqrt{(-2)^2 - 4(1)(2)}}{2(1)}$$

$$= \frac{2 \pm \sqrt{4 - 8}}{2}$$

$$= \frac{2 \pm \sqrt{-4}}{2}.$$

Write $\sqrt{-4}$ as an imaginary number: $\sqrt{-4} = \sqrt{4(-1)} = \sqrt{4}\sqrt{-1} = 2i$. Then

$$x = \frac{2 \pm 2i}{2} = \frac{2(1 \pm i)}{2} = 1 \pm i.$$

Hence the equation $x^2 - 2x + 2 = 0$ has two imaginary solutions, $1 + i$ and $1 - i$. Check as follows.

If $x = 1 + i$,

$$(1 + i)^2 - 2(1 + i) + 2 = 0$$
$$1 + 2i + i^2 - 2 - 2i + 2 = 0$$
$$1 + 2i - 1 - 2 - 2i + 2 = 0$$
$$0 = 0.$$

True

If $x = 1 - i$,

$$(1 - i)^2 - 2(1 - i) + 2 = 0$$
$$1 - 2i + i^2 - 2 + 2i + 2 = 0$$
$$1 - 2i - 1 - 2 + 2i + 2 = 0$$
$$0 = 0.$$

True

Example 2 Solve the quadratic equation $2x^2 + 6x + 5 = 0$.

Again we use the quadratic formula. Here $a = 2$, $b = 6$, and $c = 5$.

$$\begin{aligned} x &= \frac{-6 \pm \sqrt{6^2 - 4(2)(5)}}{2(2)} \\ &= \frac{-6 \pm \sqrt{36 - 40}}{4} \\ &= \frac{-6 \pm \sqrt{-4}}{4} \\ &= \frac{-6 \pm 2i}{4} \\ &= \frac{2(-3 \pm i)}{4} \\ &= \frac{-3 \pm i}{2}. \end{aligned}$$

The solutions are $\dfrac{-3 + i}{2}$ and $\dfrac{-3 - i}{2}$.

Example 3 Solve $(x + 7)^2 = -5$.

Take square roots on both sides to get

$$x + 7 = \sqrt{-5} \qquad \text{or} \qquad x + 7 = -\sqrt{-5}.$$

Rewrite $\sqrt{-5}$ using i:

$$\sqrt{-5} = \sqrt{5} \cdot \sqrt{-1} = \sqrt{5}\, i.$$

Then

$$\begin{aligned} x + 7 &= \sqrt{5}\, i \qquad \text{or} \qquad x + 7 = -\sqrt{5}\, i \\ x &= -7 + \sqrt{5}\, i \qquad \text{or} \qquad x = -7 - \sqrt{5}\, i \end{aligned}$$

EXERCISES

Solve each of the following equations.

1. $2x^2 - 2x + 1 = 0$
2. $5x^2 - 4x + 1 = 0$
3. $x^2 - 4x + 8 = 0$
4. $8x^2 - 12x + 5 = 0$
5. $2x^2 + 2x + 1 = 0$
6. $9x^2 - 12x + 8 = 0$
7. $9x^2 - 6x + 4 = 0$
8. $9x^2 + 6x + 4 = 0$
9. $-4x^2 - x = 4$
10. $12x^2 + 4x = -3$

11. $3x^2 + 5 = 4x$

12. $4x^2 + 17 = 4x$

13. $10x^2 + x + 2 = 0$

14. $x^2 + 3x + 10 = 0$

15. $4x^2 + 2x + 1 = 0$

16. $5x^2 + 4x + 8 = 0$

17. $9x^2 + 6x + 8 = 0$

18. $2x^2 + 3x + 2 = 0$

19. $8x^2 - 4x + 3 = 0$

20. $-9x^2 + 12x - 7 = 0$

21. $6x^2 + 3 = 4x$

22. $6x^2 - x + 5 = 0$

23. $9x^2 + 4x + 8 = 0$

24. $7x^2 - 8x + 6 = 0$

25. $5x^2 + 6x + 7 = 0$

26. $x^2 + 4x + 5 = 0$

27. $(x - 3)^2 = -9$

28. $(2x - 5)^2 = -16$

29. $(x - 4)^2 + 25 = 0$

30. $(3x + 16)^2 + 1 = 0$

31. $(4x - 3)^2 = -27$

32. $(2x + 1)^2 = -48$

9.6 THE DISCRIMINANT TEST

The solution of the quadratic equation

$$ax^2 + bx + c = 0 \qquad a \neq 0$$

is given by the quadratic formula

$$x = \frac{-b \pm \sqrt{b^2 - 4ac}}{2a}.$$

We have seen that the solution to a quadratic equation may be two rational numbers (occasionally, one), two irrational numbers, or two imaginary numbers. The quantity $b^2 - 4ac$ under the square-root sign in the quadratic formula is called the **discriminant**. The discriminant can be used to predict the kind of solutions that will be found *before* the actual solutions are determined.

Example 1 Find the solutions of the equation

$$6x^2 + 5x - 6 = 0.$$

We have $a = 6$, $b = 5$, and $c = -6$. The discriminant is

$$5^2 - 4(6)(-6) = 25 + 144 = 169.$$

The number 169 is a perfect square (square of an integer), as can be seen from the table of square roots inside the front cover. Since $169 = 13^2$, $\sqrt{169} = 13$. By the quadratic formula,

$$x = \frac{-5 \pm \sqrt{169}}{2(6)} = \frac{-5 \pm 13}{12}.$$

Use the plus sign first:

$$x = \frac{-5 + 13}{12} = \frac{8}{12} = \frac{2}{3}.$$

Then use the minus sign:

$$x = \frac{-5 - 13}{12} = \frac{-18}{12} = \frac{-3}{2}.$$

The solutions are two rational numbers, $\frac{2}{3}$ and $-\frac{3}{2}$.

As we have seen in this example, if the discriminant is a perfect square (not zero), there will be two solutions. If a, b, and c are all integers, then these two solutions are rational numbers.

Example 2 Find the solutions of the equation

$$2x^2 + 3x + 2 = 0.$$

Here $a = 2$, $b = 3$, and $c = 2$. The discriminant, $b^2 - 4ac$, is

$$3^2 - 4(2)(2) = 9 - 16 = -7.$$

In the quadratic formula we take the square root of the discriminant. Since $\sqrt{-7} = \sqrt{7}\,i$,

$$x = \frac{-3 \pm \sqrt{7}\,i}{4},$$

so that there are two imaginary number solutions,

$$\frac{-3 + \sqrt{7}\,i}{4} \quad \text{and} \quad \frac{-3 - \sqrt{7}\,i}{4}.$$

Thus, when the discriminant is a negative number, there are two solutions, both of which are imaginary numbers.

Example 3 Find the solutions of the equation $4x^2 - 12x + 9 = 0$.

To calculate the discriminant, let $a = 4$, $b = -12$, and $c = 9$. Thus

$$b^2 - 4ac = (-12)^2 - 4(4)(9) = 144 - 144 = 0.$$

Insert this value of the discriminant into the quadratic formula.

$$x = \frac{-(-12) \pm \sqrt{0}}{2 \cdot 4} = \frac{12}{8} = \frac{3}{2}$$

Because the discriminant, zero, has only one square root, we find only one rational number solution instead of two. In general, if the values a, b, and c are all integers, and the discriminant is zero, then there is a single solution, a rational number.

Example 4 Solve $x^2 + 2x - 5 = 0$.

We have $a = 1$, $b = 2$, and $c = -5$. The discriminant is

$$2^2 - 4(1)(-5) = 4 + 20 = 24.$$

Since the discriminant is not zero, we can find two solutions. By the quadratic formula,

$$x = \frac{-2 \pm \sqrt{24}}{2(1)}.$$

Simplify the radical: $\sqrt{24} = \sqrt{4 \cdot 6} = \sqrt{4} \cdot \sqrt{6} = 2\sqrt{6}$. Thus

$$x = \frac{-2 \pm 2\sqrt{6}}{2}.$$

Factor $-2 \pm 2\sqrt{6}$, so that

$$x = \frac{2(-1 \pm \sqrt{6})}{2} = -1 \pm \sqrt{6}.$$

The solutions, $-1 + \sqrt{6}$ and $-1 - \sqrt{6}$, are irrational numbers. In this example, where the discriminant was positive but not a perfect square, we found two irrational number solutions.

Summary In the quadratic equation

$$ax^2 + bx + c = 0, \qquad a \neq 0,$$

where a, b, and c are all integers, the kind of solutions the equation has can be determined from the discriminant, $(b^2 - 4ac)$, as follows.

1. If the discriminant is zero, then the equation has exactly one solution, and this solution is a rational number.
2. If the discriminant is a perfect square (square of an integer), then the equation has two rational number solutions.
3. If the discriminant is a positive number which is not a perfect square, then the equation has two irrational number solutions.
4. If the discriminant is negative, then the equation has two imaginary number solutions.

Example 5 Decide what kind of solutions each of the following has.

(a) $5x^2 - 8x - 6 = 0$

Here $a = 5$, $b = -8$, and $c = -6$. The discriminant is

$$(-8)^2 - 4(5)(-6) = 64 + 120 = 184.$$

As seen in the square-root table inside the front cover, 184 is not a perfect square. From statement 3 above, we see that the equation has two irrational number solutions.

(b) $4x^2 = 4x - 1$

Rewrite the equation as

$$4x^2 - 4x + 1 = 0.$$

Then $a = 4$, $b = -4$, $c = 1$, and the discriminant is

$$(-4)^2 - 4(4)(1) = 16 - 16 = 0.$$

The equation has exactly one rational number solution.

(c) $\frac{3}{4}x^2 - \frac{5}{8}x + 2 = 0.$

We cannot use the discriminant test mentioned above when any of the coefficients are not integers. We can, however, convert all the coefficients to integers by multiplying both sides by 8 to get

$$6x^2 - 5x + 16 = 0.$$

The discriminant is

$$(-5)^2 - 4(6)(16) = 25 - 384 = -359.$$

Since the discriminant is negative, the equation has two imaginary number solutions.

Example 6 Find a number b such that $16x^2 + bx + 25 = 0$ will have exactly one rational solution.

First find the discriminant.

$$b^2 - 4(16)(25) = b^2 - 1600.$$

The only way the equation can have exactly one rational solution is for the discriminant to be zero. Thus

$$b^2 - 1600 = 0$$
$$b^2 = 1600$$
$$b = 40 \text{ or } b = -40.$$

Both $16x^2 + 40x + 25 = 0$ and $16x^2 - 40x + 25 = 0$ have exactly one rational solution.

EXERCISES

In Exercises 1–24, use the discriminant to determine whether each equation has:
(a) exactly one rational solution
(b) two rational solutions
(c) two irrational solutions
(d) two imaginary solutions

Do not solve the equations.

1. $x^2 + 3x + 2 = 0$
2. $2x^2 = 1 + x$
3. $3x^2 + 2x + 1 = 0$
4. $x^2 + 4x + 3 = 0$
5. $3x^2 + x - 4 = 0$
6. $x^2 + 3x - 4 = 0$
7. $4x^2 + 3 = 2x$
8. $9x^2 + 6x + 1 = 0$
9. $4x^2 - 3 = 2x$
10. $x^2 + 6 = 6x$
11. $9x^2 + 6x = 7$
12. $4x^2 + 53 = 4x$
13. $9x^2 + 20 = 20x$
14. $16x^2 + x = 15$
15. $50x^2 + 30x = -4$
16. $x^2 + 5x = -6$
17. $\frac{3}{5}x^2 - \frac{2}{3}x = \frac{1}{6}$
18. $x^2 - \frac{3}{4}x + 4 = 0$
19. $\frac{-3}{7}x^2 + \frac{2}{3}x = \frac{1}{6}$
20. $\frac{1}{3}x^2 - 2x + 3 = 0$
21. $\frac{1}{7}x^2 = \frac{2}{7} - \frac{x}{7}$
22. $x^2 + 2x + \frac{1}{4} = 0$
23. $(3x - 1)(2x + 5) = 3x + 19$
24. $9x(x - 2) = (x + 1)(4x - 5)$

In each equation below, find a value for a, b, or c such that the equation has exactly one rational solution.

25. $ax^2 - 6x + 9 = 0$
26. $4x^2 + bx + 1 = 0$
27. $64x^2 - 48x + c = 0$
28. $4x^2 = bx - 25$
29. $14x + 1 = ax^2$
30. $x^2 + 12x = c$

9.7 GRAPHING PARABOLAS (OPTIONAL)

In Chapter 6 we graphed straight lines to represent the solutions of linear equations. Now we shall investigate the graphs of those quadratic equations in two variables which have the form $y = ax^2 + bx + c$. Perhaps the simplest such quadratic equation is

$$y = x^2$$

(which is the same as $y = 1x^2 + 0x + 0$). The graph of this equation cannot be a straight line since only linear equations, which have the form

$$ax + by + c = 0,$$

have graphs which are straight lines. However, we can graph $y = x^2$ in much the same way as we graphed straight lines, by selecting values for x and then finding the corresponding y values. For example, if $x = 2$ in the equation $y = x^2$, then

$$y = 2^2 = 4.$$

Thus the point (2, 4) belongs to the graph of $y = x^2$. (Recall that in an ordered pair such as (2, 4) the x value comes first and the y value second.) In the same way, we can complete a chart showing values of y for some values of x (which we choose arbitrarily).

Equation	*Ordered pairs*	
$y = x^2$	$(-3, 9)$	$(1, 1)$
	$(-2, 4)$	$(2, 4)$
	$(-1, 1)$	$(3, 9)$
	$(0, 0)$	

If we plot these points on a coordinate system and draw a smooth curve through them, we get the graph in Figure 9.1, called a **parabola.**

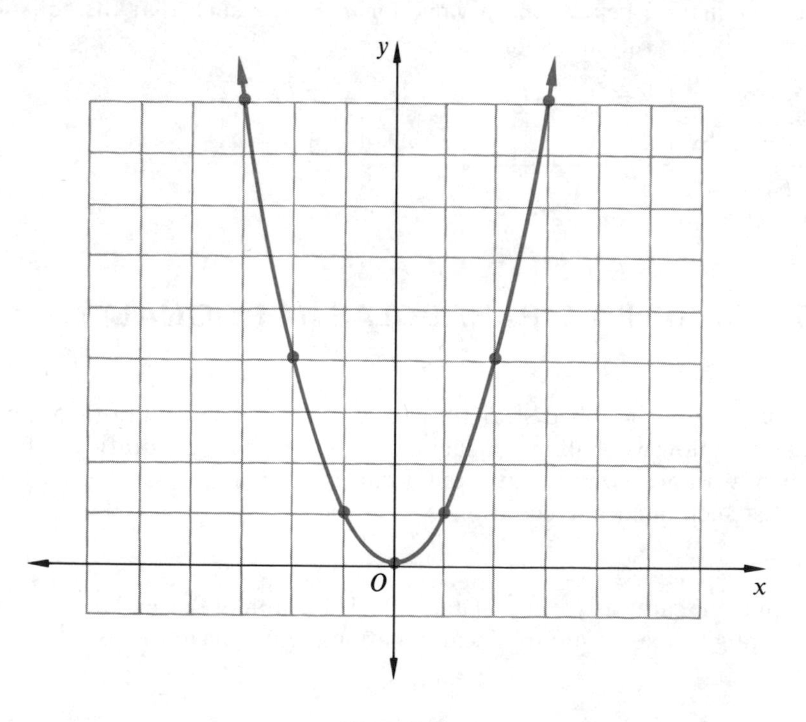

Figure 9-1

Every quadratic equation of the form

$$y = ax^2 + bx + c$$

has a graph which is a parabola. Because of its many useful properties, the parabola occurs frequently in real-life applications. If an object is thrown into the air, the path that the object follows is a parabola (discounting wind resistance). The cross-sections of radar, spotlight, and telescope reflectors also form parabolas.

Example 1 Graph the parabola $y = -x^2$.

We could select values for x and then find the corresponding y values. But note that for a given x value, the y value will be the negative of the corresponding y value of the parabola $y = x^2$ discussed above. Hence this new parabola has the same shape as the one in Figure 9.1, but is turned in the opposite direction. We say it opens downward (instead of upward), as shown in Figure 9.2.

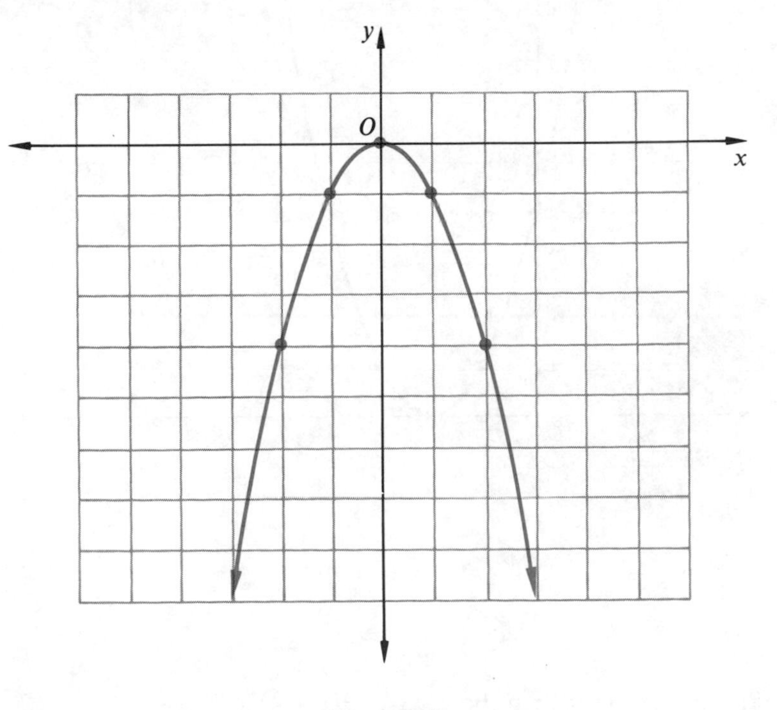

Figure 9-2

Example 2 Graph the parabola $y = (x - 2)^2$.

Again, we select values for x and find the corresponding y values. For example, if we let $x = -1$, we have

$$y = (-1 - 2)^2 = (-3)^2 = 9.$$

Calculating other ordered pairs in the same way, we get the following:

Equation	*Ordered pairs*	
$y = (x - 2)^2$	$(-1, 9)$	$(3, 1)$
	$(0, 4)$	$(4, 4)$
	$(1, 1)$	$(5, 9)$
	$(2, 0)$	

Plotting these points and joining them gives the graph of Figure 9.3. Note that the parabola of Figure 9.3 has the same shape as our original parabola $y = x^2$, but is shifted two units to the right.

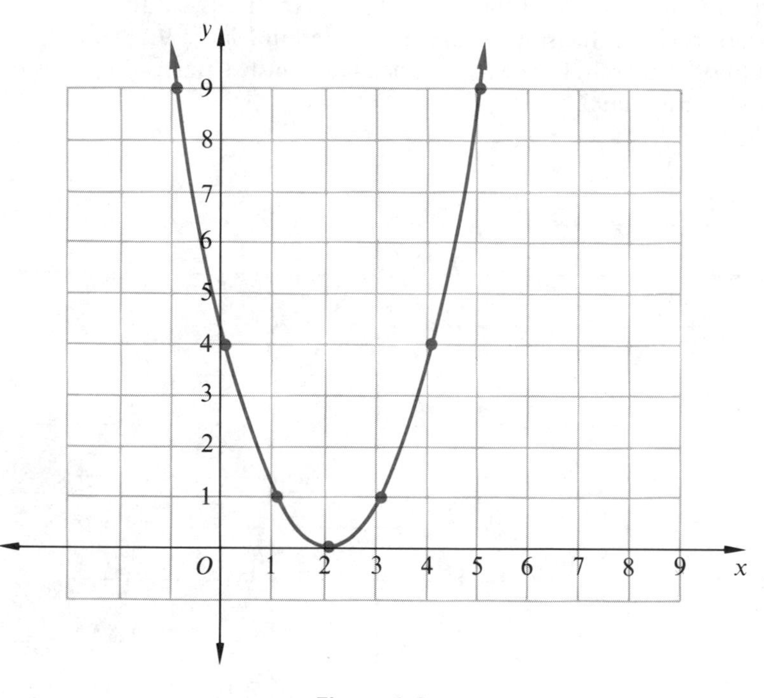

Figure 9-3

Example 3 Graph the parabola $y = -3(x + 4)^2$.

If we make a table of values and plot the points, we get the graph of Figure 9.4. The graph shown in Figure 9.4 opens downward, is shifted four units to the left, and is narrower than the graph of $y = x^2$.

Example 4 Graph the parabola $y = x^2 - 3$.

The graph is shown in Figure 9.5. Note that this time the graph is shifted three units downward, as compared to the graph of $y = x^2$.

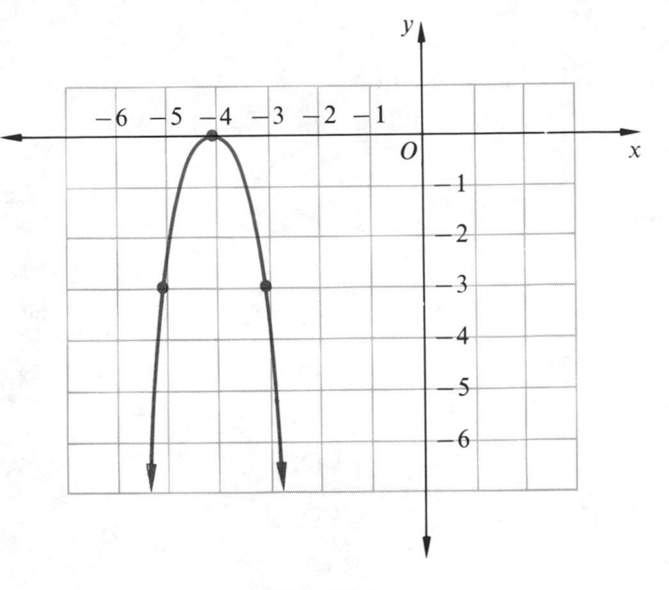

Figure 9-4

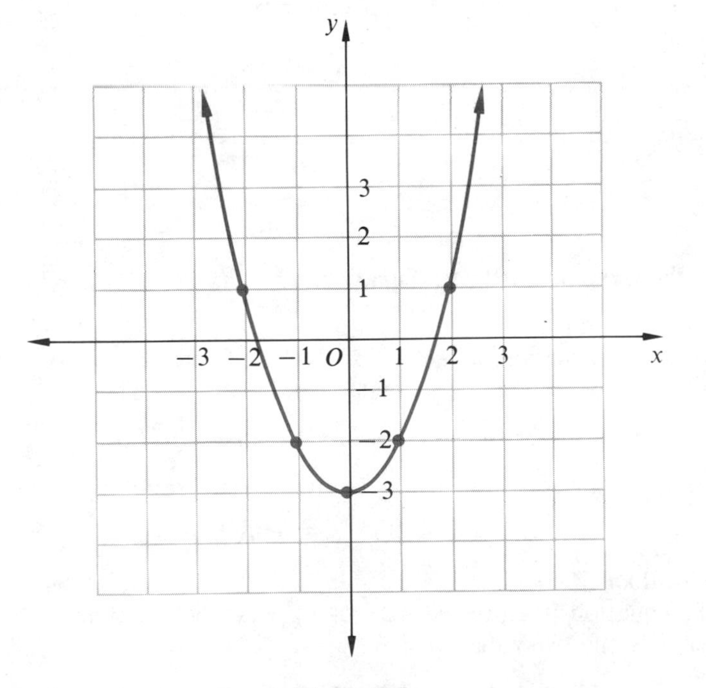

Figure 9-5

EXERCISES

Sketch the graph of each equation.

1. $y = 2x^2$
2. $y = -2x^2$
3. $y = (x + 1)^2$
4. $y = (x - 2)^2$
5. $y = -(x + 1)^2$
6. $y = -(x - 2)^2$
7. $y = x^2 + 1$
8. $y = -x^2 - 2$
9. $3y = -2x^2$
10. $2y = 3x^2$
11. $2y = -x^2 + 1$
12. $3y = -x^2 + 3$
13. $y = (2x - 1)^2$
14. $y = (8x + 5)^2$
15. $y = (x + 1)^2 + 2$
16. $y = (x - 2)^2 - 1$
17. $y = (x - 4)^2 - 1$
18. $y = (x + 3)^2 + 3$

CHAPTER 9 SUMMARY

Key Words

Completing the square
Quadratic formula
Discriminant
Imaginary numbers
Complex numbers
Parabola

Definitions

$i = \sqrt{-1}$ $\qquad$ $i^2 = -1$

Solving the General Quadratic Equation $ax^2 + bx + c = 0, a \neq 0$.

1. *Factoring Method*
   Express the equation as $ax^2 + bx + c = 0$. Factor $ax^2 + bx + c$ as the product of two binomials so that

$$ax^2 + bx + c = (mx + h)(nx + g) = 0.$$

   Set each factor equal to zero and solve the two resulting equations,

$$mx + h = 0 \qquad \text{or} \qquad nx + g = 0.$$

2. *Square Root Method*
   If the equation is expressed as $(x + d)^2 = k$, the solution can be found by solving the two equations

$$x + d = \sqrt{k} \qquad \text{or} \qquad x + d = -\sqrt{k}.$$

3. *Completing the Square*
   (a) If $a \neq 1$, multiply both sides of the equation by $1/a$.
   (b) Add $-c/a$ to both sides of the equation.
   (c) Find $(b/2a)^2$.
   (d) Add $(b/2a)^2$ to both sides of the equation.
   (e) Write the side containing the variables as a perfect square and add together any terms on the other side.
   (f) The equation should now be in a proper form for use of the square root method given above.

4. *The Quadratic Formula*
   If the equation is expressed as $ax^2 + bx + c = 0$, the solutions are given by

$$x = \frac{-b \pm \sqrt{b^2 - 4ac}}{2a}.$$

The Discriminant Test

Consider the quadratic equation $ax^2 + bx + c = 0$, where a, b, and c are integers. Calculate the quantity $b^2 - 4ac$; this quantity, the discriminant, predicts the kind of solutions.

Discriminant is	*Number of solutions*	*Kind of solutions*
zero	1	rational
positive and a perfect square	2	rational
positive and *not* a perfect square	2	irrational
negative	2	imaginary

CHAPTER 9 TEST

Solve by completing the square.

1. $2x^2 - 5x = 0$

2. $5x^2 = 2 - 9x$

Solve by the square root method.

3. $x^2 = 5$

4. $(x - 3)^2 = 49$

5. $(3x - 2)^2 = 35$

Solve by using the quadratic formula.

6. $x^2 - 8x + 16 = 0$

7. $3x^2 + 2 = 7x$

8. $4x^2 + 8x = -5$

9. $3x^2 - 5x + 1 = 0$

Use the discriminant to predict the kind of solutions for each equation.

10. $8x^2 - 10x - 3 = 0$

11. $9a^2 + 6a = -1$

12. $3m^2 + 2 = 4m$

13. $6x^2 + 2x = 5$

Simplify each as much as possible.

14. $\sqrt{-3}$

15. $\sqrt{-11}$

16. $\sqrt{-25}$

17. $\sqrt{-18}$

18. $2 + \sqrt{-32}$

19. $\sqrt{-12} + \sqrt{-75}$

20. $\dfrac{3 - 6\sqrt{2}\,i}{6}$

Draw a graph of each of the following.

21. $y = -3x^2$

22. $y = 3x^2 - 2$

The Metric System of Weights and Measures

In the United States today, length is measured in inches, feet, or miles. Weight is measured in ounces (there are two kinds of ounces), pounds, or tons (there are long tons and short tons). Volume is measured in pints, quarts, or gallons. These weights and measures make up the *English system* of measures. The United States will gradually switch to the *metric system*, which is used in almost all other countries of the world. Many industries are now working on the switch to the metric system, which will be pretty much in effect by the early 1980s. Even now, some Vega automobiles are designed with metric specifications.

The metric system was developed in France in 1789. The name comes from the basic unit of length, the meter (abbreviated m), which is a little longer than a yard.

The meter can be subdivided into smaller parts for measuring shorter distances. The most common subdivisions are millimeters, or 1/1000 meter, and centimeters, 1/100 meter. Millimeters are abbreviated mm and centimeters are abbreviated cm.

For example, a dime is about 2 mm thick. Film for instamatic cameras is 35 mm wide. Many common parts of the body can be measured with centimeters and millimeters. A man is perhaps 150 cm tall, and a woman's waist might be 66 cm around. Some manufacturers of clothing use metric sizes—a size 40 suit becomes a size 102 suit in metric.

A major advantage of the metric system of measurement is the ease in converting from one unit of measure to another. This is illustrated in the following examples.

Example Convert 28 cm to millimeters.

One centimeter is made up of 10 mm. Thus, 28 cm is

$$28 \times 10 = 280 \text{ mm.}$$

Example Convert 2 m to centimeters.

One meter is 100 centimeters. Therefore, $2 \times 100 = 200$ cm. There are 200 cm in 2 m.

Example Convert 250 mm to meters.

One meter is 1000 mm. Thus,

$$\frac{250}{1000} = .250 \text{ m.}$$

250 mm is .250 m.

Longer distances are measured in kilometers (km), or one thousand meters. A kilometer is about 5/8 mile. The table on the following page shows distances in kilometers between various cities.

	Seattle	Los Angeles	Denver	Houston	St. Louis	Chicago	Cleveland	Atlanta	Miami	Washington	New York	Boston
Seattle		1842	2167	3704	3393	3319	3868	4434	5504	4422	4673	4885
Los Angeles	1842		1825	2499	2973	3371	3894	3535	4364	4254	4690	4911
Denver	2167	1825		1651	1377	1635	2183	2254	3290	2729	4580	3200
Houston	3704	2499	1651		1278	1746	2066	1310	1957	2269	2632	3083
St. Louis	3393	2973	1377	1278		463	879	898	1968	1289	1554	1895
Chicago	3319	3371	1635	1746	463		552	1175	2188	1146	1352	1569
Cleveland	3868	3894	2183	2066	879	552		1104	2097	565	816	1017
Atlanta	4434	3535	2254	1310	898	1175	1104		1070	1014	1376	1718
Miami	5504	4364	3290	1957	1968	2188	2097	1070		1778	2140	2481
Washington	4422	4254	2729	2269	1289	1146	565	1014	1778		368	703
New York	4673	4690	4580	2632	1554	1352	816	1376	2140	368		348
Boston	4885	4911	3200	3083	1895	1569	1017	1718	2481	703	348	

Weights in the metric system are based on the *gram* (gm). A nickel weighs about 5 gm, for example. Since a gram is such a small weight, milligrams (1/1000 gm) and centigrams (1/100 gm) are mainly used to measure very small weights in science. A *kilogram* (kg), or 1000 grams, is about 2.2 pounds. Kilograms (abbreviated kg) are sometimes called kilos.

Volume is measured in *liters* (l). A liter is about a quart. Milliliters (1/1000 liter), centiliters (1/100 liter), and kiloliters (1000 liters) are sometimes used, but mainly in science.

In summary, the metric system uses the following four common prefixes:

Prefix	Definition	Example
milli	1/1000	1 millimeter = 1/1000 meter
centi	1/100	1 centiliter = 1/100 liter
deci	1/10	1 decigram = 1/10 gram
kilo	1000	1 kilogram = 1000 grams

Eventually, people will think in the metric system as easily as they think now in the English system. To help you "think metric," you should get in the habit of estimating in the metric system. As an aid in doing this, use the *approximate* conversion table on the next page.

Example Convert 2 yards to meters.

Yards to meters is an English to metric conversion. From the table, yards can be converted to meters by multiplying by .9144. Thus, 2 yards are $2 \times .9144 = 1.8288$ m.

Example Convert 690 grams to pounds.

From the table, we need to multiply by .00220.

$$690 \times .00220 = 1.518 \text{ pounds.}$$

Exercises

Make the following conversions within the metric system.

1. 20 m to mm
2. 7 cm to mm
3. 80 mm to cm
4. 500 mm to cm
5. 5200 m to km
6. 15,000 m to km
7. 7.8 km to m
8. 6 kg to gm
9. 1.92 kg to gm
10. 8000 gm to kg
11. 9 l to ml
12. 2.98 l to ml
13. 57,000 ml to l
14. 800 ml to l

Use the table below to make conversions as indicated.

15. 12 yards to m
16. 20 m to yards
17. 50 ft to m
18. 832 ft to m
19. 300 lbs to kg
20. 5 lbs to gm
21. 850 gm to lbs
22. 9600 gm to lbs
23. 5000 in. to m
24. 39 in. to m

Think in metric.

25. Find your height in cm.
26. Find your height in mm.
27. Find the length of your longest finger in cm.
28. What are the dimensions of the cover of this book in mm?
29. Give your Levi size (waist and length) in cm.
30. What is your weight in kg?
31. What is the distance in km between the two largest cities in your state?
32. How many liters in a six-pack of 12-ounce cans of beer?

METRIC TO ENGLISH			ENGLISH TO METRIC		
From	to	Multiply by	From	to	Multiply by
meters	yards	1.094	yards	meters	.9144
meters	feet	3.281	feet	meters	.3048
meters	inches	39.37	inches	meters	.0254
kilometers	miles	.6214	miles	kilometers	1.6093
grams	pounds	.00220	pounds	grams	454
kilograms	pounds	2.20	pounds	kilograms	.454
liters	quarts	1.057	quarts	liters	.946
liters	gallons	.264	gallons	liters	3.785

Answers to Odd-numbered Exercises

SECTION 1.1

1. T **3.** F **5.** T **7.** T **9.** F **11.** F **13.** T **15.** F **17.** F **19.** T
21. F **23.** F **25.** T **27.** F **29.** T **31.** T **33.** F **35.** T
37. $10 - (7 - 3) = 6$ **39.** $(3 \cdot 5) + 7 = 22$ (or, use no parentheses at all)
41. $(3 \cdot 5) - 4 = 11$ (or, no parentheses) **43.** $3(5 + 2)4 = 84$ **45.** $3(5 - 2)4 = 36$
47. $(100 \div 20) \div 5 = 1$ **49.** $100 \div (20 \div 5) = 25$

SECTION 1.2

1. 12 **3.** 15 **5.** 14 **7.** 6 **9.** 43 **11.** 24 **13.** 10 **15.** 2
17. 10/12 or 5/6 **19.** yes **21.** no **23.** yes **25.** yes **27.** no

29. 0 2 4 6 **31.** 1 2 3 4 5

33. 5 6 7 **35.** 3 4 5

37. no solutions **39.** 2 3 4 5 6

41. 0 1 2 3 4 5 **43.** 1 2 3 4 5

45. no solutions **47.** 1 2 3 4 5 6 7

49. $2x$ **51.** $6 + x$ **53.** $8 - x$ **55.** $8 + 3x$ or $3x + 8$ **57.** $15 - 2x$ **59.** 4
61. 0, 2, 4, 6, 8 **63.** 0, 2 **65.** 4, 6, 8, 10 **67.** 2

SECTION 1.3

1. -8, negative **3.** 9 **5.** 2 **7.** -15, negative **9.** -5, negative
11. -4, negative **13.** -5 **15.** -12 **17.** -8 **19.** 3 **21.** $|-3|$ **23.** $-|-6|$
25. T **27.** T **29.** F **31.** T **33.** T **35.** F **37.** T **39.** F **41.** T

43. -6 -5 -4 -3 -2 -1 0 1 2 3

45. -6 -5 -4 -3 -2 -1 0 1 2 3 4

47. -4 -3 -2 -1 0 1 2 3

49. 3, -3 **51.** -1, 0, 1 **53.** -1, 0, 1 **55.** -4, -3, -2, 2, 3, 4 **57.** 0
59. all numbers **61.** no solutions **63.** all numbers **65.** no solutions

SECTION 1.4

1. 2 **3.** -2 **5.** -8 **7.** -11 **9.** -12 **11.** 4 **13.** 12 **15.** 5 **17.** 2
19. -9 **21.** 13 **23.** -11 **25.** -8 **27.** 0 **29.** -20 **31.** T **33.** F
35. T **37.** T **39.** T **41.** F **43.** T **45.** -2 **47.** -3 **49.** -2
51. -2 **53.** 2 **55.** $9 **57.** -135 feet, or 135 feet below sea level **59.** 90

SECTION 1.5

1. -3 **3.** -1 **5.** -8 **7.** -14 **9.** 9 **11.** 17 **13.** -4 **15.** 4 **17.** 4
19. 1 **21.** 10 **23.** -5 **25.** 11 **27.** -5 **29.** -10 **31.** -18 **33.** -5
35. 2 **37.** -16 **39.** -12 **41.** $12 - (-6) = 18$ **43.** $-25 - (-4) = -21$
45. $-24 - (-27) = 3$ **47.** $-5 - 10 = -15$ **49.** $14{,}494 - (-282) = 14{,}776$

SECTION 1.6

1. commutative **3.** associative **5.** commutative **7.** associative **9.** closure **11.** inverse **13.** identity **15.** commutative **17.** associative **19.** inverse **21.** 8 **23.** 2 **25.** 4 **27.** 6 **29.** 6 **31.** yes, no **33.** yes, yes **35.** yes, no **37.** no, no **39.** associative, inverse, identity **41.** associative, commutative, associative, inverse, identity

SECTION 1.7

1. 12 **3.** -12 **5.** 5 **7.** 44 **9.** 120 **11.** -48 **13.** -30 **15.** -65 **17.** 0 **19.** -165 **21.** -36 **23.** 12 **25.** 5 **27.** 12 **29.** 18 **31.** -14 **33.** -10 **35.** 45 **37.** 12 **39.** 16 **41.** 17 **43.** -28 **45.** 12 **47.** -360 **49.** 0 **51.** -2 **53.** 0 **55.** -2 **57.** -2 **59.** -2

SECTION 1.8

1. 1/9 **3.** $-1/4$ **5.** 3/2 **7.** $-10/9$ **9.** no reciprocal for 0 **11.** $k+9$ **13.** m **15.** $3r+3m$ **17.** 1 **19.** 0 **21.** $-5+5=0$ **23.** $-3r-6$ **25.** 9 **27.** $k+[5+(-6)]=k-1$ **29.** $4z+(2r+3k)$ **31.** $5m+10$ **33.** $-4r-8$ **35.** $-8k+16$ **37.** $-9a-27$ **39.** $4r+32$ **41.** $-16+2k$ **43.** $10r+12m$ **45.** $-12x+16y$ **47.** $5(8+9)$ **49.** $7(2+8)$ **51.** $9(p+q)$ **53.** $5(7z+8w)$ **55.** commutative **57.** associative **59.** inverse **61.** associative **63.** inverse **65.** inverse **67.** distributive

SECTION 1.9

1. -2 **3.** -3 **5.** -6 **7.** -5 **9.** 2 **11.** 15 **13.** 36 **15.** 0 **17.** -4 **19.** -10 **21.** 5 **23.** -4 **25.** -60 **27.** -6 **29.** 2 **31.** 4 **33.** 3 **35.** -3 **37.** 8 **39.** -2 **41.** -1 **43.** no such number
45. 36, 18, 12, 9, 6, 4, 3, 2, 1, -1, -2, -3, -4, -6, -9, -12, -18, -36
47. 25, 5, 1, -1, -5, -25
49. 40, 20, 10, 8, 5, 4, 2, 1, -1, -2, -4, -5, -8, -10, -20, -40
51. 17, 1, -1, -17 **53.** 29, 1, -1, -29 **55.** -8 **57.** -6 **59.** 0 **61.** 8 **63.** -4 **65.** $6x=-42$; -7 **67.** $x/5=15$; 75 **69.** $x/3=-9$; -27 **71.** $x/2=-6$; -12 **73.** $6/(x+1)=3$; 1

SECTION 1.10

1. real, rational, integer, whole number **3.** real, rational **5.** real, rational, integer **7.** real, irrational **9.** real, rational, integer, whole number **11.** real, rational, integer **13.** real, irrational **15.** not a real number **17.** real, rational, integer, whole number **19.** not a real number **21.** 3/5, 2/3, 5/9 **23.** -6, 2/3, -5 **25.** π, $-\sqrt{2}$, $\sqrt{3}$ **27.** T **29.** F **31.** F **33.** T **35.** F

Chapter 1 Test

1. 4 **2.** -9 **3.** -1 **4.** 32 **5.** -8 **6.** -4 **7.** 10 **8.** 6 **9.** -72 **10.** -3 **11.** -9 **12.** -1 **13.** 2 **14.** not a number **15.** 0 **16.** -2 **17.** -13 **18.** -1 **19.** -3

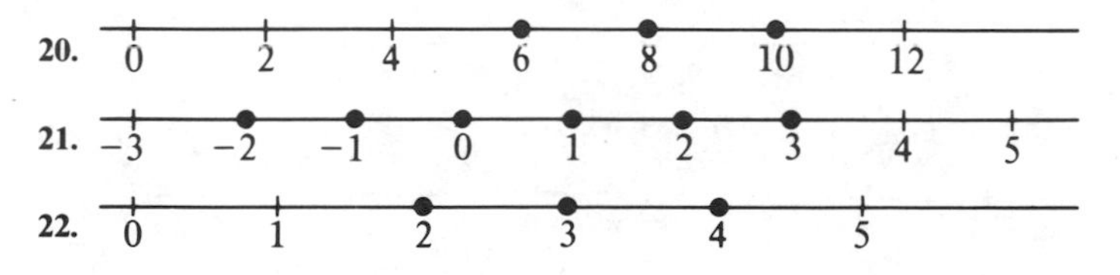

23. -5 **24.** $-|-8|$ **25.** 0 **26.** 3 **27.** -9 **28.** $|4|$ **29.** H, J **30.** B
31. F, G **32.** C, D **33.** A, I **34.** E **35.** 0/2 (or 0), 15/3 (or 5), 12
36. -5, 0/2, 15/3, 12, -6 **37.** -5, 0/2, 9/4, 15/3, 12, -6 **38.** $\sqrt{3}$
39. -5, 0/2, $\sqrt{3}$, 9/4, 15/3, 12, -6 **40.** 3/0

SECTION 2.1

1. yes **3.** no **5.** yes **7.** yes **9.** no **11.** $7k + 15$ **13.** $m - 1$ **15.** $-4y$
17. $2x + 6$ **19.** $20 - 7m$ **21.** 2 **23.** 8 **25.** 6 **27.** 19 **29.** -4 **31.** 5
33. $5x + 3x + 1 = 49; 6$ **35.** $x + (x + 1) = 5; 2$

SECTION 2.2

1. 10 **3.** -2 **5.** 10 **7.** -8 **9.** -5 **11.** -2 **13.** 4 **15.** -5
17. -11 **19.** -6 **21.** 7 **23.** 8 **25.** -10 **27.** -9 **29.** 0 **31.** 17
33. 9

SECTION 2.3

1. 5 **3.** 25 **5.** -8 **7.** -7 **9.** -4 **11.** -9 **13.** 0 **15.** -6 **17.** 4
19. 0 **21.** 4 **23.** 8 **25.** 20 **27.** 32 **29.** 49 **31.** 9 **33.** 8 **35.** -80
37. $x/4 = 6; 24$ **39.** $x/4 = 62; 248$

SECTION 2.4

1. $3k - 18$ **3.** $30t + 66$ **5.** $-3n - 15$ **7.** $-6x + 8$ **9.** $4r + 14$ **11.** 21
13. $12k - 13$ **15.** $13k - 21$ **17.** 2 **19.** -4 **21.** 9 **23.** -2 **25.** 18
27. -1 **29.** -4 **31.** -3 **33.** 8 **35.** -2 **37.** 6 **39.** -5 **41.** 0
43. 5 **45.** 0 **47.** $-1/5$ **49.** $-5/7$ **51.** $3(x - 17) = 102; 51$
53. $8 - 3(x + 4) = 2; -2$

SECTION 2.5

1. 8 **3.** 6 **5.** $20\frac{1}{2}$ inches **7.** 48 **9.** 36 **11.** 613 **13.** 70 **15.** 800 feet
17. 36 quart containers

SECTION 2.6

1. 128 **3.** 36 **5.** 8 **7.** 14 **9.** 21 **11.** 4 **13.** 1.5 **15.** $l = A/w$
17. $t = d/r$ **19.** $h = V/lw$ **21.** $t = I/(pr)$ **23.** $b = 2A/h$ **25.** $w = (P - 2l)/2$
27. $b = 2A/h - B$ or $b = (2A - Bh)/h$ **29.** 10 **31.** 37.68 **33.** 24 **35.** 385 miles
37. 4 **39.** 2 cm by 14 cm **41.** 3 **43.** 2 hours **45.** 3
47. 2500 at $5\frac{1}{2}\%$, 7500 at $6\frac{1}{2}\%$ **49.** 25 mph **51.** $172

SECTION 2.7

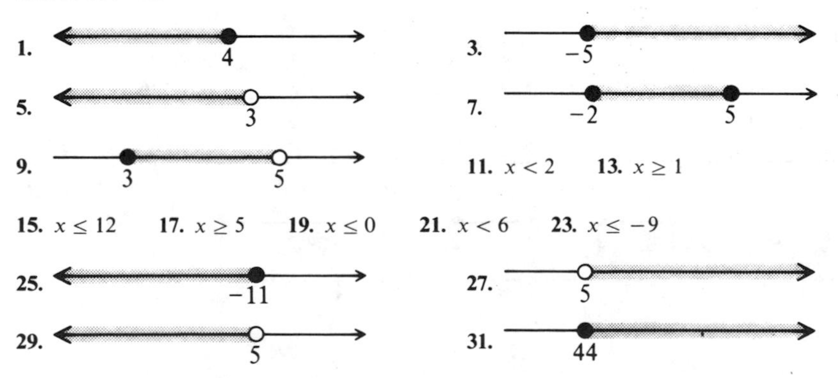

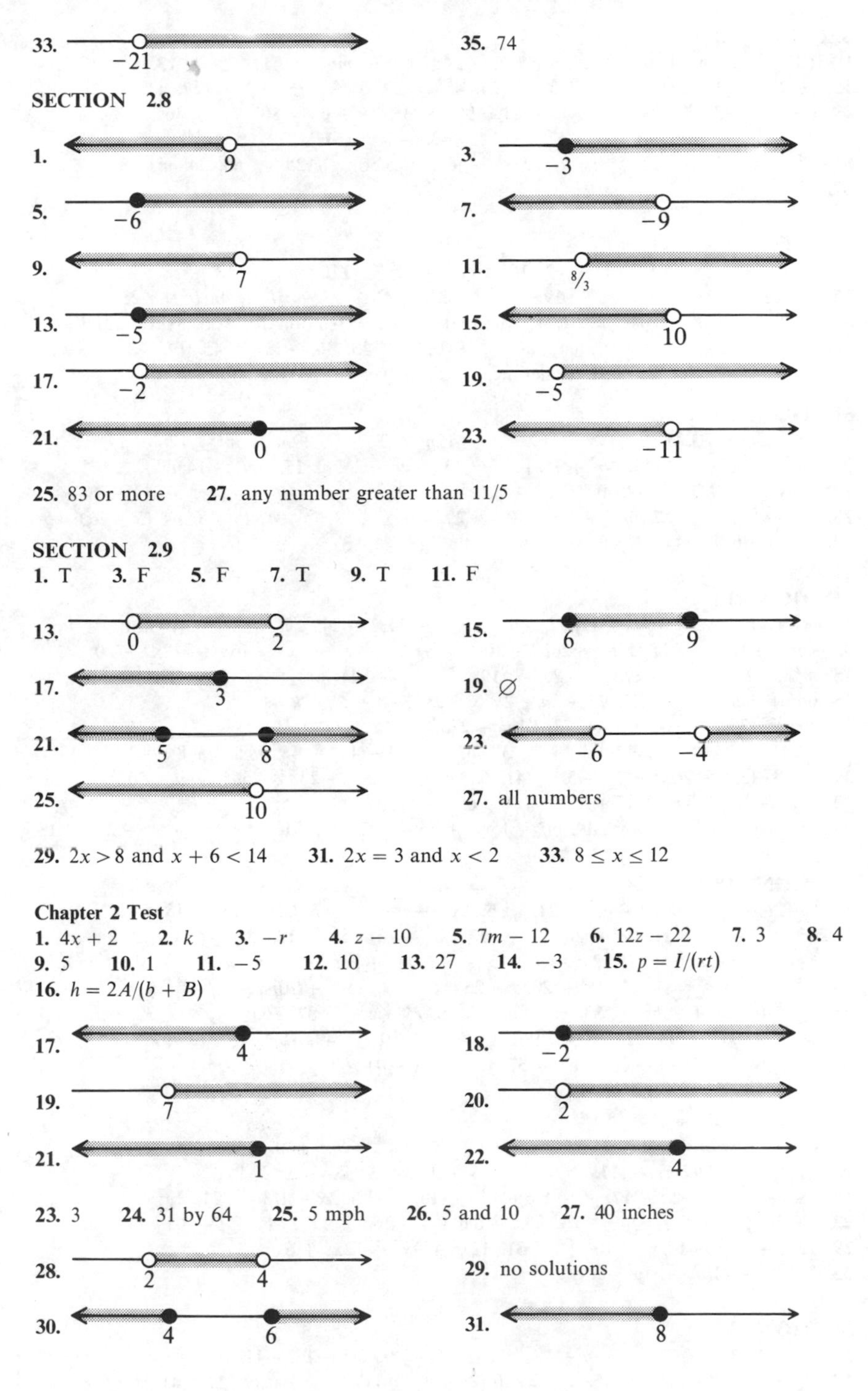

33. -21 **35.** 74

SECTION 2.8

1. 9 **3.** -3

5. -6 **7.** -9

9. 7 **11.** $8/3$

13. -5 **15.** 10

17. -2 **19.** -5

21. 0 **23.** -11

25. 83 or more **27.** any number greater than 11/5

SECTION 2.9

1. T **3.** F **5.** F **7.** T **9.** T **11.** F

13. 0, 2 **15.** 6, 9

17. 3 **19.** $\varnothing$

21. 5, 8 **23.** -6, -4

25. 10 **27.** all numbers

29. $2x > 8$ and $x + 6 < 14$ **31.** $2x = 3$ and $x < 2$ **33.** $8 \leq x \leq 12$

Chapter 2 Test

1. $4x + 2$ **2.** k **3.** $-r$ **4.** $z - 10$ **5.** $7m - 12$ **6.** $12z - 22$ **7.** 3 **8.** 4
9. 5 **10.** 1 **11.** -5 **12.** 10 **13.** 27 **14.** -3 **15.** $p = I/(rt)$
16. $h = 2A/(b + B)$

17. 4 **18.** -2

19. 7 **20.** 2

21. 1 **22.** 4

23. 3 **24.** 31 by 64 **25.** 5 mph **26.** 5 and 10 **27.** 40 inches

28. 2, 4 **29.** no solutions

30. 4, 6 **31.** 8

SECTION 3.1

1. 5; 12 **3.** $3m$; 4 **5.** 125; 3 **7.** -24; 2 **9.** m; 2 **11.** 3^5 **13.** 5^4
15. $(-2)^5$ **17.** $1/4^5$ **19.** $1/3^4$ **21.** 4^5 **23.** 9^8 **25.** 3^{11} **27.** 4^6
29. $(-3)^5$ **31.** $(-2)^9$ **33.** $9 + 81 = 90$ **35.** $16 + 64 = 80$ **37.** 36 **39.** 2
41. 2 **43.** 4 **45.** 1/16 **47.** -14 **49.** -1 **51.** x^9 **53.** r^{11} **55.** y^5
57. a^3 **59.** k^6 **61.** 5^3m^3 or $125m^3$ **63.** $a^3/5^3$ or $a^3/125$ **65.** $243m^5n^5/32$
67. x^5 **69.** $1/b^2$

SECTION 3.2

1. -9 **3.** -3 **5.** -27 **7.** 15 **9.** -135 **11.** $8m^5$ **13.** $-r^5$
15. can't simplify **17.** x^5 **19.** $-p^7$ **21.** $7y^2$ **23.** $8a^3b^6 + 4a^3b^2$ **25.** 2
27. $23c^4$ **29.** $1 + 64m^6$ **31.** 4; binomial **33.** 9; trinomial **35.** 8; trinomial
37. 5; monomial **39.** 0; monomial **41.** 0, 6 **43.** 36, -12 **45.** 19, -2
47. 11, 5 **49.** -4, 2 **51.** sometimes **53.** never **55.** sometimes

SECTION 3.3

1. $5m^2 + 3m$ **3.** $4x^4 - 4x^2$ **5.** $-n^5 - 12n^3 - 2$ **7.** $12m^3 + m^2 + 12m - 14$
9. $15m^2 - 3m + 4$ **11.** $4b^2 + 7b + 8$ **13.** $-r^2 - 2r$ **15.** $5m^2 - 14m$
17. $-6s^2 + 5s + 1$ **19.** $4s + 2s^2$ **21.** $4x^3 + 2x^2 + 5x$ **23.** $-11y^4 + 8y^2 + 3y$
25. $a^4 - a^2 + 1$ **27.** $5m^2 + 8m - 10$ **29.** $4 + x^2 > 8$ **31.** $(5 + x^2) + (3 - 2x) \neq 5$
33. $P(x) + (6x^2 - 4x + 2) = (3x^2 - 6x + 1)$; $P(x) = -3x^2 - 2x - 1$

SECTION 3.4

1. $6m^2 + 4m$ **3.** $-6p^4 + 12p^3$ **5.** $-16z^2 - 24z^3 - 24z^4$ **7.** $6y + 4y^2 + 10y^5$
9. $-6m^5 + 9m^6$ **11.** $10b^6 + 20b^5 - 10b^4$ **13.** $-4x^7 - 8x^6 + 16x^5 + 12x^4 - 20x^3$
15. $m^2 + 12m + 35$ **17.** $x^2 - 25$ **19.** $t^2 - 16$ **21.** $6p^2 - p - 5$ **23.** $16m^2 - 9$
25. $6b^2 + 46b - 16$ **27.** $24 + 2a - 2a^2$ **29.** $-8 + 6k - k^2$
31. $12x^3 + 26x^2 + 10x + 1$ **33.** $81a^3 + 27a^2 + 11a + 2$
35. $20m^4 - m^3 - 8m^2 - 17m - 15$ **37.** $6x^6 - 3x^5 - 4x^4 + 4x^3 - 5x^2 + 8x - 3$
39. $5x^4 - 13x^3 + 20x^2 + 7x + 5$ **41.** $6a^6 + 3a^5 + 2a^4 + 23a^3 + 5a^2 + 4a + 20$
43. $2m^4 - 7m^3 + 3m^2 + 17m - 15$ **45.** $6x^3 - 11x^2 - 12x + 5$
47. $6y^3 - y^2 - 27y - 20$ **49.** $m^3 - 15m^2 + 75m - 125$ **51.** $x^4 + 5x^3 + x^2 - 21x - 18$

SECTION 3.5

1. $r^2 + 2r - 3$ **3.** $x^2 - 10x + 21$ **5.** $6x^2 + x - 2$ **7.** $6z^2 - 13z - 15$
9. $2a^2 + 9a + 4$ **11.** $8r^2 + 2r - 3$ **13.** $6a^2 + 8a - 8$ **15.** $20x^2 - 9x - 20$
17. $-12 + 5r + 2r^2$ **19.** $2a^2 - a - 15$ **21.** $m^2 + 2mn + n^2$ **23.** $25 + 10x + x^2$
25. $x^2 + 4xy + 4y^2$ **27.** $4z^2 - 20zx + 25x^2$ **29.** $25p^2 + 20pq + 4q^2$
31. $16a^2 + 40ab + 25b^2$ **33.** $m^2 - n^2$ **35.** $r^2 - z^2$ **37.** $36a^2 - p^2$ **39.** $4m^2 - 25$
41. $49y^2 - 100$ **43.** 324 **45.** 961 **47.** 10,201 **49.** 1599 **51.** 9999
53. 999,999 **55.** $(3 + x)^2 = 5$ **57.** $(3 + x)(x - 4) > 7$

SECTION 3.6

1. $30m^3 - 10m$ **3.** $60m^5 - 30m^2 + 40m$ **5.** $3m^4 - 2m^2 + m$ **7.** $4m^2 - 2m + 3$
9. $m/2 + 1/2 + 1/(2m)$ **11.** $x^3 + 3x^2 + x + 2$ **13.** $4x^3 - x^2 + 1$
15. $9x^2 - 3x^3 + 6x^4$ **17.** $12 + 8x + x^2$ **19.** $x^2/3 + 2x - 1/3$ **21.** $3y^2 - 2y$
23. $10p^3 - 5p^2 + 3p - 3/p$ **25.** $9m^2 - 6m + 4 - 2/m$ **27.** $4y^3 - 2 + 3/y$
29. $12/x - 6/x^2 + 14/x^3 - 10/x^4$ **31.** $12x^5 + 9x^4 - 12x^3 + 6x^2$
33. $-63y^4 - 21y^3 - 35y^2 + 14y$

SECTION 3.7

1. $x + 2$ **3.** $2y - 5$ **5.** $p - 4$ **7.** $r - 5$ **9.** $6m - 1$ **11.** $a - 7$
13. $x + 2 + 1/(2x + 1)$ **15.** $a - 2 + 6/(2a + 1)$ **17.** $d - 3 + 17/(2d + 4)$
19. $x^2 - x + 2$ **21.** $4k^3 - k + 2$ **23.** $3y + 1$ **25.** $x^2 + 1 + (-6x + 2)/(x^2 - 2)$
27. $x^2 - x + 1$ **29.** $x^2 - 1$

SECTION 3.8

1. 1/27 **3.** 1/25 **5.** 1/9 **7.** 1/36 **9.** 1/7 **11.** 32 **13.** 2 **15.** 27/8 **17.** 5/6 **19.** 9/20 **21.** 1/3 **23.** 9 **25.** 1/9 **27.** 8^3 **29.** $1/4^9$ **31.** $1/5^9$ **33.** m^5 **35.** $1/m$ **37.** $1/r$ **39.** a^7 **41.** $9/x^{10}$ **43.** $81/y^{10}$ **45.** b/a **47.** $25/m^2$ **49.** $125/(27x^2)$ **51.** $a^7/(2b^5)$

SECTION 3.9

1. 6.835×10^9 **3.** 8.36×10^{12} **5.** 2.15×10^2 **7.** 2.5×10^4 **9.** 3.5×10^{-2} **11.** 1.01×10^{-2} **13.** 1.2×10^{-5} **15.** 2.7×10^{-4} **17.** 1.14×10^{-10} **19.** 8,100,000,000 **21.** 9,132,000 **23.** 324,000,000 **25.** .00032 **27.** .041 **29.** .0000078 **31.** 800,000 **33.** .000004 **35.** 3,000,000 **37.** .05 **39.** 4×10^{-4}; 8×10^{-4} **41.** 3.68×10^{15} **43.** 1000; .06102 **45.** 35,000 **47.** .0000186

Chapter 3 Test

1. 12 **2.** 12 **3.** $4^5 = 1024$ **4.** $1/6^2 = 1/36$ **5.** $(-2)^5 = -32$ **6.** $2^4 = 16$ **7.** $2^6 = 64$ **8.** $2^2/3^2 = 4/9$ **9.** $1/8^2 = 1/64$ **10.** $8^1 = 8$ **11.** 1/3 **12.** $1/(8x^6y^9)$ **13.** $-x^2 + 6x$; 2; (b) **14.** $m^4 + 11m^3 - m^2$; 4; (a) **15.** $3x^3 - 4x^2 + 2x - 1$; 3; (d) **16.** 7; 0; (c) **17.** $8m^5 - 3m^2 + 4m^9$; 9; (a) **18.** $-x^2$; 2; (c) **19.** $10x^3 - 2x^2 - 8x$ **20.** $x^5 + 6x^3 - x^2 - 2x + 12$ **21.** $3y^2 - 2y - 2$ **22.** $-40x^{10}$ **23.** $6m^5 + 12m^4 - 18m^3 + 42m^2$ **24.** $r^2 - 3r - 10$ **25.** $6t^2 - t - 12$ **26.** $4p^2 + 20p + 25$ **27.** $x^2 - 64$ **28.** $k^5 + 4k^4 + 2k^3 - 4k^2 - 15k$ **29.** $3r^3 - 2r^2 + 5r - 3$ **30.** $3x^4 + 18x^3 + 27x^2 + 60x + 60$ **31.** 6×10^6 **32.** 2.45×10^8 **33.** 2.5×10^{-5} **34.** 1.04×10^{-2} **35.** 48,000 **36.** 291,000,000 **37.** .0645 **38.** .0000103 **39.** .02 **40.** 2 **41.** 1×10^{-6}

SECTION 4.1

1. 2 **3.** x **5.** $3m^2$ **7.** $2z^4$ **9.** xy^2 **11.** xy^2 **13.** $7x^3y^2$ **15.** $12(x + 2)$ **17.** $3(1 + 12d)$ **19.** $9a(a - 2)$ **21.** $5y^5(13y^4 - 7)$ **23.** $11p^4(11p - 3)$ **25.** $11(z^2 - 11)$ **27.** $9m^2(1 + 10m)$ **29.** $19y^2p^2(y + 2p)$ **31.** $6x^2y(3y^2 - 4x^2)$ **33.** $13y^3(y^3 + 2y^2 - 3)$ **35.** $8a(2a^2 + a + 3)$ **37.** $9qp^3(5q^3p^2 - 4p^3 + 9q)$ **39.** $ab^3(a^2b^3 - ab^4 + 1)$ **41.** $5z^3a^3(25z^2 - 12za^2 + 17a)$ **43.** $11y^4(3y^5 - 4y^9 + 7 + y)$

SECTION 4.2

1. $x + 3$ **3.** $r + 8$ **5.** $t - 12$ **7.** $x - 8$ **9.** $m + 6$ **11.** $p - 1$ **13.** $x + 3$ **15.** $(x + 5)(x + 1)$ **17.** $(a + 4)(a + 5)$ **19.** $(x - 1)(x - 7)$ **21.** cannot be factored **23.** $(y - 2)(y - 4)$ **25.** $(s - 5)(s + 7)$ **27.** cannot be factored **29.** $(b - 3)(b - 8)$ **31.** $(y + 3)(y - 7)$ **33.** cannot be factored **35.** $(z + 5)(z - 8)$ **37.** $3m(m + 3)(m + 1)$ **39.** $6(a + 2)(a - 10)$ **41.** $3j(j - 4)(j - 6)$ **43.** $3x^2(x + 5)(x - 6)$ **45.** $(x + 3a)(x + a)$ **47.** $(y + 5b)(y - 6b)$ **49.** $(x - 5y)(x + 6y)$ **51.** $(r - s)(r - s)$ **53.** $(p + 2q)(p - 5q)$

SECTION 4.3

1. $x - 1$ **3.** $b - 3$ **5.** $4y - 3$ **7.** $5x + 4$ **9.** $m + 10$ **11.** $(2x + 1)(x - 3)$ **13.** $(3x + 7)(x + 1)$ **15.** $(4r - 3)(r + 1)$ **17.** $(3m - 1)(5m + 2)$ **19.** $(2m - 3)(4m + 1)$ **21.** $(5a + 3)(a - 2)$ **23.** $(3r - 5)(r + 2)$ **25.** $(y + 17)(4y + 1)$ **27.** $(19x + 2)(2x + 1)$ **29.** $(2x + 3)(5x - 2)$ **31.** $(2w + 5)(3w + 2)$ **33.** $(2q + 3)(3q + 7)$ **35.** $(5m - 4)(2m - 3)$ **37.** $(4k - 5)(2k + 3)$ **39.** $(5m - 8)(2m + 3)$ **41.** $(3 - 2x)(1 - 4x)$ **43.** $(3 - 8m)(2 + 5m)$ **45.** $2m(4 - m)(5 + m)$ **47.** $2a^2(4a - 1)(3a + 2)$ **49.** $4z^3(z - 1)(8z + 3)$ **51.** $(4p - 3q)(3p + 4q)$ **53.** $(5a + 2b)(5a + 3b)$ **55.** $(3a - 5b)(2a + b)$

SECTION 4.4

1. $(x+4)(x-4)$ **3.** $(p+2)(p-2)$ **5.** $(m+n)(m-n)$ **7.** $(a+b)(a-b)$
9. $(3m+1)(3m-1)$ **11.** $(5m+4)(5m-4)$ **13.** $(2a+9)(2a-9)$
15. $(11r+12)(11r-12)$ **17.** $16(2z+1)(2z-1)$ **19.** $4(3t+2)(3t-2)$
21. $(5a+4r)(5a-4r)$ **23.** $(a^2+1)(a+1)(a-1)$ **25.** $(m^2+9)(m+3)(m-3)$
27. $(a+2)^2$ **29.** $(x+7)^2$ **31.** $(2y-5)^2$ **33.** not a perfect square **35.** $(4a-5b)^2$
37. $25(2m+1)^2$ **39.** $(7x+2y)^2$ **41.** $(2c+3d)^2$ **43.** $(5h-2y)^2$

SECTION 4.5

1. 2, −4 **3.** −5/3, 1/2 **5.** 0, −1, 1 **7.** 0, 1, −3, 2 **9.** −2, −3 **11.** −1, 6
13. −7, 4 **15.** 3, −8 **17.** −1, 3 **19.** −1, −2 **21.** −4 **23.** −2/5
25. 5/2, −2 **27.** −4/3, 1/2 **29.** −5/2, 1/3 **31.** 2, 5 **33.** 11/2, −12
35. −1, 3 **37.** 2 **39.** 5/4, −5/4 **41.** 2, −2 **43.** 5/2, 1/3, 5 **45.** −7/2, −1, 3
47. 1/2, −3, −1 **49.** 0, 1 **51.** 0, 2, −2 **53.** 0, 2, 4 **55.** 0, 4, −5

SECTION 4.6

1. 4 and 5 or −1 and 0 **3.** 12, 14, and 16; or −2, 0, and 2 **5.** 4 and 8
7. −9 and −5 or 6 and 10 **9.** 4 by 8 **11.** 3 inches **13.** 6 inches; 16 square inches
15. 2 feet **17.** $1\frac{1}{2}$ feet **19.** 15 feet by 15 feet

SECTION 4.7

1. 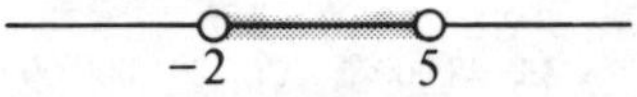

3. 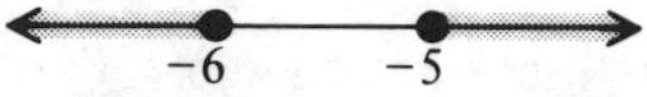

5. 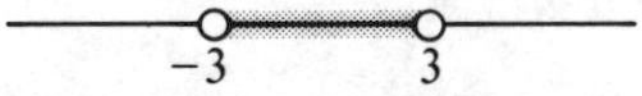

7. 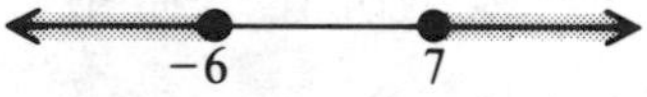

9. 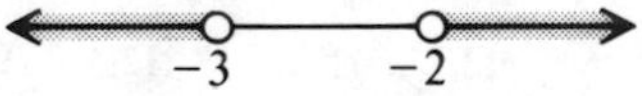

11. 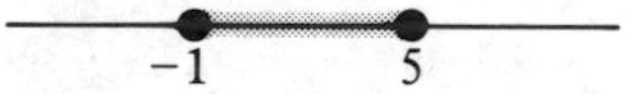

13. ∅

15. 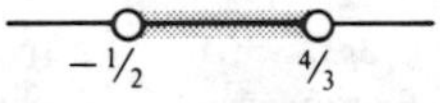

17. 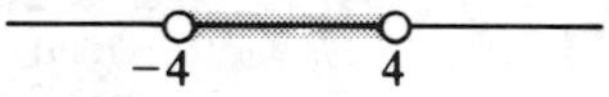

19. 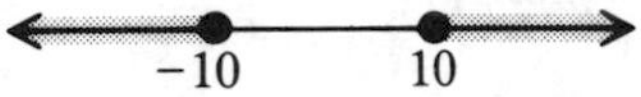

21. 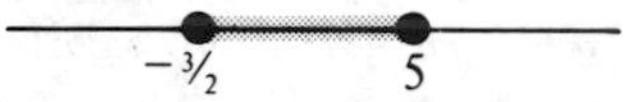

23. 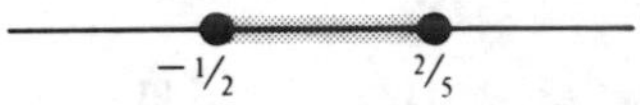

25. 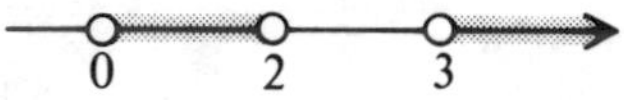

Chapter 4 Test

1. $m^2(8m-1)$ **2.** $6m(n+2m)$ **3.** $4(2m+n)(2m-n)$ **4.** $-3pq(p-3-2q)$
5. $(x+5)(x+6)$ **6.** $(x+7)(x-1)$ **7.** $(2y+3)(y-5)$ **8.** $(2x+3)^2$
9. $(x+5)(x-5)$ **10.** $4xt(x-2t)(x-t)$ **11.** −2, −1 **12.** 2, −2 **13.** 1/3, −2
14. −1, 2, 3 **15.** −4, 0, 4 **16.** −1, 8

17. 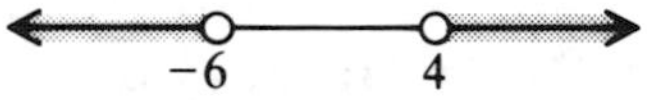

18. 8 and 17 or −7 and 2

19. 3 by 5 **20.** 2 **21.** 1/2 foot

SECTION 5.1

1. −27/20 **3.** 1/5 **5.** 3 **7.** 3/10 **9.** 1/9 **11.** 3/20 **13.** 5/12 **15.** 13/3
17. −28 **19.** −15/32 **21.** 21/4 **23.** 1 **25.** 13/30, 12/25 **27.** 7/9, 4/5, 6/7
29. 2/7, 3/10, 1/3, 3/8 **31.** 5/12, 1/2, 8/15, 4/7 **33.** 10/9 **35.** −9/10 **37.** 1/6

SECTION 5.2
1. 1/3 **3.** 7/10 **5.** 10/9 **7.** 19/22 **9.** −2/15 **11.** 8/15 **13.** 75/8
15. 31/6 **17.** 17/12 **19.** 49/30 **21.** 13/28 **23.** 1/2 **25.** 7/40 **27.** 103/300
29. 21/26 **31.** −11/5 **33.** 2/15 **35.** 5/3 **37.** 11/24 **39.** 9/8

SECTION 5.3
1. −3/2 **3.** −14/3 **5.** no value **7.** 2/3 **9.** $2k$ **11.** $-4y^3/3$ **13.** $4m/(3p)$
15. $3/[-4(y+2)]$ **17.** $(x-1)/(x+1)$ **19.** $m-n$ **21.** $m/2$ **23.** $2(2r+s)$
25. $(m-2)/(m+3)$ **27.** $(x+4)/(x+1)$ **29.** $(2y+1)/(y+1)$ **31.** 1
33. $-(x+1)$ **35.** −1

SECTION 5.4
1. $3m/4$ **3.** 3/32 **5.** $2a^4$ **7.** 1/4 **9.** 1/6 **11.** $6/(a+b)$ **13.** 2 **15.** 2/9
17. 3/10 **19.** $2r/3$ **21.** $(y+4)(y-3)$ **23.** $18/[(m-1)(m+2)]$ **25.** −7/8
27. −1 **29.** $(k+2)/(k+3)$ **31.** $(n+4)/(n-4)$ **33.** 1 **35.** $(m-3)/(2m-3)$
37. $(p-q)/(p+q)$ **39.** $(x+2)(x+1)$ **41.** $10/(x+10)$

SECTION 5.5
1. $7/p$ **3.** $-3/k$ **5.** 1 **7.** $m+n$ **9.** m **11.** 1 **13.** $(6+m)/(2m)$
15. $(18+3m)/(2m)$ **17.** $(3y-5)/(5y)$ **19.** $m/3$ **21.** 7/24 **23.** $(6-2y)/y^2$
25. $(9p+8)/(2p^2)$ **27.** $(7m+4n)/6$ **29.** $(-y-3x)/(x^2y)$
31. $(2m^2+4m+4)/[m(m+2)]$ **33.** $(4x+24)/[(x-2)(x+2)]$ **35.** $x/[2(x+y)]$
37. $3/[(m+1)(m-1)(m+2)]$ **39.** $(m-2)/[(m+3)(m-3)]$ **41.** $3/(m-2)$
43. $-3/(y-3)$ **45.** $8m/(m+2n)$ **47.** $(2x^2+6xy+8y^2)/[(x+y)^2(x+3y)]$
49. $-(a^2+5ab)/[(a+b)^2(a-b)]$

SECTION 5.6
1. 1/2 **3.** 12 **5.** 1/4 **7.** 2/5 **9.** 24 **11.** 2 **13.** −7 **15.** 1 **17.** 2
19. 5 **21.** −6 **23.** −8 **25.** 8 **27.** 5 **29.** 2 **31.** 6, 4
33. any real number except −1 **35.** 1, −24 **37.** 1/2, −6

SECTION 5.7
1. 9 **3.** 12/17 **5.** −3/4 and 9/4 **7.** 2/3 or −3 **9.** 2/3 or 1 **11.** $260
13. $66\frac{2}{3}\%$ **15.** 200 **17.** $11.52 **19.** $7.14 **21.** 5.56%
23. After the raise he earns, say, $100 + $10 = $110. A 10% cut reduces this to $110 − $11 = $99.
25. 900 miles **27.** 110 mph **29.** 84/19 hours **31.** 20 days **33.** 100/11 minutes

SECTION 5.8
1. 3/2 **3.** 7/5 **5.** −5/6 **7.** $1/x$ **9.** r/p **11.** $(m^2+1)/(3-m^2)$ **13.** $2/y$
15. x/y **17.** $(x+y)^2/(xy)$ **19.** q **21.** $-n/2$ **23.** 2/3

Chapter 5 Test
1. 9/16, 7/12, 5/8 **2.** 25/36 **3.** 36/25 **4.** x **5.** mn^2 **6.** $-1/x$
7. $11/[6(a+1)]$ **8.** $2/(t+1)$ **9.** $(2a+7)/[(a+3)(a-2)(a+4)]$
10. $(3m-2)/(3m+2)$ **11.** $(a+3)/(a+4)$ **12.** 2 **13.** $3(1+x)/x$ **14.** 3/2
15. −2/3, 4 **16.** 160 **17.** 1/4 or 1/2 **18.** 9/5 **19.** 20/9 hours **20.** 1/5 hours

SECTION 6.1
1. yes **3.** yes **5.** no **7.** yes **9.** yes **11.** no **13.** 7; 1; −1 **15.** 2; 8; 17
17. 9; 3; −15 **19.** 2; 3; −3 **21.** −3; 5; −5 **23.** −4; −4; −4 **25.** 3; 3; 3
27. −9; −9; −9 **29.** 2; 5/2; −5/4 **31.** −5/4; 5/6; 7/4

SECTION 6.2

1. (2, 5) **3.** (−5, 5) **5.** (7, 3)

7. (6, 1) **9.** (3, 5)

11. (−2, 4) **13.** (−3, 5)

15. (4, 0) **17.** (−2, 0)

19. (0, 3) **21.** (0, −5)

23. I **25.** II **27.** III **29.** II **31.** IV **33.** none

35. (0, 6); (2, 10); (−3, 0); (−2, 2)

37. (0, 3); (10, −3); (5, 0); (−5, 6)

39. (0, 0); (−2, −6); (4, 12); (−1, −3)

41. (3, 2); (3, 5); (3, 0); (3, −3)

43. (5, −2); (0, −2); (−3, −2); (−2, −2)

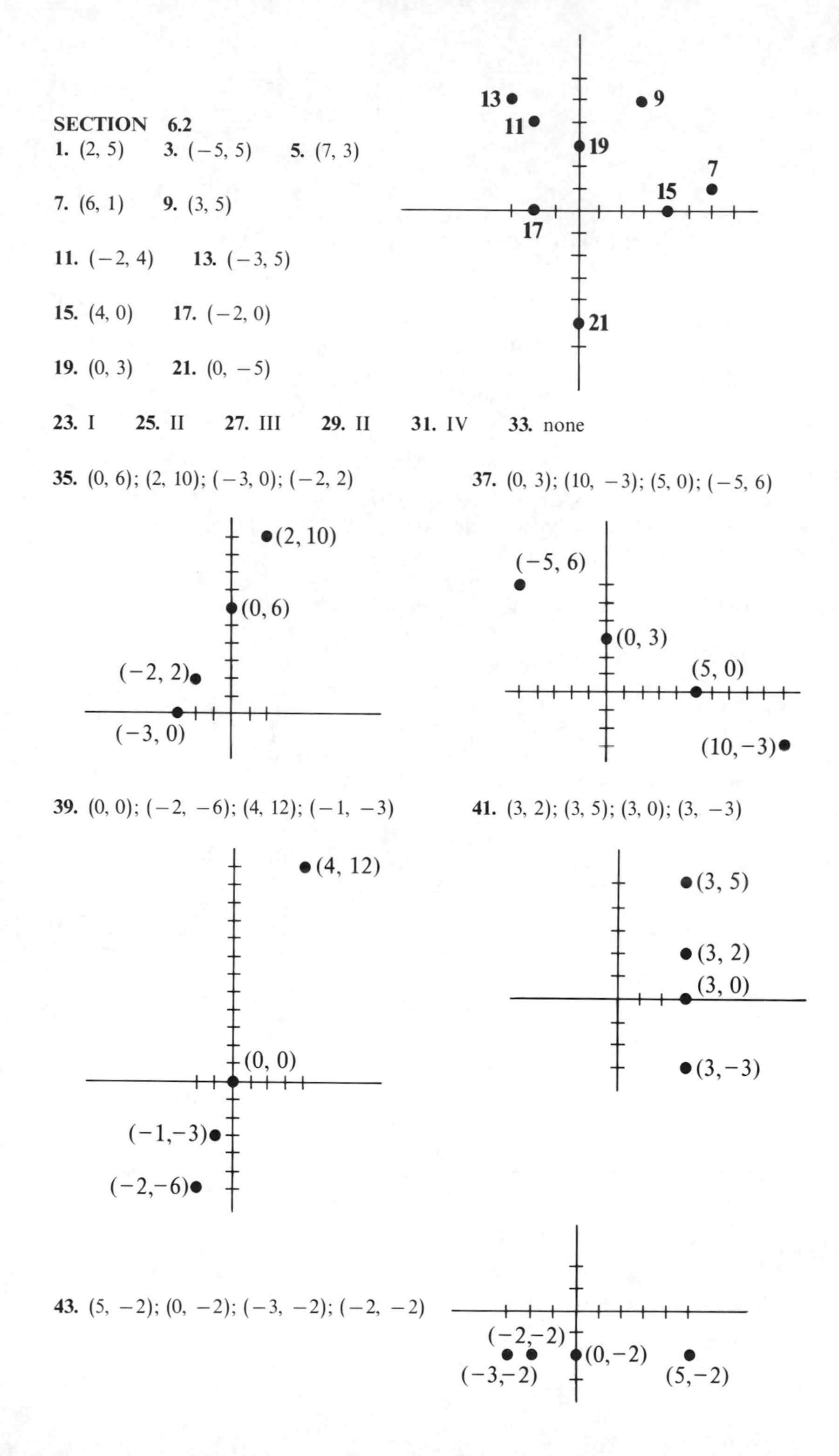

SECTION 6.3

1. 5; 5; 3

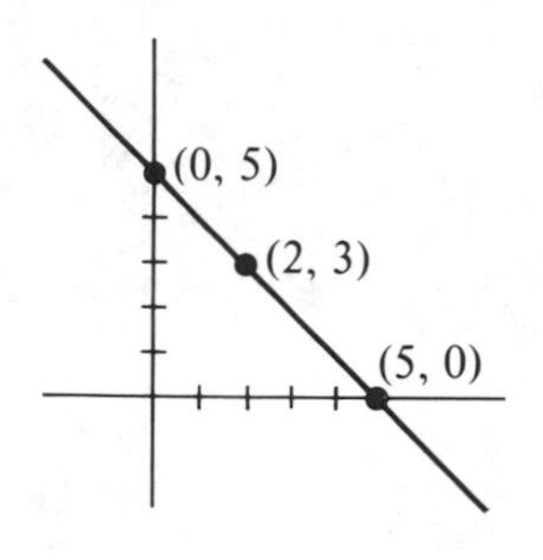

3. 4; −4; 2

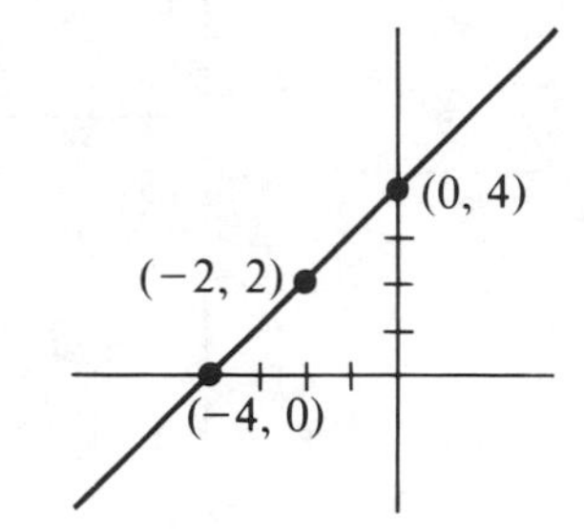

5. −6; 2; 3

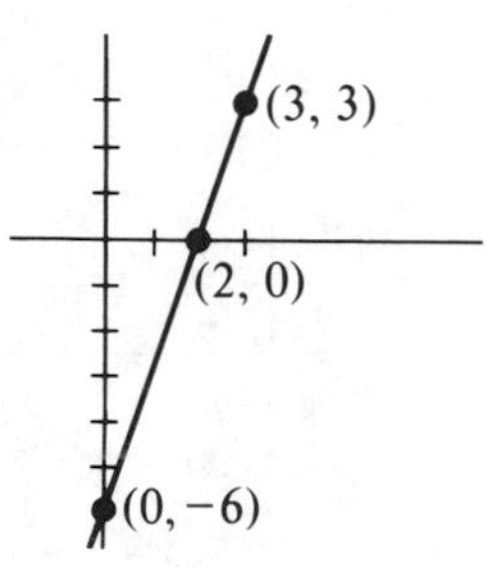

7. 4; 10; 2

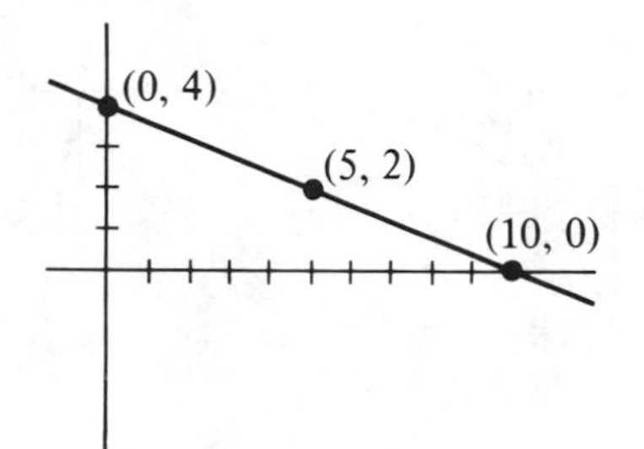

9. −5; −5; −5

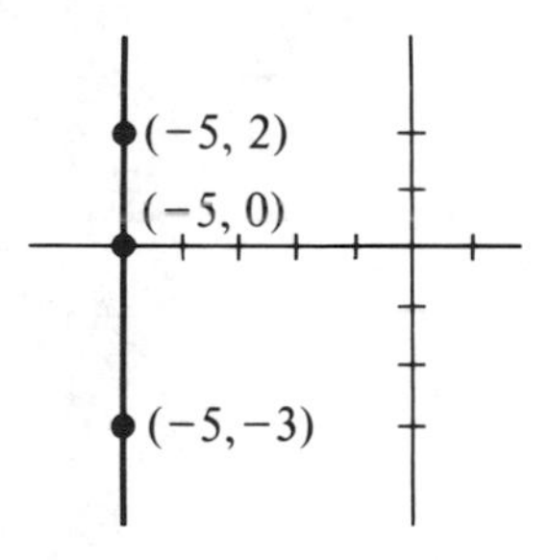

11.

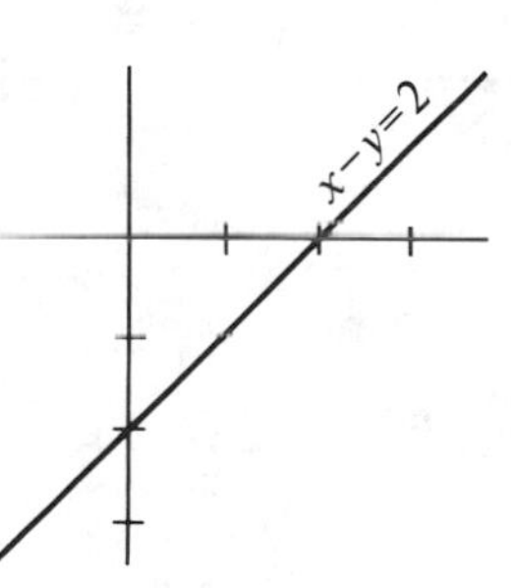

13.

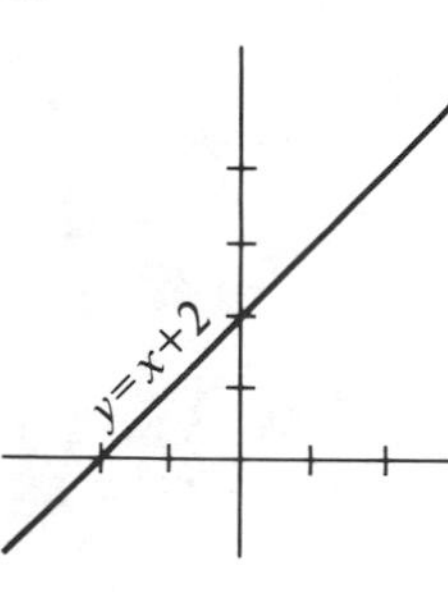

15.

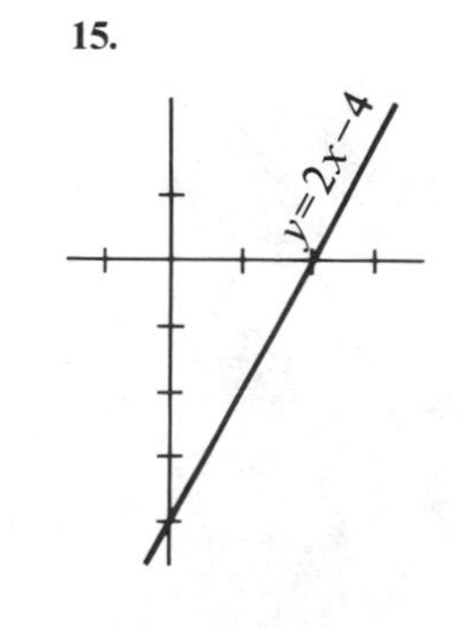

17.

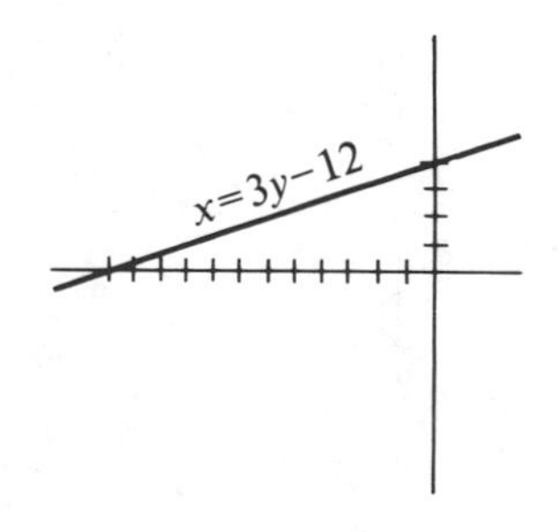

19.

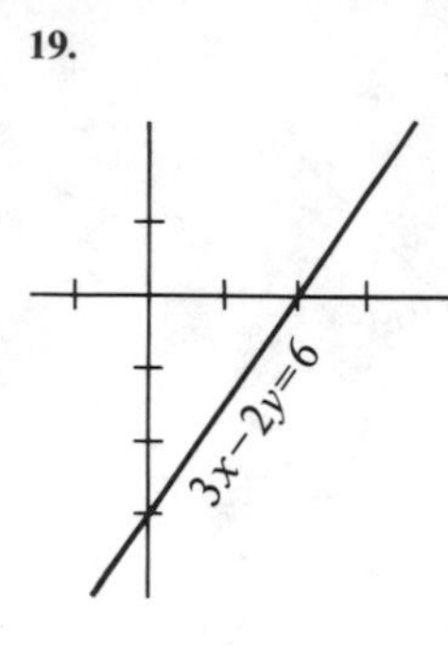

21.

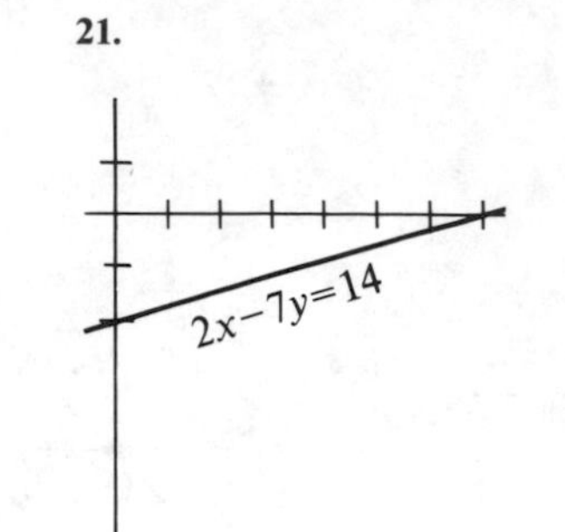

23.

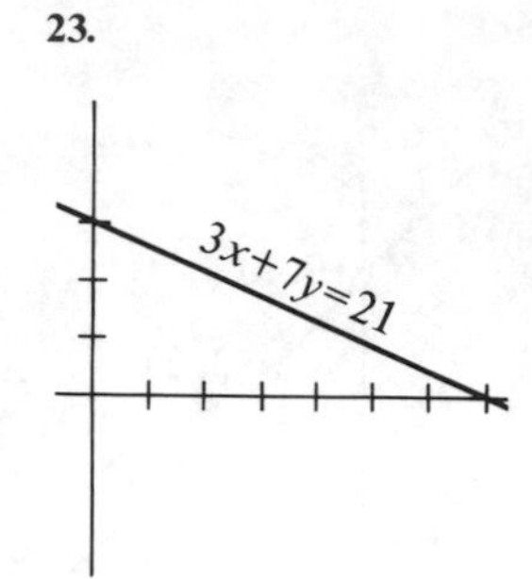

25.

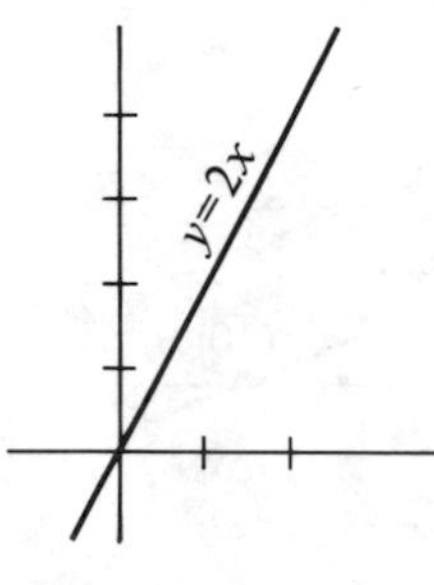

27.

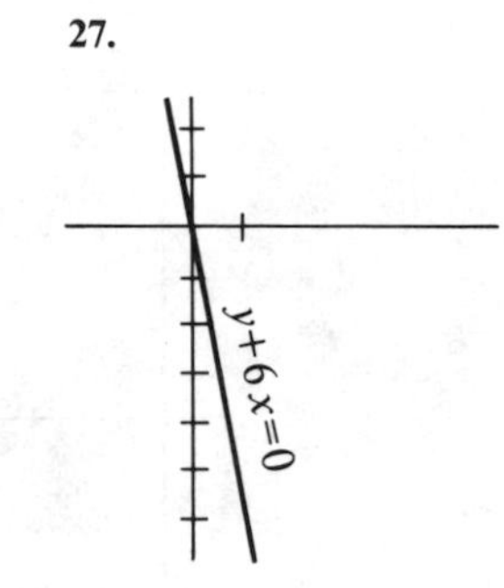

29.

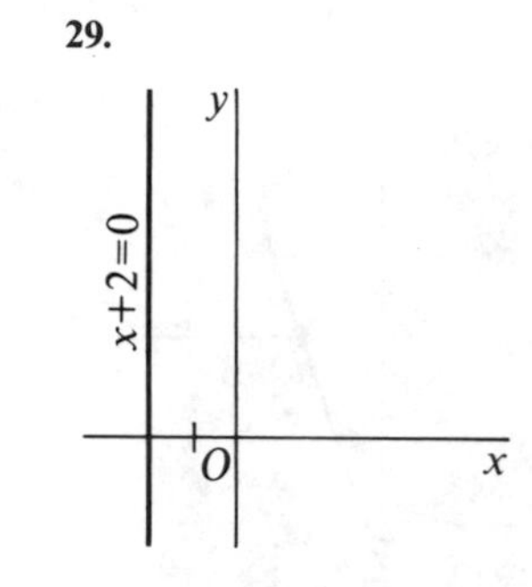

31.

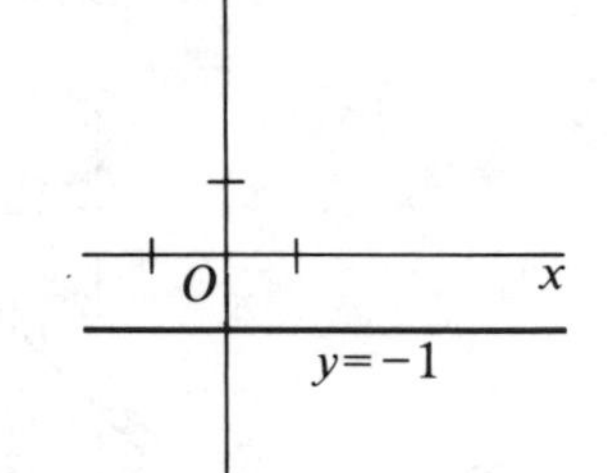

33.

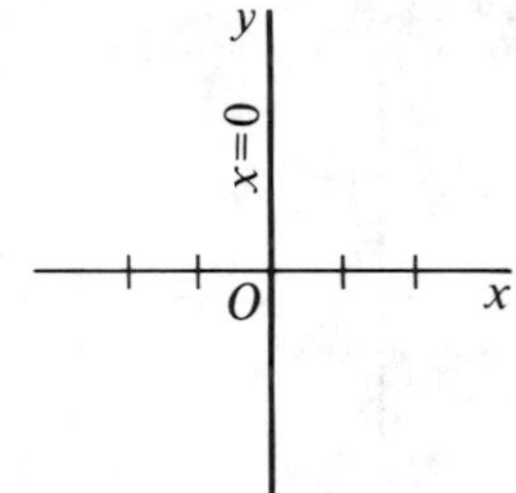

35.

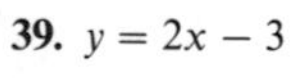

37. $x = y + 2$

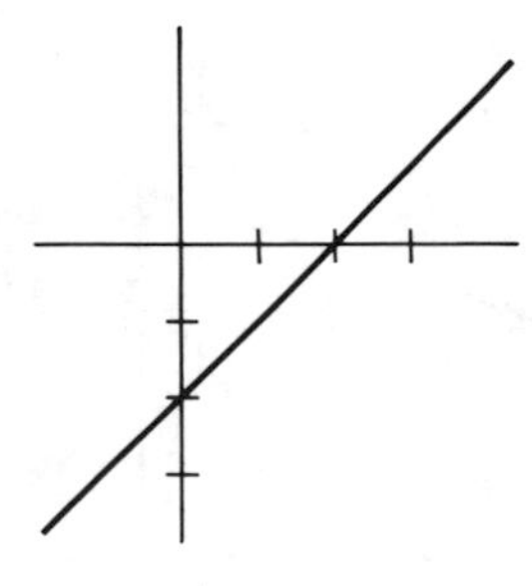

39. $y = 2x - 3$

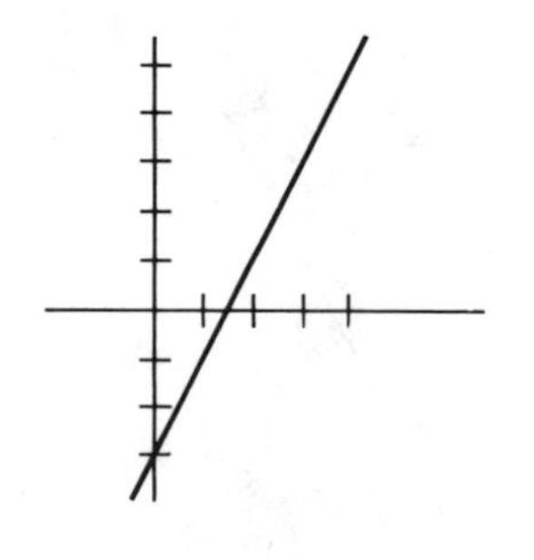

41. $3 + y = 2x - 4$
$y = 2x - 7$

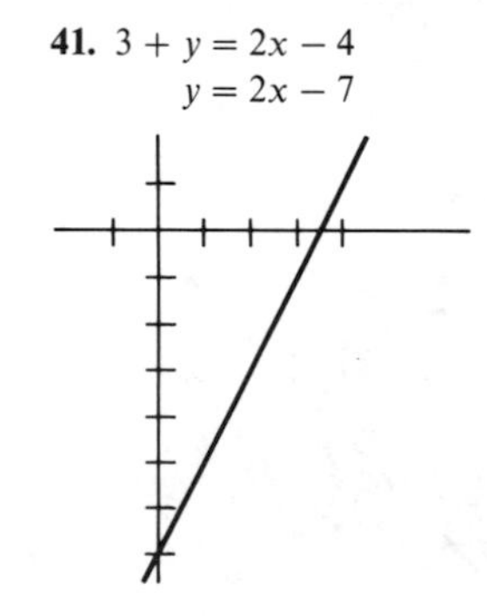

43. (a) 121 (f)

(b) 132

(c) 154

(d) 165

(e) 176

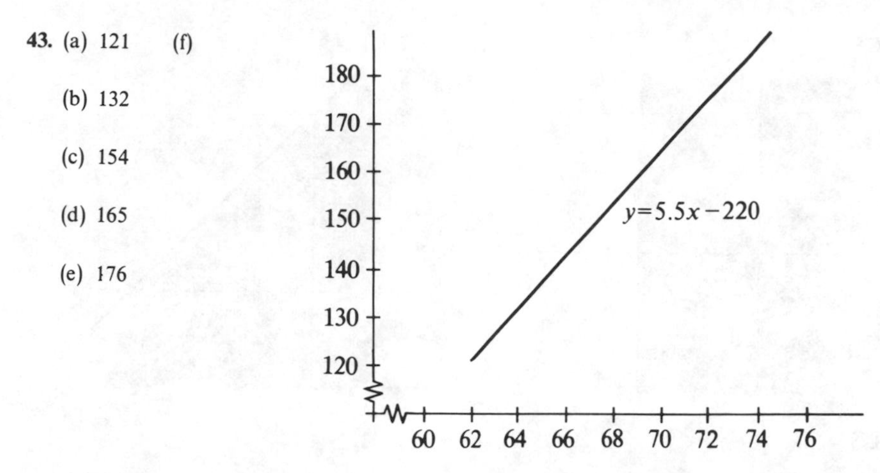

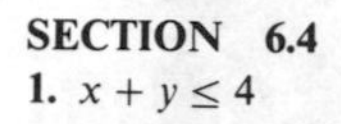

SECTION 6.4

1. $x + y \leq 4$

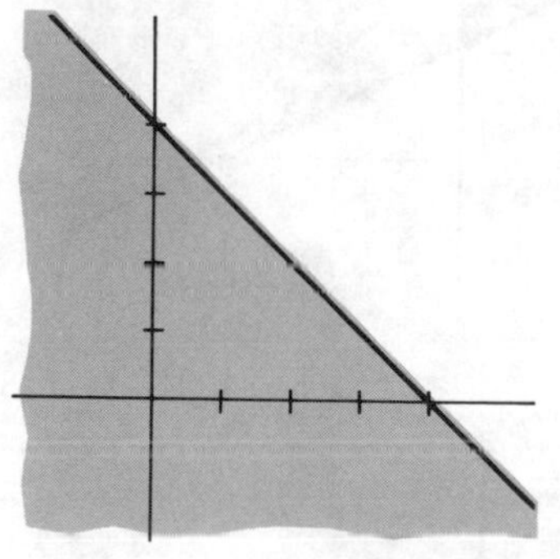

3. $x + 2y \leq 7$

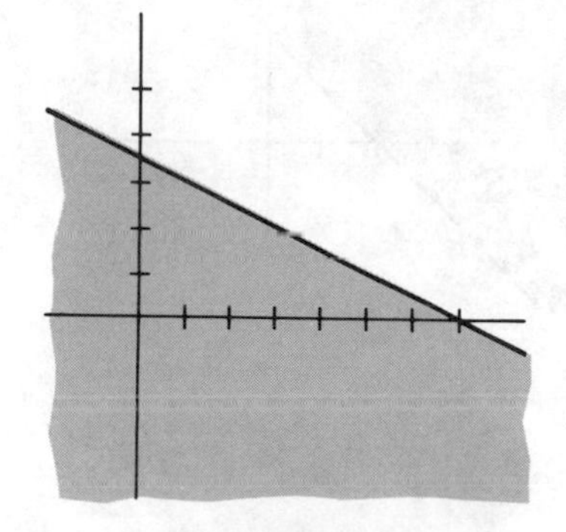

5. $-3x + 4y < 12$

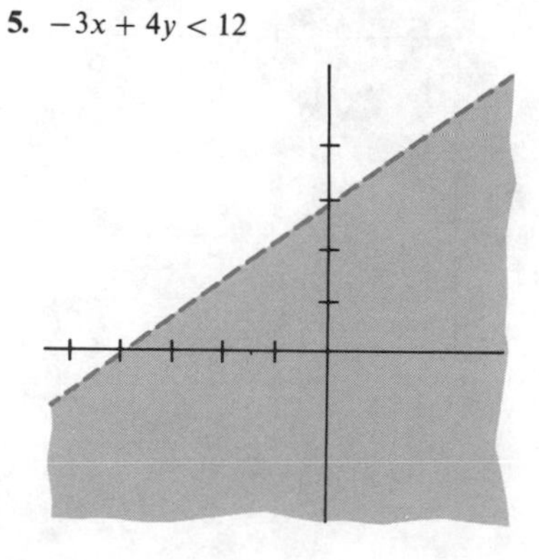

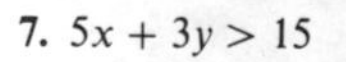

7. $5x + 3y > 15$

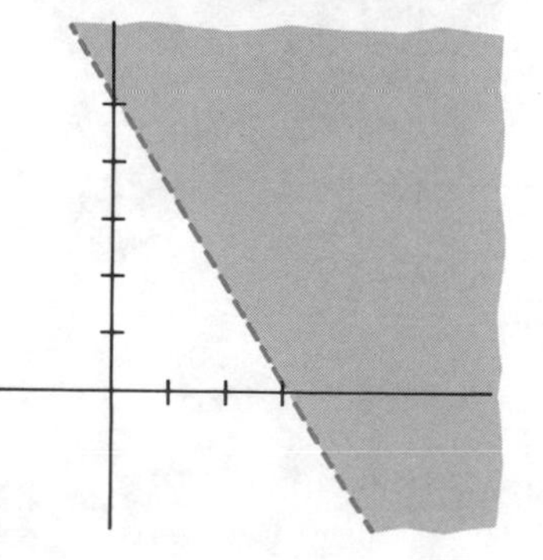

9. $x < 4$

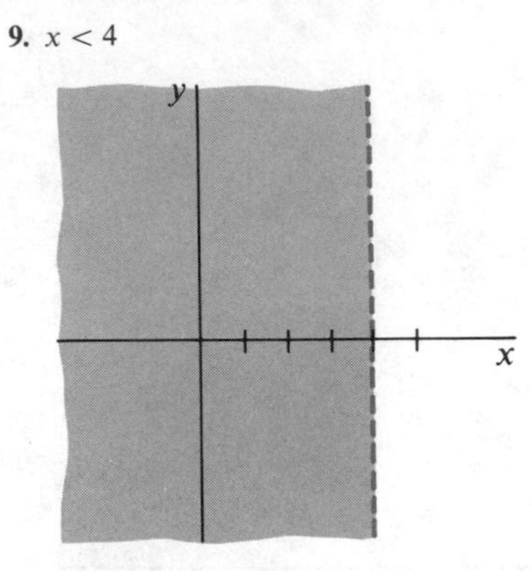

11. $x + y \leq 8$

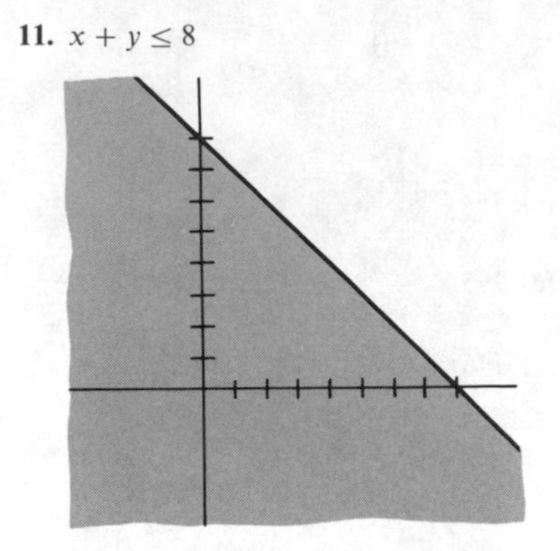

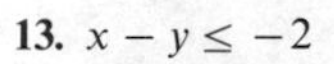
13. $x - y \leq -2$

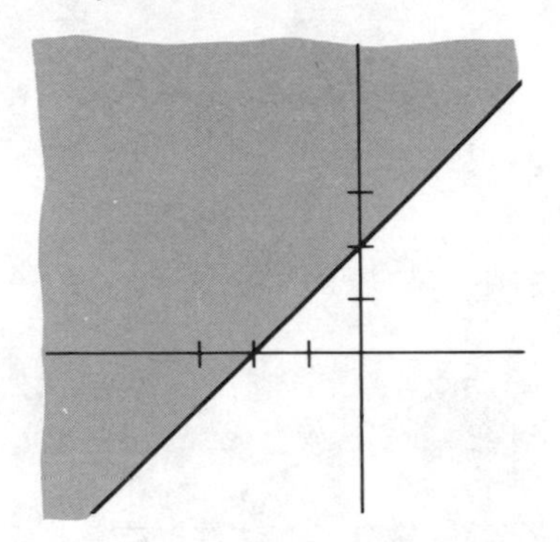

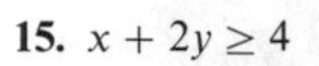
15. $x + 2y \geq 4$

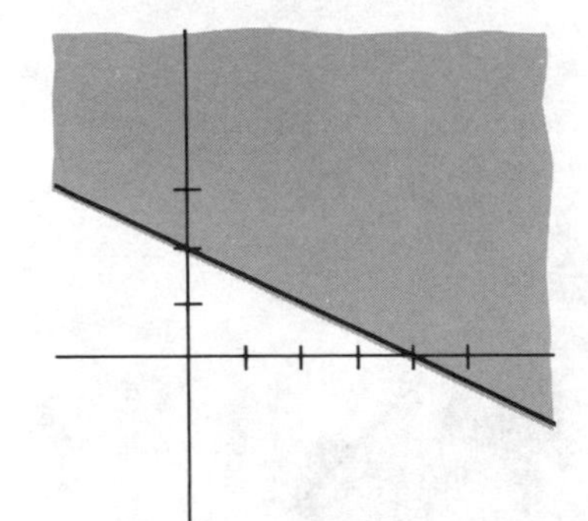

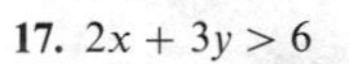
17. $2x + 3y > 6$

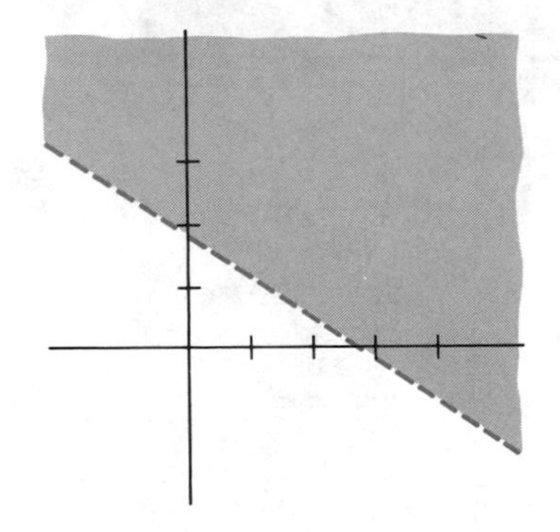

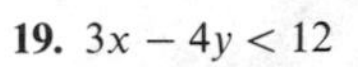
19. $3x - 4y < 12$

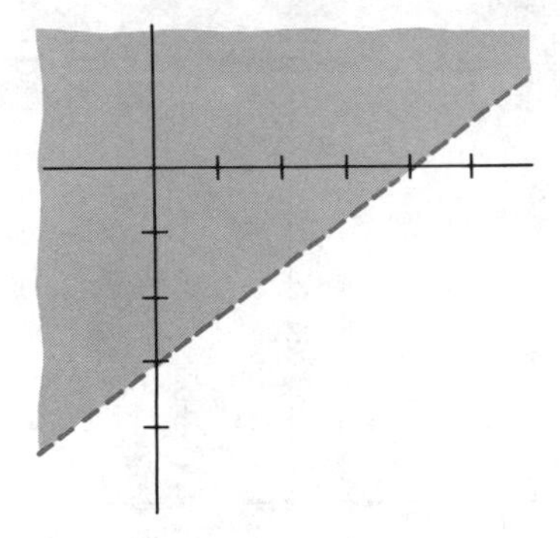

21. $3x + 7y \geq 21$

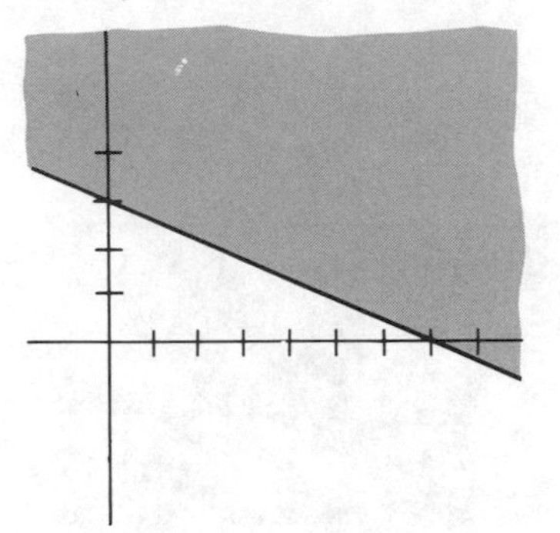

23. $4x - 5y \geq 20$

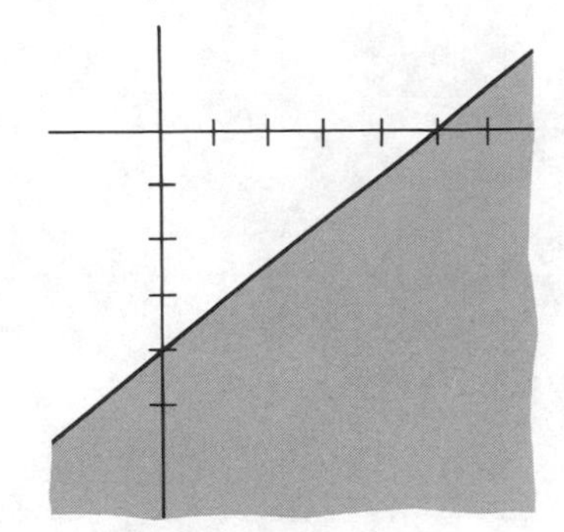

25. $4x + 7y \leq 14$

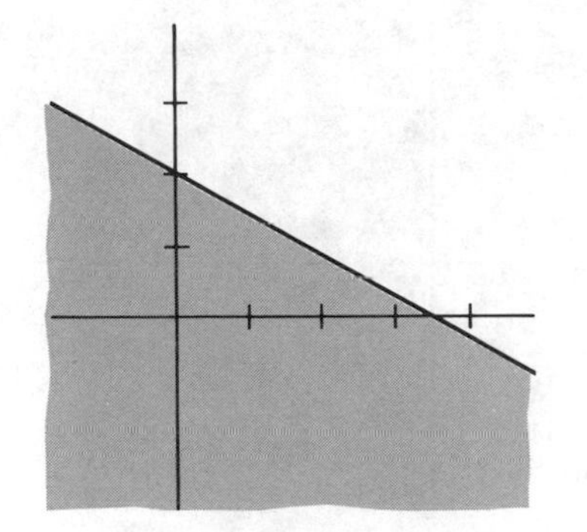

27. $x \leq 3y$

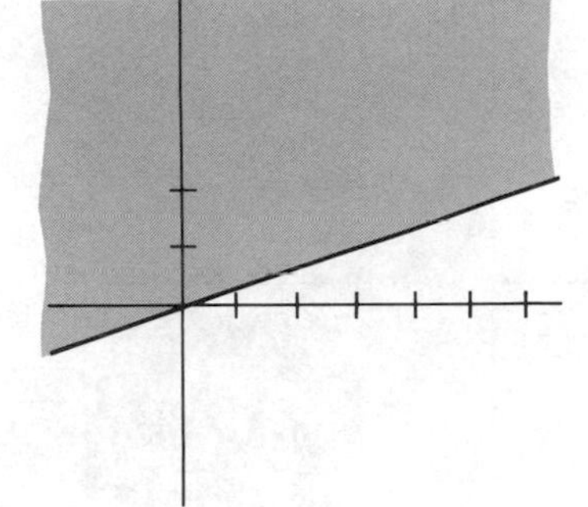

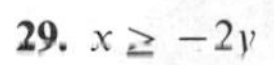

29. $x \geq -2y$

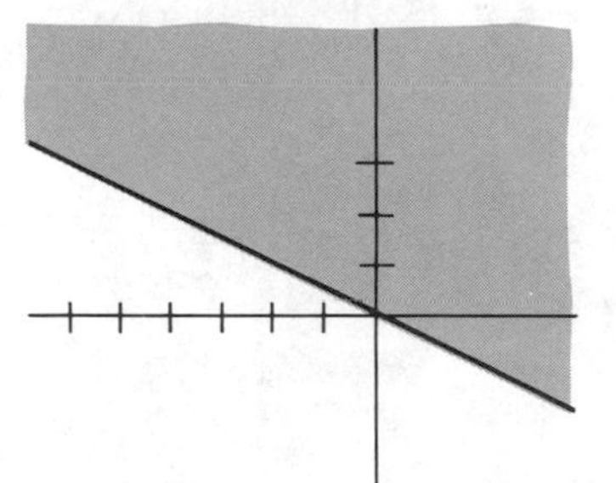

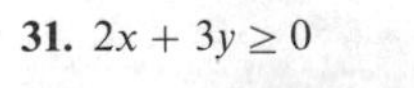

31. $2x + 3y \geq 0$

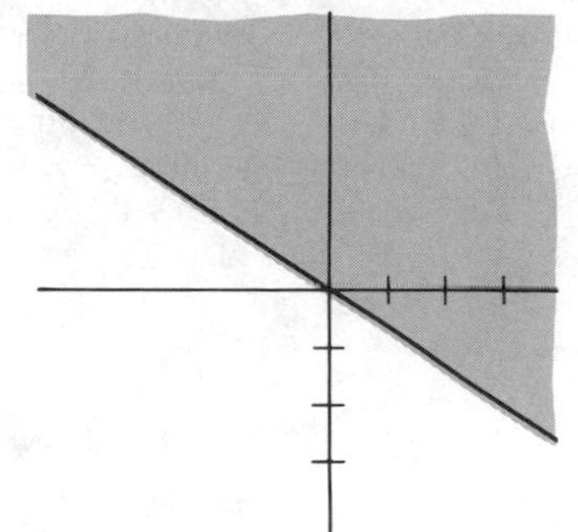

33. $x < 4$

35. $y \leq 2$

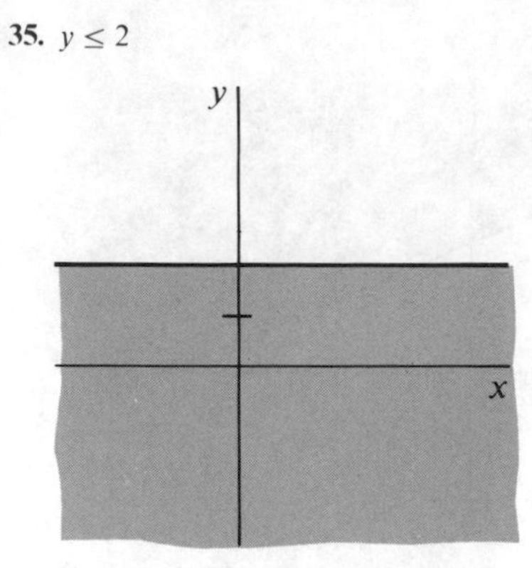

37. $x \geq -2$

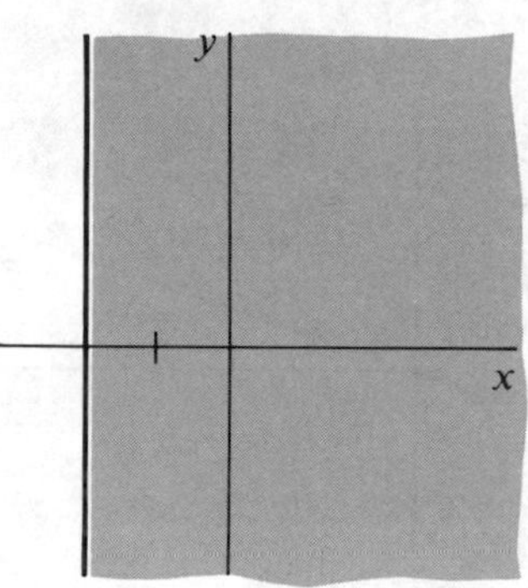

39. $x \geq 0$

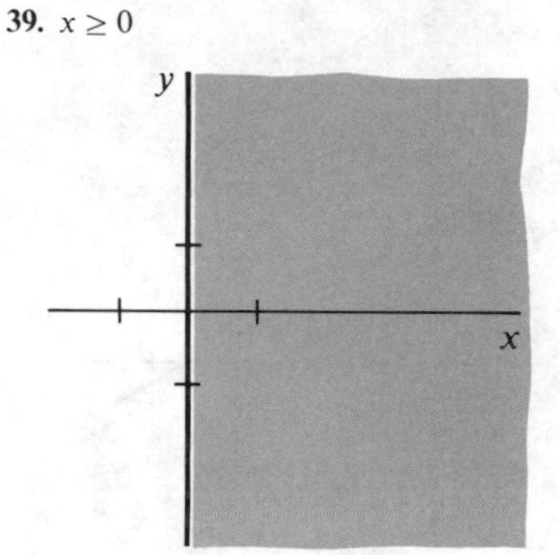

SECTION 6.5

1. yes **3.** yes **5.** yes **7.** no **9.** no **11.** no **13.** all reals; all reals **15.** all reals; all reals **17.** all reals; $y \geq 4$ **19.** all reals; $y \geq 0$ **21.** all reals; $y \leq 0$ **23.** all reals; $y \geq 2$ **25.** all reals; $y \geq 0$ **27.** (a) 8 (b) 2 (c) -7 **29.** (a) 2 (b) 4 (c) 7 **31.** (a) -12 (b) -4 (c) 8 **33.** (a) 6 (b) 2 (c) 11 **35.** (a) 1 (b) 9 (c) 36 **37.** (a) -16 (b) -4 (c) -1 **39.** (a) -4 (b) -2 (c) -1 **41.** (a) 0 (b) 3 (c) 8 **43.** (a) 9 (b) 24 (c) -11 **45.** (a) 1 (b) -62 (c) -7 **47.** function **49.** not a function **51.** function **53.** not a function

Chapter 6 Test

1. -6; -16; 4

2. -6; 10; -3

3. -4; -4; -4

4.

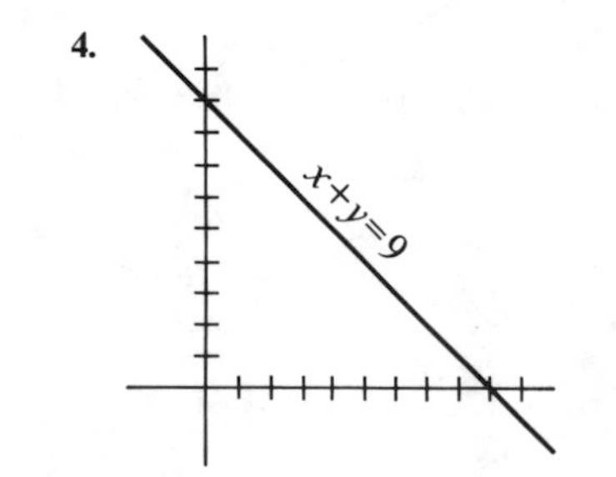

5.

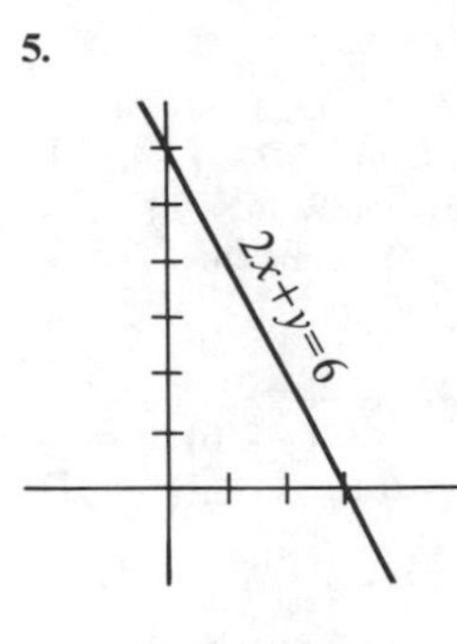

6.

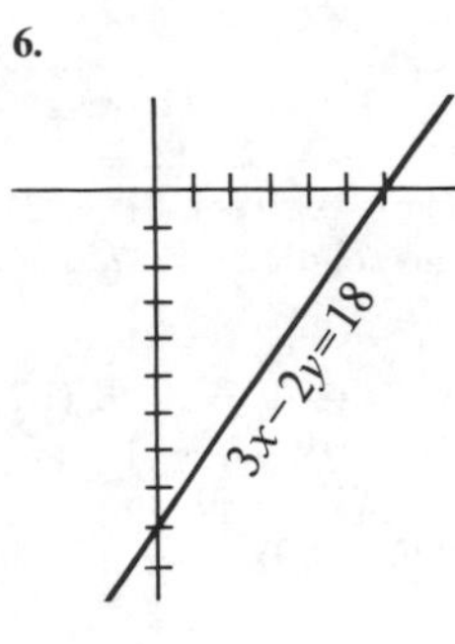

7.

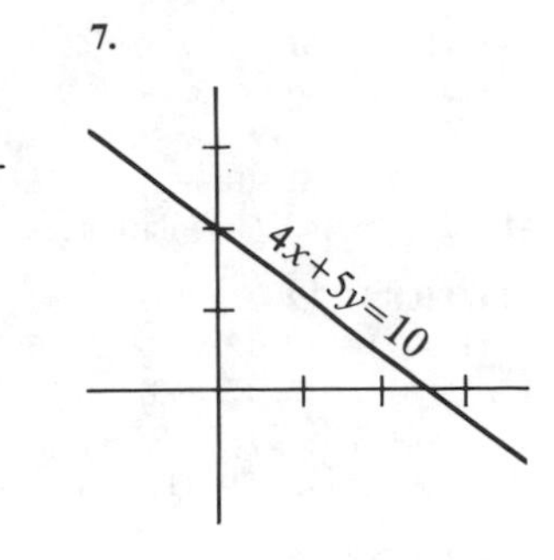

8. **9.** **10.**

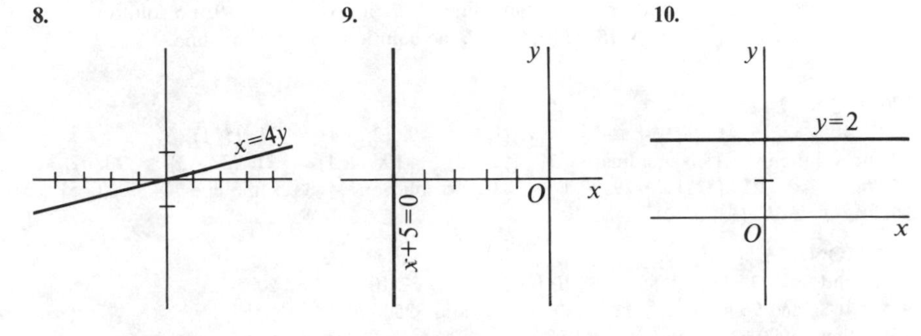

11.

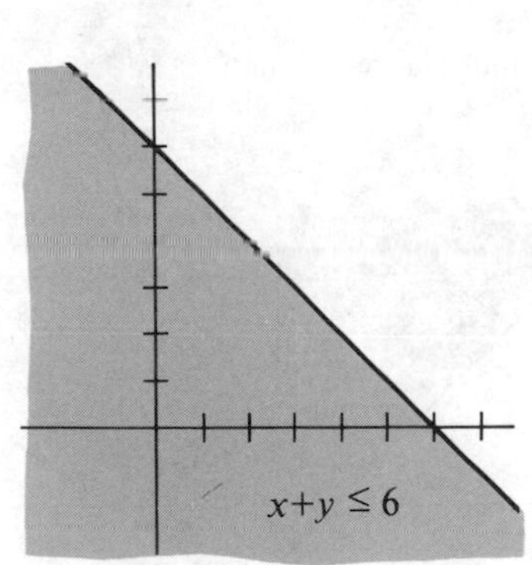

12.

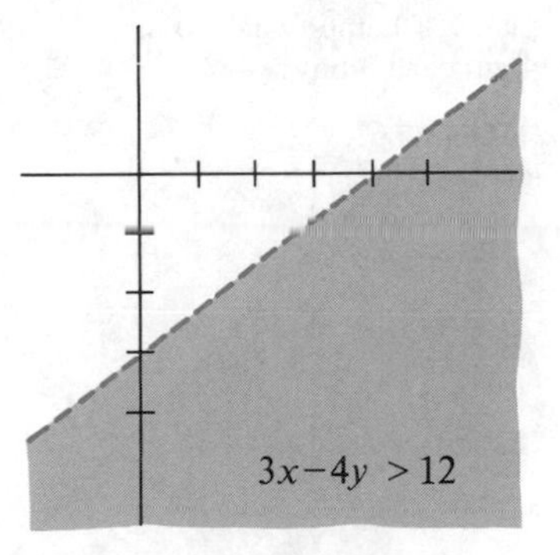

13.

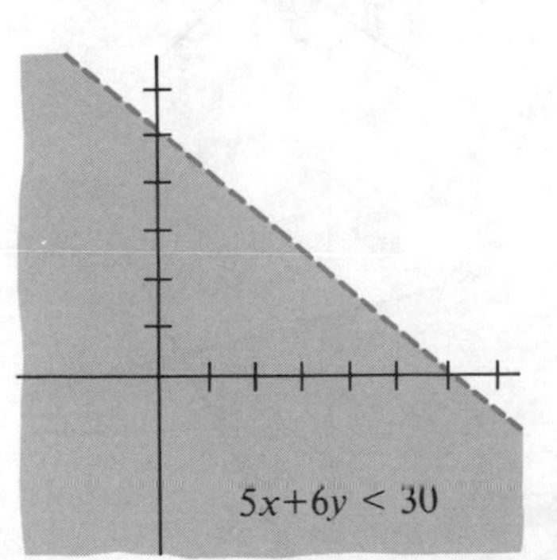

14.

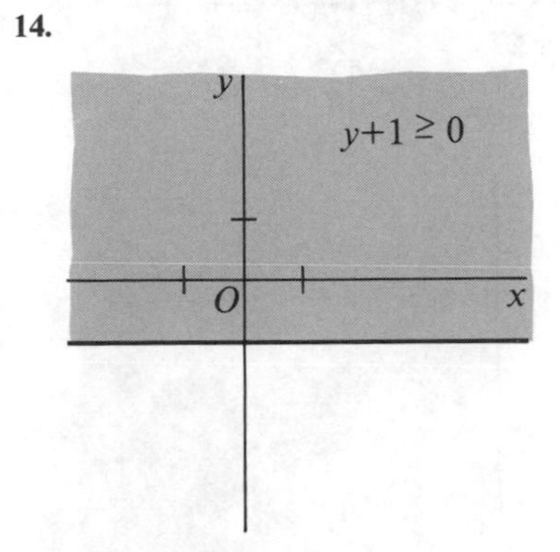

15. function **16.** function **17.** function **18.** not a function **19.** function
20. not a function **21.** -2 **22.** 22 **23.** -20 **24.** 22 **25.** 6

SECTION 7.1

1. yes **3.** no **5.** no **7.** yes **9.** no **11.** yes **13.** (5, 3) **15.** (4, 8) **17.** (4, −2) **19.** (1, −3) **21.** (3, −2) **23.** (2, −1) **25.** (−8, 6) **27.** (−4, −1) **29.** (5, 0) **31.** (0, −2) **33.** (5, 2) **35.** (−1, 3) **37.** (2, −3) **39.** (6, −2) **41.** (2, 7) **43.** no solution **45.** no solution **47.** same line **49.** same line

SECTION 7.2

1. (1, −2) **3.** (2, 0) **5.** (6, 2) **7.** (−2, 6) **9.** (1/2, 2) **11.** (3/2, −2) **13.** (3, 0) **15.** (2, −3) **17.** (4, 4) **19.** (2, −5) **21.** (4, 9) **23.** (−4, 0) **25.** (4, −3) **27.** (−9, −11) **29.** (6, 3) **31.** (6, −5) **33.** (−6, 0) **35.** (1/2, 2/3) **37.** (3/8, 5/6) **39.** (11, 15) **41.** (22/9, 8/9)

SECTION 7.3

1. no solution **3.** no solution **5.** same line **7.** no solution **9.** no solution **11.** same line **13.** (0, 0) **15.** (3, 2) **17.** no solution **19.** same line **21.** same line

SECTION 7.4

1. (2, 4) **3.** (6, 4) **5.** (8, −1) **7.** (1, 5) **9.** (2, −4) **11.** (5, 1) **13.** no solution **15.** same line **17.** (5, −3) **19.** (4, 1) **21.** (2, −3) **23.** (2, 8) **25.** (4, −6) **27.** (3, 2) **29.** (7, 0) **31.** no solution **33.** same line **35.** (6, 5) **37.** (0, 3) **39.** (18, −12)

SECTION 7.5

1. 43 and 9 **3.** 72 and 24 **5.** 10 by 20 **7.** 22 10's and 63 20's
9. 74 at 8¢ and 96 at 10¢ **11.** \$7500 at 5% and \$2500 at 7%
13. 4 liters of 90% and 16 liters of 75%
15. 30 barrels of \$40 olives and 20 barrels of \$60 olives
17. plane, 470 mph; wind, 70 mph **19.** John, $3\frac{1}{4}$ mph; Harriet, $2\frac{3}{4}$ mph
21. 4 girls and 3 boys

SECTION 7.6

1.

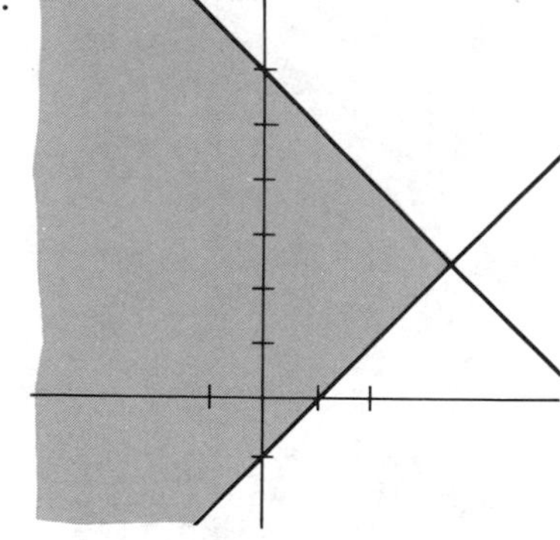

3.

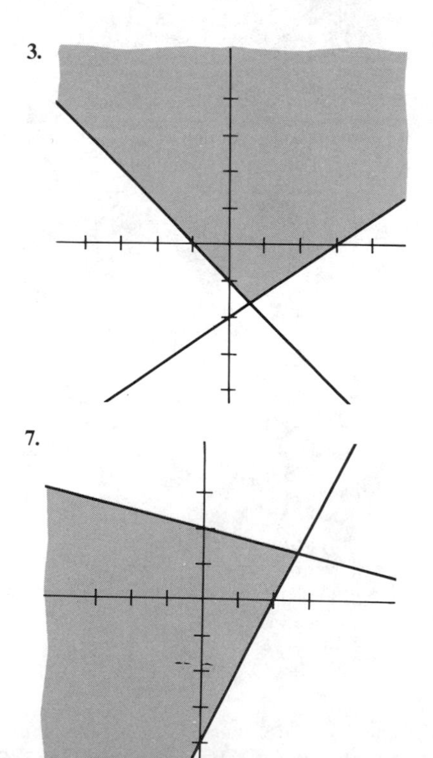

5.

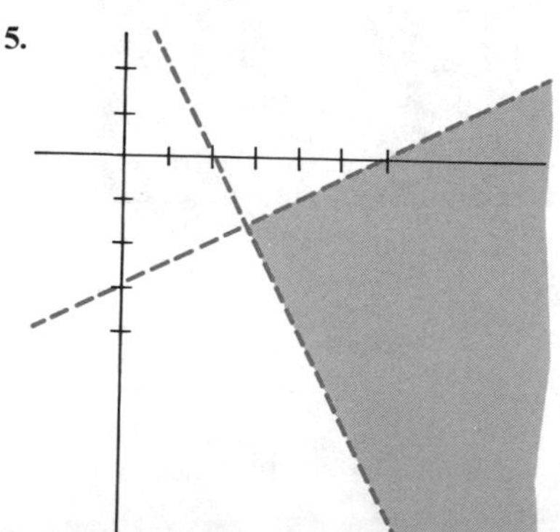

7.

9.

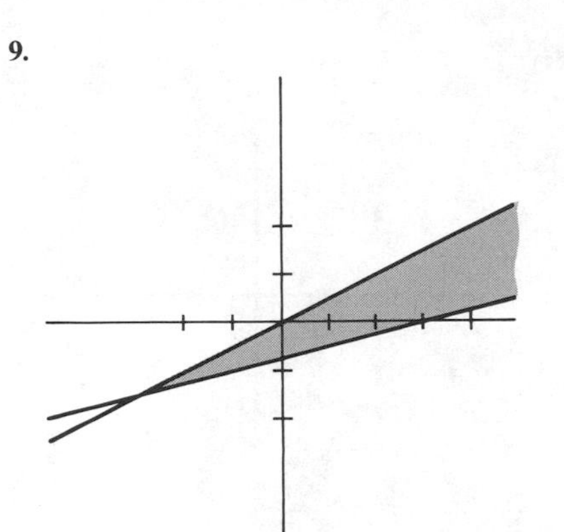

11.

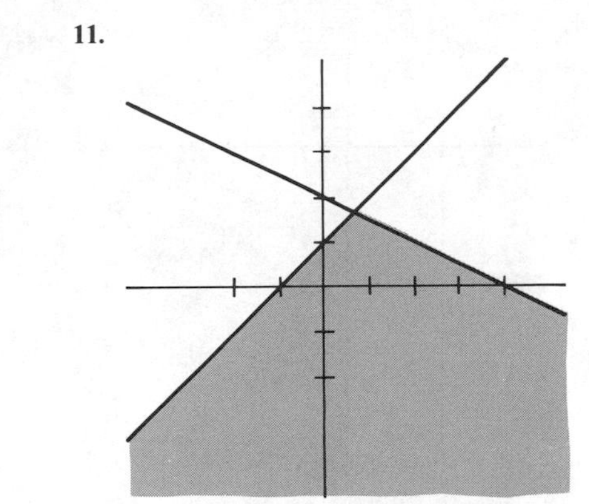

13.

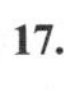

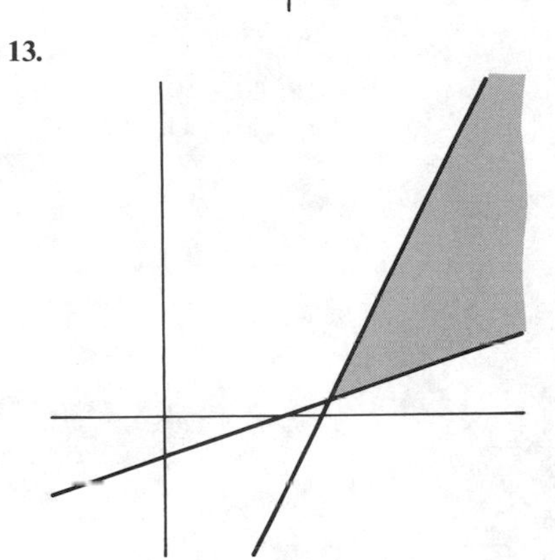

15.

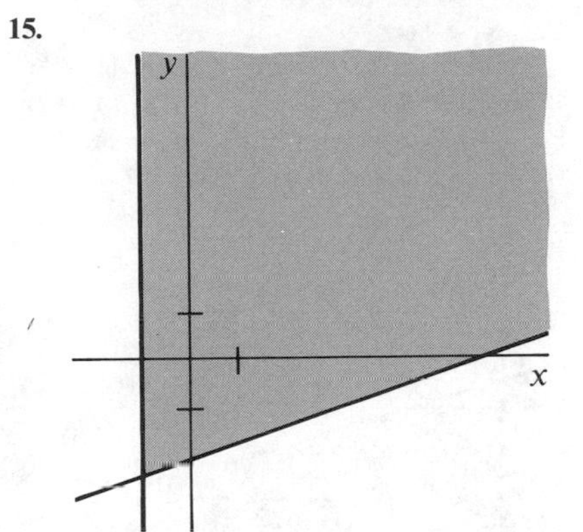

17.

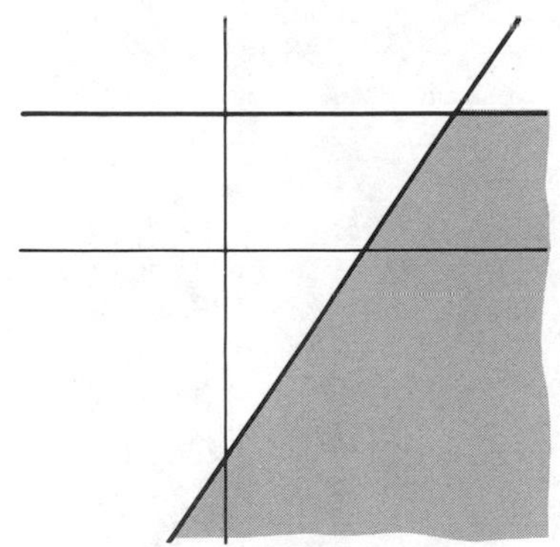

19.

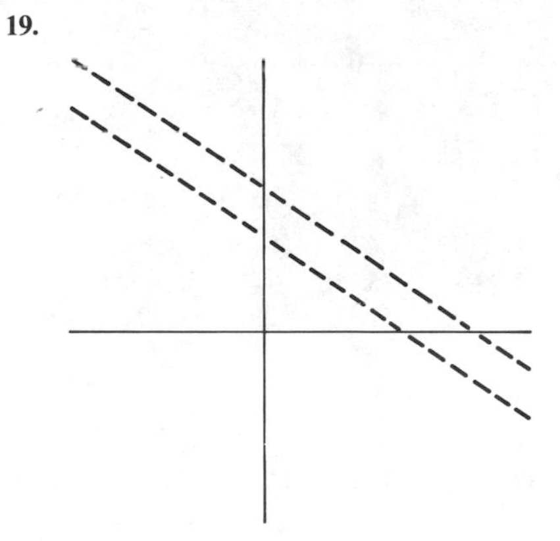

21. 23.

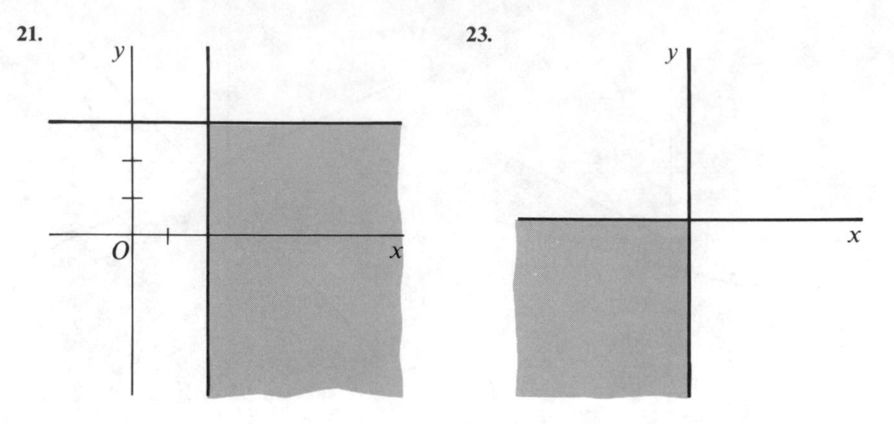

Chapter 7 Test

1. (4, −3) **2.** (6, −5) **3.** (4, 1) **4.** (6, 5) **5.** (8, −2) **6.** (2, −5)
7. parallel lines; no solution **8.** (0, −2) **9.** (2, 2) **10.** (3/2, −2) **11.** (4/3, 6)
12. (3, −5) **13.** (−6, 8) **14.** same line **15.** 12 and 27
16. 2 at \$2.50 and 4 at \$3.75 **17.** 45 mph and 60 mph

18. 19. 20.

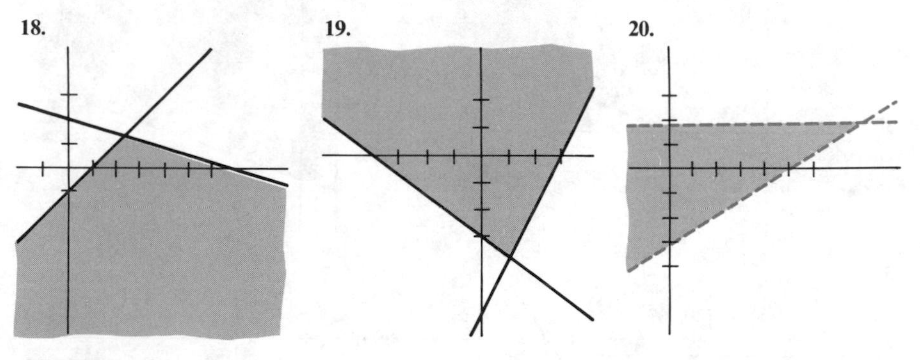

SECTION 8.1

1. 3, −3 **3.** 11, −11 **5.** 20, −20 **7.** 91, −91 **9.** $\sqrt{15}$, $-\sqrt{15}$
11. $\sqrt{72}$, $-\sqrt{72}$ **13.** no such real number **15.** 5 **17.** 5, −5 **19.** 1 **21.** 8
23. −9 **25.** 16 **27.** 25 **29.** −35 **31.** −41 **33.** no such real number
35. 3 **37.** −1 **39.** 3 **41.** 1 **43.** .5 **45.** 1.1 **47.** 6.5 **49.** .04

SECTION 8.2

1. $3\sqrt{3}$ **3.** $2\sqrt{7}$ **5.** $3\sqrt{2}$ **7.** $4\sqrt{3}$ **9.** $5\sqrt{5}$ **11.** $10\sqrt{7}$ **13.** 4 **15.** 6
17. 21 **19.** $\sqrt{21}$ **21.** $15\sqrt{3}$ **23.** 40 **25.** 36 **27.** 60 **29.** $7\sqrt{3}$
31. $15\sqrt{3}$ **33.** $20\sqrt{3}$ **35.** $10\sqrt{10}$ **37.** 10/3 **39.** 6/7 **41.** $\sqrt{5}/4$
43. $\sqrt{30}/7$ **45.** 2/5 **47.** 4/25 **49.** 5 **51.** 4 **53.** $3\sqrt{5}$ **55.** $5\sqrt{10}$ **57.** y
59. $\sqrt{xz}$ **61.** x **63.** x^2 **65.** xy^2 **67.** $x\sqrt{x}$ **69.** x^2 **71.** ab^2

SECTION 8.3

1. $7\sqrt{3}$ **3.** $-5\sqrt{7}$ **5.** $2\sqrt{6}$ **7.** $3\sqrt{17}$ **9.** $4\sqrt{7}$ **11.** $12\sqrt{2}$ **13.** $7\sqrt{5}$
15. $11\sqrt{2}$ **17.** $3\sqrt{3}$ **19.** $20\sqrt{2}$ **21.** $-13\sqrt{2}$ **23.** $19\sqrt{7}$ **25.** $-12\sqrt{5}$
27. $6\sqrt{2} + 7\sqrt{3}$ **29.** $-16\sqrt{2} + 8\sqrt{3}$ **31.** $20\sqrt{2} + 6\sqrt{3} + 15\sqrt{5}$ **33.** $2\sqrt{2}$

35. $3\sqrt{3} - 2\sqrt{5}$ **37.** $5\sqrt{3}$ **39.** $3\sqrt{21} - \sqrt{7}$ **41.** $6\sqrt{x}$ **43.** $13\sqrt{a}$
45. $15x\sqrt{3}$ **47.** $2x\sqrt{2}$ **49.** 0

SECTION 8.4
1. $6\sqrt{5}/5$ **3.** $\sqrt{5}$ **5.** $3\sqrt{7}/7$ **7.** $8\sqrt{15}/5$ **9.** $\sqrt{30}/2$ **11.** $8\sqrt{3}/9$
13. $3\sqrt{2}/10$ **15.** $\sqrt{2}$ **17.** $9\sqrt{2}/8$ **19.** 2 **21.** $\sqrt{2}$ **23.** $2\sqrt{30}/3$ **25.** $\sqrt{2}/2$
27. $\sqrt{70}/7$ **29.** $3\sqrt{5}/5$ **31.** $\sqrt{14}$ **33.** $\sqrt{15}/10$ **35.** $\sqrt[3]{2}/2$ **37.** $\sqrt[3]{12}/2$
39. $\sqrt[3]{12}/3$ **41.** $\sqrt{5x}/x$ **43.** $2r\sqrt{s}$ **45.** $2a\sqrt{3ab}$ **47.** $xz^2\sqrt{6y}/(3y)$ **49.** $\sqrt[3]{r}/r$

SECTION 8.5
1. $27\sqrt{5}$ **3.** $21\sqrt{2}$ **5.** -4 **7.** $\sqrt{15} + \sqrt{35}$ **9.** $2\sqrt{10} + 10$ **11.** $-4\sqrt{7}$
13. $21 - \sqrt{6}$ **15.** $87 + 9\sqrt{21}$ **17.** $34 + 24\sqrt{2}$ **19.** $37 - 12\sqrt{7}$ **21.** 7
23. -4 **25.** 1 **27.** 16 **29.** 20 **31.** 2 **33.** $(3 - \sqrt{2})/7$ **35.** $-10 + 5\sqrt{5}$
37. $-2 - \sqrt{11}$ **39.** $-3\sqrt{2} + 6$ **41.** $(-3\sqrt{5} - 15)/4$ **43.** $6 - 2\sqrt{3}$
45. $(9 - \sqrt{3})/26$ **47.** $(\sqrt{66} - \sqrt{11} - 3\sqrt{2} + \sqrt{3})/8$ **49.** $(6\sqrt{2} + 2\sqrt{3} + 3\sqrt{6} + 3)/3$

SECTION 8.6
1. 3.606 **3.** 7.141 **5.** 12.247 **7.** 24.083 **9.** 27.386 **11.** 76.16 **13.** 37.42
15. 616.4 **17.** 7280 **19.** 118.32 **21.** 281.07 **23.** .2828 **25.** .7874
27. .07348 **29.** .04472 **31.** .12649 **33.** $2\sqrt{7} = 5.292$ **35.** $2\sqrt{3} = 3.464$
37. $2\sqrt{3}/3 = 1.155$ **39.** $4\sqrt{6}/3 = 3.265$ **41.** $(3\sqrt{5} + 5)/5 = 2.342$
43. $(1 - \sqrt{5})/-2 = .618$ **45.** $-3 + 3\sqrt{2} = 1.242$ **47.** $(3 - \sqrt{5})/2 = .382$
49. $15\sqrt{2} - 20 = 1.21$

SECTION 8.7
1. 4 **3.** 1 **5.** 7 **7.** -7 **9.** -5 **11.** no solution **13.** 9 **15.** 16 **17.** 6
19. 7 **21.** -4 **23.** 12 **25.** 5 **27.** 0, -1 **29.** 5 **31.** 3 **33.** 9 **35.** 9
37. 21 **39.** 8 **41.** (a) 90 (b) 60 (c) 60

Chapter 8 Test
1. 10, -10 **2.** 25, -25 **3.** -3 **4.** 2, -2 **5.** none **6.** 8 **7.** 14
8. $5\sqrt{2}$ **9.** $8\sqrt{2}$ **10.** $5\sqrt{3}$ **11.** $8\sqrt{2}$ **12.** $11\sqrt{6}$ **13.** $-7\sqrt{3x}$
14. $4xy\sqrt{2y}$ **15.** -1 **16.** $11 + 2\sqrt{30}$ **17.** $4\sqrt{3}/3$ **18.** $\sqrt{3}$ **19.** $3 + \sqrt{7}$
20. $(-8 - 2\sqrt{3})/13$ **21.** 19.748 **22.** 86.02 **23.** 308.22 **24.** 2.683 **25.** 14.928
26. 23 **27.** 37 **28.** 4 **29.** -5 **30.** -1, -2

SECTION 9.1
1. 2, 3 **3.** -2, -10 **5.** 1/2, -3 **7.** $-3/2$, -1 **9.** 5/2, -3 **11.** 6, -2
13. $-4 + \sqrt{10}$, $-4 - \sqrt{10}$ **15.** $1 + 4\sqrt{2}$, $1 - 4\sqrt{2}$ **17.** 2, -1 **19.** $-2/3$, $-8/3$
21. 13/6, $-3/2$ **23.** $(5 + \sqrt{30})/2$, $(5 - \sqrt{30})/2$ **25.** $(1 + 3\sqrt{2})/3$, $(1 - 3\sqrt{2})/3$
27. $(5 + 7\sqrt{2})/2$, $(5 - 7\sqrt{2})/2$ **29.** $(-4 + 2\sqrt{2})/3$, $(-4 - 2\sqrt{2})/3$
31. about 1/2 second

SECTION 9.2
1. 1 **3.** 81 **5.** 81/4 **7.** 49 **9.** 25/4 **11.** -1, -3
13. $-1 + \sqrt{6}$, $-1 - \sqrt{6}$ **15.** -2, -4 **17.** $3 + 2\sqrt{2}$, $3 - 2\sqrt{2}$
19. $(-3 + \sqrt{17})/2$, $(-3 - \sqrt{17})/2$ **21.** no real number solutions
23. no real number solutions **25.** $3 + \sqrt{5}$, $3 - \sqrt{5}$ **27.** $(2 + \sqrt{14})/2$, $(2 - \sqrt{14})/2$

29. 1, $-1/3$ **31.** $5+\sqrt{17}, 5-\sqrt{17}$ **33.** $(3+2\sqrt{6})/3, (3-2\sqrt{6})/3$
35. $-2+\sqrt{3}, -2-\sqrt{3}$

SECTION 9.3

1. $3, 4, -8$ **3.** $-8, -2, -3$ **5.** $2, -3, 2$ **7.** $1, 0, -2$ **9.** $3, -8, 0$ **11.** $1, 1, -12$
13. $9, 9, -26$ **15.** $-1+\sqrt{3}, -1-\sqrt{3}$ **17.** -2
19. $(7+\sqrt{41})/4, (7-\sqrt{41})/4$ **21.** $(-1+\sqrt{5})/2, (-1-\sqrt{5})/2$ **23.** $1, -13$
25. $(-6+\sqrt{26})/2, (-6-\sqrt{26})/2$ **27.** $1/5, -1$ **29.** $5/2, -1$
31. $(-1+\sqrt{73})/6, (-1-\sqrt{73})/6$ **33.** 1 **35.** $2\sqrt{5}, -2\sqrt{5}$
37. $(6+2\sqrt{6})/3, (6-2\sqrt{6})/3$ **39.** $(2+\sqrt{22})/6, (2-\sqrt{22})/6$

SECTION 9.4

1. always **3.** always **5.** never **7.** always **9.** $2i$ **11.** $4i$ **13.** $-7i$
15. $2\sqrt{3}\,i$ **17.** $7\sqrt{2}\,i$ **19.** $-3\sqrt{3}\,i$ **21.** $1+6i$ **23.** $2-2\sqrt{2}\,i$
25. $2\sqrt{3}\,i$ **27.** $7\sqrt{3}\,i$ **29.** $1+i$ **31.** $1+\sqrt{2}\,i$ **33.** $(1-2i)/2$

SECTION 9.5

1. $(1+i)/2, (1-i)/2$ **3.** $2+2i, 2-2i$
5. $(-1+i)/2, (-1-i)/2$ **7.** $(1+\sqrt{3}\,i)/3, (1-\sqrt{3}\,i)/3$
9. $(-1+3\sqrt{7}\,i)/8, (-1-3\sqrt{7}\,i)/8$ **11.** $(2+\sqrt{11}\,i)/3, (2-\sqrt{11}\,i)/3$
13. $(-1+\sqrt{79}\,i)/20, (-1-\sqrt{79}\,i)/20$ **15.** $(-1+\sqrt{3}\,i)/4, (-1-\sqrt{3}\,i)/4$
17. $(-1+\sqrt{7}\,i)/3, (-1-\sqrt{7}\,i)/3$ **19.** $(1+\sqrt{5}\,i)/4, (1-\sqrt{5}\,i)/4$
21. $(2+\sqrt{14}\,i)/6, (2-\sqrt{14}\,i)/6$ **23.** $(-2+2\sqrt{17}\,i)/9, (-2-2\sqrt{17}\,i)/9$
25. $(-3+\sqrt{26}\,i)/5, (-3-\sqrt{26}\,i)/5$ **27.** $3+3i, 3-3i$ **29.** $4+5i, 4-5i$
31. $(3+3\sqrt{3}\,i)/4, (3-3\sqrt{3}\,i)/4$

SECTION 9.6

1. (b) **3.** (d) **5.** (b) **7.** (d) **9.** (c) **11.** (c) **13.** (d) **15.** (b)
17. (c) **19.** (c) **21.** (b) **23.** (b) **25.** 1 **27.** 9 **29.** -49

SECTION 9.7

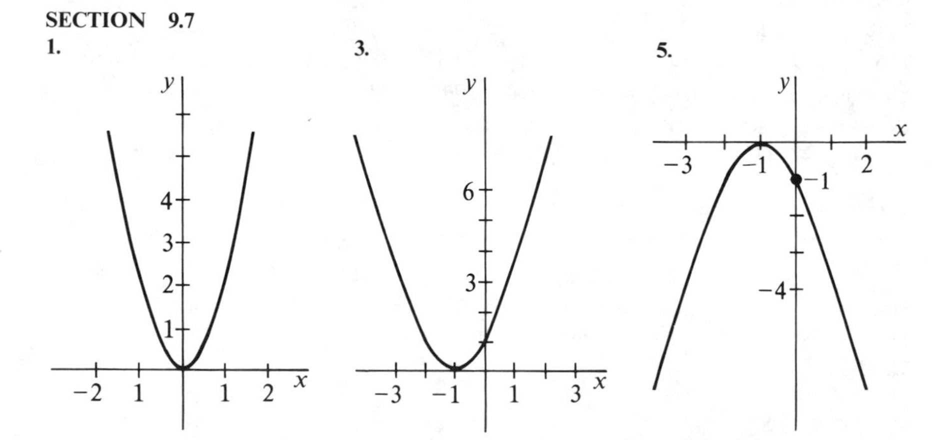

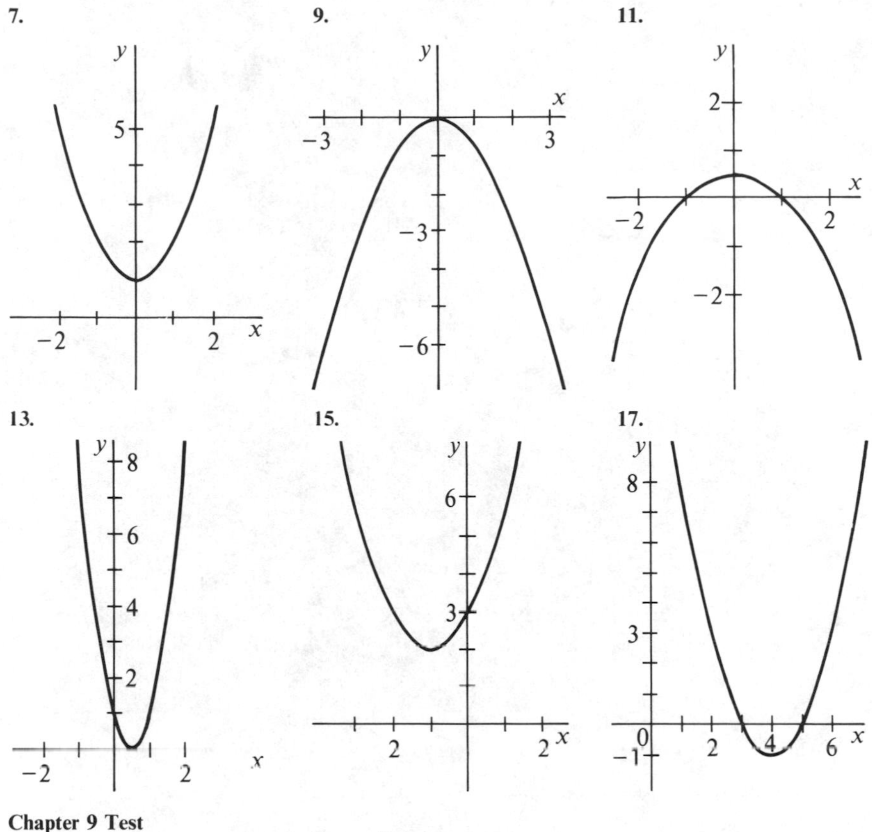

Chapter 9 Test

1. 0, 5/2 **2.** 1/5, -2 **3.** $\sqrt{5}$, $-\sqrt{5}$ **4.** 10, -4
5. $(2+\sqrt{35})/3$, $(2-\sqrt{35})/3$ **6.** 4 **7.** 2, 1/3 **8.** $(-2+i)/2$, $(-2-i)/2$
9. $(5+\sqrt{13})/6$, $(5-\sqrt{13})/6$ **10.** 2 rational **11.** 1 rational
12. 2 imaginary **13.** 2 irrational **14.** $\sqrt{3}\,i$ **15.** $\sqrt{11}\,i$ **16.** $5i$ **17.** $3\sqrt{2}\,i$
18. $2+4\sqrt{2}\,i$ **19.** $7\sqrt{3}\,i$ **20.** $(1-2\sqrt{2}\,i)/2$
21. **22.**

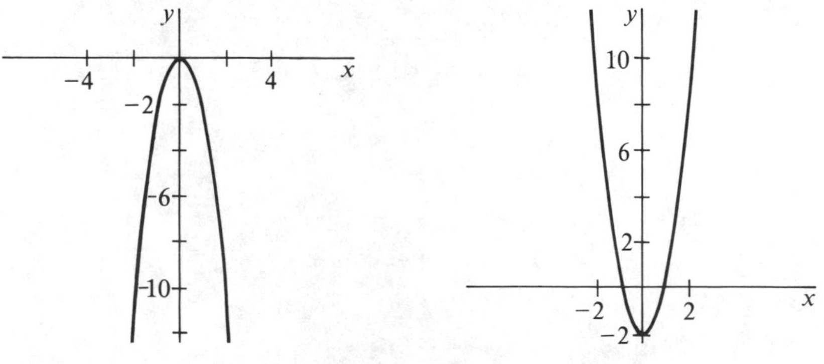

APPENDIX ON METRIC SYSTEM

1. 20,000 **3.** 8 **5.** 5.2 **7.** 7800 **9.** 1920 **11.** 9000 **13.** 57
15. 10.9728 **17.** 15.24 **19.** 136.2 **21.** 1.87 **23.** 127

INDEX

We would appreciate it if you would take a few minutes to answer these questions. Then cut the page out, fold it, seal it, and mail it. No postage is required.

Which chapters did you cover?
(circle) 1 2 3 4 5 6 7 8 9 All ______

Which helped most?
Explanations ______ Examples ______ Exercises ______ All three ______

Does the book have enough worked-out examples? Yes ______ No ______

enough exercises? Yes ______ No ______

Were the answers in the back of the book helpful? Yes ______ No ______

Did you use the *Study Guide?*
Yes ______ No ______ Did not know of it ______

If YES, was the *Study Guide* helpful?
Yes ______ For some topics ______ No ______

How was your course taught? Regular class ______ Self paced ______

For you, was the course elective ______ required by ____________

Do you plan to take more mathematics courses? Yes ______ No ______

If YES, which ones?

Intermediate algebra ______ Geometry ______ Math for elementary teachers ______

Business math ______ Nursing (or allied health) math ______
Technical math ______

Introduction to math (survey) ______ College algebra ______
Data processing ______

Other ____________

How much algebra did you have before this course? None ______

Terms in high school (circle) 1 2 3 4
Courses in college 1 2 3

If you had algebra before, how long ago?
Last 2 years ______ 3–5 years ______ 5 years or more ______

What is your major or your career goal? ____________ Your age? ______

What did you like most about the book?

What did you like least about the book?

FOLD HERE

FOLD HERE

First Class
PERMIT NO. 282
Glenview, Il.

BUSINESS REPLY MAIL
No postage necessary if mailed in United States

Postage will be paid by
SCOTT, FORESMAN AND COMPANY
College Division Attn: Lial/Miller
1900 East Lake Avenue
Glenview, Illinois 60025

Formulas

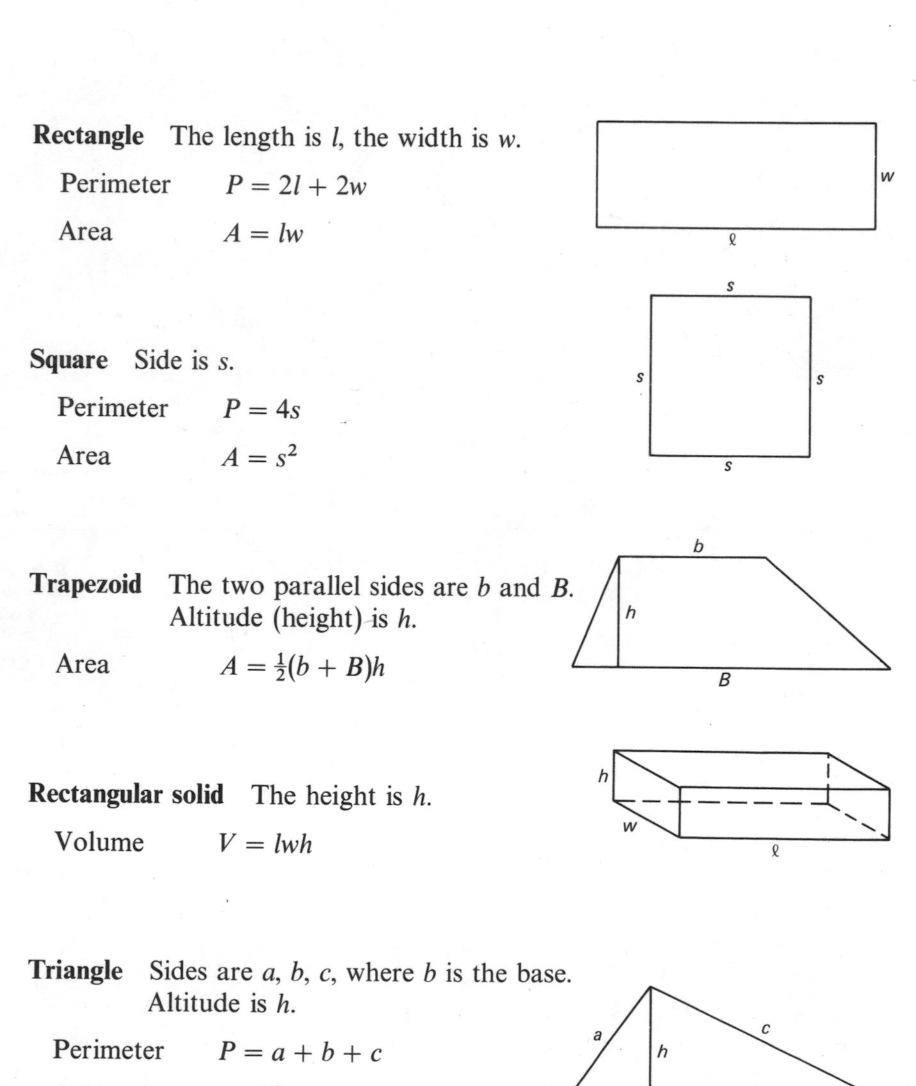

Rectangle The length is l, the width is w.

Perimeter $P = 2l + 2w$

Area $A = lw$

Square Side is s.

Perimeter $P = 4s$

Area $A = s^2$

Trapezoid The two parallel sides are b and B. Altitude (height) is h.

Area $A = \frac{1}{2}(b + B)h$

Rectangular solid The height is h.

Volume $V = lwh$

Triangle Sides are a, b, c, where b is the base. Altitude is h.

Perimeter $P = a + b + c$

Area $A = \frac{1}{2}bh$